PROGRESS IN CLINICAL AND BIOLOGICAL RESEARCH

Series Editors
Nathan Back
George J. Brewer

Vincent P. Eijsvoogel
Robert Grover
Kurt Hirschhorn

Seymour S. Kety
Sidney Udenfriend
Jonathan W. Uhr

Vol 1: **Erythrocyte Structure and Function,** George J. Brewer, *Editor*
Vol 2: **Preventability of Perinatal Injury,** Karlis Adamsons and Howard A. Fox, *Editors*
Vol 3: **Infections of the Fetus and the Newborn Infant,** Saul Krugman and Anne A. Gershon, *Editors*
Vol 4: **Conflicts in Childhood Cancer: An Evaluation of Current Management,** Lucius F. Sinks and John O. Godden, *Editors*
Vol 5: **Trace Components of Plasma: Isolation and Clinical Significance,** G.A. Jamieson and T.J. Greenwalt, *Editors*
Vol 6: **Prostatic Disease,** H. Marberger, H. Haschek, H.K.A. Schirmer, J.A.C. Colston, and E. Witkin, *Editors*
Vol 7: **Blood Pressure, Edema and Proteinuria in Pregnancy,** Emanuel A. Friedman, *Editor*
Vol 8: **Cell Surface Receptors,** Garth L. Nicolson, Michael A. Raftery, Martin Rodbell, and C. Fred Fox, *Editors*
Vol 9: **Membranes and Neoplasia: New Approaches and Strategies,** Vincent T. Marchesi, *Editor*
Vol 10: **Diabetes and Other Endocrine Disorders During Pregnancy and in the Newborn,** Maria I. New and Robert H. Fiser, *Editors*
Vol 11: **Clinical Uses of Frozen-Thawed Red Blood Cells,** John A. Griep, *Editor*
Vol 12: **Breast Cancer,** Albert C.W. Montague, Geary L. Stonesifer, Jr., and Edward F. Lewison, *Editors*
Vol 13: **The Granulocyte: Function and Clinical Utilization,** Tibor J. Greenwalt and G.A. Jamieson, *Editors*
Vol 14: **Zinc Metabolism: Current Aspects in Health and Disease,** George J. Brewer and Ananda S. Prasad, *Editors*
Vol 15: **Cellular Neurobiology,** Zach Hall, Regis Kelly, and C. Fred Fox, *Editors*
Vol 16: **HLA and Malignancy,** Gerald P. Murphy, *Editor*
Vol 17: **Cell Shape and Surface Architecture,** Jean Paul Revel, Ulf Henning, and C. Fred Fox, *Editors*
Vol 18: **Tay-Sachs Disease: Screening and Prevention,** Michael M. Kaback, *Editor*
Vol 19: **Blood Substitutes and Plasma Expanders,** G.A. Jamieson and T.J. Greenwalt, *Editors*
Vol 20: **Erythrocyte Membranes: Recent Clinical and Experimental Advances,** Walter C. Kruckeberg, John W. Eaton, and George J. Brewer, *Editors*
Vol 21: **The Red Cell,** George J. Brewer, *Editor*
Vol 22: **Molecular Aspects of Membrane Transport,** Dale Oxender and C. Fred Fox, *Editors*
Vol 23: **Cell Surface Carbohydrates and Biological Recognition,** Vincent T. Marchesi, Victor Ginsburg, Phillips W. Robbins, and C. Fred Fox, *Editors*
Vol 24: **Twin Research,** Proceedings of the Second International Congress on Twin Studies
Walter E. Nance, *Editor*
Published in 3 Volumes:
    Part A: Psychology and Methodology
    Part B: Biology and Epidemiology
    Part C: Clinical Studies

Vol 25:	**Recent Advances in Clinical Oncology,** Tapan A. Hazra and Michael C. Beachley, *Editors*
Vol 26:	**Origin and Natural History of Cell Lines,** Claudio Barigozzi, *Editor*
Vol 27:	**Membrane Mechanisms of Drugs of Abuse,** Charles W. Sharp and Leo G. Abood, *Editors*
Vol 28:	**The Blood Platelet in Transfusion Therapy,** Tibor J. Greenwalt and G.A. Jamieson, *Editors*
Vol 29:	**Biomedical Applications of the Horseshoe Crab (Limulidae),** Elias Cohen, *Editor-in-Chief*
Vol 30:	**Normal and Abnormal Red Cell Membranes,** Samuel E. Lux, Vincent T. Marchesi, and C. Fred Fox, *Editors*
Vol 31:	**Transmembrane Signaling,** Mark Bitensky, R. John Collier, Donald F. Steiner, and C. Fred Fox, *Editors*
Vol 32:	**Genetic Analysis of Common Diseases: Applications to Predictive Factors in Coronary Disease,** Charles F. Sing and Mark Skolnick, *Editors*
Vol 33:	**Prostate Cancer and Hormone Receptors,** Gerald P. Murphy and Avery A. Sandberg, *Editors*
Vol 34:	**The Management of Genetic Disorders,** Constantine J. Papadatos and Christos S. Bartsocas, *Editors*
Vol 35:	**Antibiotics and Hospitals,** Carlo Grassi and Giuseppe Ostino, *Editors*
Vol 36:	**Drug and Chemical Risks to the Fetus and Newborn,** Richard H. Schwarz and Sumner J. Yaffe, *Editors*
Vol 37:	**Models for Prostate Cancer,** Gerald P. Murphy, *Editor*
Vol 38:	**Ethics, Humanism, and Medicine,** Marc D. Basson, *Editor*
Vol 39:	**Neurochemistry and Clinical Neurology,** Leontino Battistin, George Hashim, and Abel Lajtha, *Editors*
Vol 40:	**Biological Recognition and Assembly,** David S. Eisenberg, James A. Lake, and C. Fred Fox, *Editors*
Vol 41:	**Tumor Cell Surfaces and Malignancy,** Richard O. Hynes and C. Fred Fox, *Editors*
Vol 42:	**Membranes, Receptors, and the Immune Response: 80 Years After Ehrlich's Side Chain Theory,** Edward P. Cohen and Heinz Köhler, *Editors*
Vol 43:	**Immunobiology of the Erythrocyte,** S. Gerald Sandler, Jacob Nusbacher, and Moses S. Schanfield, *Editors*
Vol 44:	**Perinatal Medicine Today,** Bruce K. Young, *Editor*
Vol 45:	**Mammalian Genetics and Cancer: The Jackson Laboratory Fiftieth Anniversary Symposium,** Elizabeth S. Russell, *Editor*
Vol 46:	**Etiology of Cleft Lip and Cleft Palate,** Michael Melnick, David Bixler, and Edward D. Shields, *Editors*
Vol 47:	**New Developments With Human and Veterinary Vaccines,** A. Mizrahi, I. Hertman, M.A. Klingberg, and A. Kohn, *Editors*
Vol 48:	**Cloning of Human Tumor Stem Cells,** Sydney E. Salmon, *Editor*
Vol 49:	**Myelin: Chemistry and Biology,** George A. Hashim, *Editor*
Vol 50:	**Rights and Responsibilities in Modern Medicine: The Second Volume in a Series on Ethics, Humanism, and Medicine,** Marc D. Basson, *Editor*
Vol 51:	**The Function of Red Blood Cells: Erythrocyte Pathobiology,** Donald F. H. Wallach, *Editor*
Vol 52:	**Conduction Velocity Distributions: A Population Approach to Electrophysiology of Nerve,** Leslie J. Dorfman, Kenneth L. Cummins, and Larry J. Leifer, *Editors*
Vol 53:	**Cancer Among Black Populations,** Curtis Mettlin and Gerald P. Murphy, *Editors*

MAMMALIAN GENETICS AND CANCER
The Jackson Laboratory Fiftieth Anniversary Symposium

MAMMALIAN GENETICS AND CANCER
The Jackson Laboratory
Fiftieth Anniversary Symposium

Editor
ELIZABETH S. RUSSELL

Senior Staff Scientist
The Jackson Laboratory

Alan R. Liss, Inc. • New York

Address all Inquiries to the Publisher
Alan R. Liss, Inc., 150 Fifth Avenue, New York, NY 10011

Copyright © 1981 Alan R. Liss, Inc.

Printed in the United States of America.

Under the conditions stated below the owner of copyright for this book hereby grants permission to users to make photocopy reproductions of any part or all of its contents for personal or internal organizational use, or for personal or internal use of specific clients. This consent is given on the condition that the copier pay the stated per copy fee through the Copyright Clearance Center, Incorporated, 21 Congress Street, Salem, MA 01970, as listed in the most current issue of "Permissions to Photocopy" (Publisher's Fee List, distributed by CCC, Inc.) for copying beyond that permitted by sections 107 or 108 of the US copyright Law. This consent does not extend to other kinds of copying, such as copying for general distribution, for advertising or promotional purposes, for creating new collective works, or for resale.

Library of Congress Cataloging in Publication Data
Main entry under title:

Mammalian genetics and cancer.

(Progress in clinical and biological research; v. 45)
Includes index.
1. Cancer—Genetic aspects—Congresses. 2. Mammals—
Genetics—Congresses. I. Russell, Elizabeth Shull, 1913–
II. Jackson Laboratory, Bar Harbor, Me. III. Series:
[DNLM: 1. Mammals—Genetics—Congresses. 2.
Hereditary diseases—Congresses. 3. Immunogenetics—
Congresses. 4. Neoplasms—Etiology—Congresses.
5. Cell differentiation—Congresses. W1 PR668E v. 45 /
QZ 50 M265 1979]
RC268.4.M35   616.99'4071   80-27531
ISBN 0-8451-0045-9

Contents

CONTRIBUTORS ix

PREFACE: A Century of Mammalian Genetics and Cancer,
1929–2029, A View at Midpassage
Elizabeth S. Russell xi

SESSION I. GENE AND CHROMOSOME ORGANIZATION,
Margaret C. Green, Chairman

    Introduction to Session I 3

    Foundation for the Future: Formal Genetics of the Mouse
    Eva M. Eicher 7

    The Organization and Evolution of Cloned Globin Genes
    Philip Leder, David A. Konkel, Yutaka Nishioka, Aya Leder,
    Dean H. Hamer, and Marian Kaehler 51

    Gene and Chromosome Organization: Chairman's Summary
    Margaret C. Green 67

SESSION II. ANALYSIS OF MAMMALIAN DIFFERENTIATION,
James D. Ebert, Chairman

    Introduction to Session II 73

    Experimental Chimaeras and the Study of Differentiation
    Virginia E. Papaioannou 77

    Genetic Influences on Teratocarcinogenesis and
    Parthenogenesis
    Leroy C. Stevens 93

    Experimental Manipulation of the Mammalian Embryo:
    Biological and Genetic Consequences
    Karl Illmensee 105

SESSION III. INHERITED DISEASES OF MOUSE AND MAN,
Victor A. McKusick, Chairman

    Introduction to Session III 123

    The Last Twenty Years: An Overview of Advances in Medical
    Genetics
    Victor A. McKusick 127

Inherited Obesity-Diabetes Syndromes in the Mouse
D.L. Coleman 145

Hemolytic Anemias Due to Abnormalities in Red Cell Spectrin:
A Brief Review
Samuel E. Lux, Lawrence C. Wolfe, Barbara Pease, Mary Beth
Tomaselli, Kathryn M. John, and Seldon E. Bernstein 159

Garrod's Legacy to the Nations of Mice and Men
Charles R. Scriver 169

SESSION IV. IMMUNOGENETICS, Dorothea Bennett, Chairman

Introduction to Session IV 195

Mouse Histocompatibility Genetics and Tumor Immunology
George Klein 197

The Major Histocompatibility Gene Clusters of Man and Mouse
Walter F. Bodmer 213

The Future of Immunogenetics
George D. Snell 241

Immunogenetics: Chairman's Summary
Dorothea Bennett 273

SESSION V. THE ETIOLOGY OF CANCER, Richmond T. Prehn, Chairman

Introduction to Session V 277

Development and Utilization of Inbred Strains of Mice for
Cancer Research
Walter E. Heston 279

Murine Leukemia Viruses as Chromosomal Genes of the Mouse
Wallace P. Rowe, Janet W. Hartley, and Christine A. Kozak 291

Abelson Murine Leukemia Virus-Induced Transformation of
Immature Lymphoid Cells
David Baltimore 297

SUMMARY OVERVIEW

A Century of Mammalian Genetics and Cancer: Where Are We
at Midpassage?
James F. Crow 309

INDEX 325

Contributors

David Baltimore, Center for Cancer Research, Massachusetts Institute of Technology, Cambridge, Massachusetts 02139 [297]

Dorothea Bennett, Sloan Kettering Institute for Cancer Research, New York, New York 10021 [273]

Seldon E. Bernstein, The Jackson Laboratory, Bar Harbor, Maine 04609 [159]

Walter F. Bodmer, Imperial Cancer Research Fund, PO Box 123, Lincoln's Inn Fields, London WC2A 3PX, England [213]

D.L. Coleman, The Jackson Laboratory, Bar Harbor, Maine 04609 [145]

James F. Crow, Department of Genetics, University of Wisconsin, Madison, Wisconsin 53706 [309]

Eva M. Eicher, The Jackson Laboratory, Bar Harbor, Maine 04609 [7]

Margaret C. Green, The Jackson Laboratory, Bar Harbor, Maine 04609 [67]

Dean H. Hamer, Laboratory of Molecular Genetics, National Institute of Child Health and Human Development, National Institutes of Health, Bethesda, Maryland 20205 [51]

Janet W. Hartley, Laboratory of Viral Diseases, National Institute of Allergy and Infectious Diseases, National Institutes of Health, Bethesda, Maryland 20205 [291]

Walter E. Heston, National Cancer Institute, retired [279]

Karl Illmensee, Department of Animal Biology, University of Geneva, CH-1224 Geneva, Switzerland [105]

Kathryn M. John, Department of Medicine, Childrens Hospital Medical Center, Boston, Massachusetts 02115 [159]

Marian Kaehler, Laboratory of Molecular Genetics, National Institute of Child Health and Human Development, National Institutes of Health, Bethesda, Maryland 20205 [51]

George Klein, Department of Tumor Biology, Karolinska Institutet, S-10401 Stockholm, Sweden [197]

David A. Konkel, Laboratory of Molecular Genetics, National Institute of Child Health and Human Development, National Institutes of Health, Bethesda, Maryland 20205 [51]

The number in brackets following a contributor's affiliation is the opening page number of that author's chapter.

x / Contributors

Christine A. Kozak, Laboratory of Viral Diseases, National Institute of Allergy and Infectious Diseases, National Institutes of Health, Bethesda, Maryland 20205 [291]

Aya Leder, Laboratory of Molecular Genetics, National Institute of Child Health and Human Development, National Institutes of Health, Bethesda, Maryland 20205 [51]

Philip Leder, Laboratory of Molecular Genetics, National Institute of Child Health and Human Development, National Institutes of Health, Bethesda, Maryland 20205 [51]

Samuel E. Lux, Department of Medicine, Childrens Hospital Medical Center, Boston, Massachusetts 02115 [159]

Victor A. McKusick, Department of Medicine, Johns Hopkins Hospital, Baltimore, Maryland 21205 [127]

Yutaka Nishioka, Laboratory of Molecular Genetics, National Institute of Child Health and Human Development, National Institutes of Health, Bethesda, Maryland 20205 [51]

Virginia E. Papaioannou, Sir William Dunn School of Pathology, University of Oxford, Oxford, UK OX1 3RE [77]

Barbara Pease, Department of Medicine, Childrens Hospital Medical Center, Boston, Massachusetts 02115 [159]

Wallace P. Rowe, Laboratory of Viral Diseases, National Institute of Allergy and Infectious Diseases, National Institutes of Health, Bethesda, Maryland 20205 [291]

Elizabeth S. Russell, The Jackson Laboratory, Bar Harbor, Maine 04609 [xi]

Charles R. Scriver, Biochemical Genetics Laboratory, Montreal Children's Hospital, Montreal, Quebec, Canada H3H1P3 [169]

George D. Snell, The Jackson Laboratory, Bar Harbor, Maine 04609 [241]

Leroy C. Stevens, The Jackson Laboratory, Bar Harbor, Maine 04609 [93]

Mary Beth Tomaselli, Department of Medicine, Childrens Hospital Medical Center, Boston, Massachusetts 02115 [159]

Lawrence C. Wolfe, Department of Medicine, Childrens Hospital Medical Center, Boston, Massachusetts 02115 [159]

Preface: A Century of Mammalian Genetics and Cancer, 1929-2029, A View at Midpassage

This volume is designed to provide (in five exciting current topics in mammalian biomedical research) a "View in Midpassage" between the status of those fields in 1929, when The Jackson Laboratory was founded, and anticipated accomplishments within the same areas achieved by that future year, 2029. The contents of this book are the proceedings of a scientific symposium held in July 1979 to celebrate the 50th birthday of The Jackson Laboratory. This anniversary came at a most propitious moment in the progress of research on mammalian genetics and cancer. Although experimental mammalian research expanded steadily and produced important results between 1929 and 1979, "center stage" was dominated by the discovery of DNA and the rise of molecular biology. In the early days of molecular genetics, critical questions could be attacked most expeditiously using microbial and viral test systems. Molecular biology focused almost exclusively on prokaryotes, while research on differentiated eukaryotes concentrated on the visible cellular and whole organism level. Now the center of excitement appears to be shifting to analysis of eukaryotic systems including control of development, differentiation, and function. By using in vitro systems, man himself has become an important experimental organism. But more and more of this new research is based on work with experimental mammals, and the five research areas discussed in this symposium are especially "hot" topics within an overall growing research field.

The five diverse topics selected for this symposium seem at first consideration somewhat unrelated: gene and chromosome organization; analysis of mammalian development; inherited diseases of mouse and man; immunogenetics; and the etiology of cancer. Actually, they tend to be mutually interdependent, so that advances in one topic lead to new possibilities in another. Also, as becomes apparent in reading papers from each session, advances in research with experimental mammals lead to advances in the understanding of human conditions; and, conversely, results of research on humans often open fruitful new avenues for experimental mammalian research.

Why was this particular set of topics selected for this symposium? None of them is brand new. Considerable attention, resulting in marked progress, has been devoted to each of these topics during the past 50 years. These topics are not old; it seems highly probable that each will continue to attract considerable interest over much of the next 50 years. It is almost a foregone conclusion that findings from basic research related to these topics (as well as to many others) will long continue to find new applications beneficial to human welfare.

Why were these five topics singled out? All have two characteristics in common. Especially rapid advances in recent years caused these to be "hot" topics in 1979; and the Jackson Laboratory feels that research at this institution has contributed materially to the long-term development of each of these fields. We wanted to celebrate! The chairman selected for each session is a long-standing friend of the Jackson Laboratory. Dr. Margaret C. Green, now Senior Staff Scientist Emeritus, has been a major force in expanding knowledge of the formal genetics of the mouse. Dr. James D. Ebert, renowned developmental biologist, was Chairman of the Board of Scientific Overseers of the Jackson Laboratory. Victor McKusick, medical geneticist from Johns Hopkins School of Medicine, has established and guided for the past 20 years an annual series of two-week Short Courses in Genetics. At first, all courses consisted of combined Johns Hopkins-Jackson Laboratory courses in Medical Genetics; later, these sessions were alternated with Jackson Laboratory Short Courses in Experimental Mammalian Genetics. Dr. Dorothea Bennett, an important mouse developmental geneticist with special interests in immunology and cell surfaces, is a member of the Jackson Laboratory Board of Scientific Overseers. Dr. Richmond Prehn, current Director of The Jackson Laboratory, did an excellent job as Chairman of Session V – The Etiology of Cancer. His introductory comments stressed the contributions of his early mentor, Dr. Howard B. Andervont, one of the organizing members of the National Cancer Institute; one of the first Jackson Laboratory summer investigators; and, for many years, a greatly valued member of the Laboratory's Board of Scientific Overseers. We regret that Dr. Andervont, originally scheduled to serve as Chairman of Session V, was unable to come to the Symposium. Dr. James F. Crow, Jackson Laboratory Overseer and a delightfully insightful geneticist whose own work centers on Drosophila, gave us an excellent overview of 50 years of developments in the whole science of genetics.

We also were very proud to have as our honored guest world-famous geneticist Sewall Wright, friend from Bussey Institute days of Clarence Cook Little, founder of The Jackson Laboratory; postdoctoral mentor of Earl L. Green, the Jackson Laboratory's second director; and doctoral sponsor of former Jackson Laboratory staff members W. L. Russell, J. P. Scott, and W. K. Silvers, and of one current staff member – myself.

In the short introduction to each session, I have attempted to orient the unfamiliar reader to the impact of each paper. Most of the papers deal with very recent advances and are presented by leading scientists in the field, four of them currently working at the Jackson Laboratory. Eight others have a "Jackson Laboratory history." Walter Heston was an early staff member; David Baltimore and Virginia Papaioannou have been Jackson Lab "summer students"; Victor McKusick, Charles Scriver, and Philip Leder have been leading lights in Genetics Short Courses; Samuel Lux and Karl Illmensee carry on collaborative research with Jackson Laboratory investigators; and Philip Leder has recently become one of our Scientific Overseers.

It is not surprising that, even though the major emphasis in this Symposium is on "Science Today," several papers (at least one per session) include historical perspectives, with gratifying reference to the role of the Jackson Laboratory. Since this occasion was the Laboratory's birthday party, I hope you will forgive me for closing with a few remarks on Jackson Laboratory history.

Some of you may recall the crash of 1929, which ushered in the Great Depression. A fine time to start a private research laboratory, particularly when there were no federal and few private foundation sources of research support! The first great accomplishment of the Jackson Laboratory was survival through its first ten years. Major research accomplishments were characterization of inbred strains of mice and demonstration of their value in research, especially through the delineation of extrachromosomal influences on the development of mouse mammary cancer. In its second decade (1939–1949), The Jackson Laboratory was still tiny and struggling, but its influence expanded because of further analysis of maternal influences on mammary cancer, demonstration of genetic influence on incidence of other types of cancer, and the beginning of advance in knowledge of the formal genetics of the mouse. To these we should add two activities whose effect was to focus attention on the research importance of genetically-controlled mice: supply of inbred and F_1 hybrid mice to outside investigators; and publication of the first edition of "The Biology of the Laboratory Mouse."

The third decade of the laboratory's history (1949–1959) was a time of rapid growth, helped greatly by the beginning of "Big Science," with support for research endeavors from the federal government and from the National Institutes of Health. The Jackson Laboratory was engaged in research on mouse immunogenetics, teratology, behavior, and genetic diseases, always utilizing genetically controlled stocks. Great advances were made in the knowledge of the formal genetics of the mouse. Studies in rabbit genetics were undertaken, and behavior-genetics research concentrated largely on dogs. During this decade Earl L. Green became the second director of The Jackson Laboratory (1956), and the 25th Anniversary was celebrated (1954) with an enthusiastic symposium. It is interesting that Sewall Wright summarized that conference, presenting "Patterns of Mammalian Gene Action." This third decade of rapid growth also was a time when the Jackson Laboratory recognized and assumed its responsibility for the genetics of the mouse by working for the Mouse News Letter and for standardized genetic nomenclature; by establishment of the breeding expansion system involving foundation stocks, pedigreed expansion stocks, and animal resources stocks; and by attention to further characterization of a wider variety of inbred lines and organization of published information on genetically controlled mice through development of a subject-strain bibliography.

The Jackson Laboratory's accomplishments in the fourth and fifth decades are too extensive to be covered in this capsule presentation. Because our selection of five topics necessarily omitted many important facets of Jackson Laboratory contributions, it may be desirable to mention a few other highlights here. Radiation effects were studied extensively in the fourth decade, and an important start was made in biochemical genetics and the study of the genetic control of metabolism. We put forth the greatly expanded second edition of "The Biology of the Laboratory Mouse" (1966), with Earl L. Green as editor. The very large animal resources stocks were utilized for an organized study of mouse spontaneous-mutation rates. Many research advances were also made in our "five topics" in the fourth decade and, of course, even more in the past ten years. We cannot leave this capsule history without mentioning development in the fifth decade of especially valuable new genetic tools. The principles and first establishment of recombinant inbred strains and congenic and coisogenic lines stem from Jackson Laboratory geneticists, and these become more important research tools with every passing day.

It is a great pleasure to have been a part of this Symposium, celebrating research developments in mammalian genetics and cancer as well as the 50th Anniversary of The Jackson Laboratory. We are very grateful to the American Cancer Society and to the National Foundation for their generous support, which helped to make possible this joyous occasion.

> Elizabeth S. Russell
> The Jackson Laboratory
> January 1981

SESSION I. GENE AND CHROMOSOME ORGANIZATION

MARGARET C. GREEN, Chairman

Introduction to Session I

During the period 1929–1979, a great deal has been learned about the genetic makeup and the cytogenetics of the 20 pairs of mouse chromosomes. This information, which has come to have elegant detail, still belongs in the breeding-experiment and light-microscope portion of a broad range of possible information about the mouse genome. A great deal has also been learned about the "midrange" organizational level, about relations between structure and function of many of the named genes recognized in the mouse genome. Very recently, great advances have been made in understanding the *very* fine structure and regulation of function of certain favorable mouse genes. Session I deals only with the two ends of this spectrum — with what is often called formal genetics and with what is called molecular genetics. We leave the entire midspectrum for other sessions.

Session I starts by considering the history of increase in knowledge of the positions of named genes on the genetic linkage map, and its coordination with the cytogenetics map. Dr. Eva Eicher also includes fascinating predictions regarding future linkage studies. This is followed by Dr. Philip Leder's paper, molecular to the nth degree, dealing with extremely fine internal structure of mouse globin genes. The "middle range" of information about mouse genes is not at all covered in this session, where genes are either points on a linkage map or cloned DNA sequences!

So we plunge, in the first paper by Dr. Eicher, into a working cytogeneticist's excellent depiction of the growth, the present state, and the probable future course of studies of mouse formal genetics. I recommend that any nongeneticist readers who might shrink from contemplation of linkage look closely at the six successive versions of the mouse genetic linkage map, as known in 1935, 1945, 1954, 1971, 1975, and 1979, and follow in the text the evolving pattern of acquisition of new technologies and discovery of new mutants that made this rapid progress possible. The first genetic linkage in the mouse was discovered in 1915, but thereafter for a long period additions came very slowly, because the base of information to link to was so small. Prior to 1971, some acceleration resulted from simple increase in the number of recognized mutants already on the map (more "handles" available). Analysis from translocations helped somewhat in earlier studies, but the discovery of chromosome banding and the resultant ease in establishing relations between cytologic segments in chromosome aberrations provided a great boost. Other exciting new techniques appeared in the 1970s, including DNA hybridization in situ; discovery of many new varying genetic loci through isozyme electrophoresis; establishment of synteny through somatic cell hybridization; and analysis of distribution of genes in families of recombinant inbred strains.

The mouse genetic linkage map has now become so crowded that it is difficult to fit new mutants into their proper "slots" in exact order. Thus, one problem for the future is what Dr. Eicher calls "information integration and retrieval."

In mammalian material, no one can tackle the huge segregation tests necessary to line up *all* loci in exact order, with intergenic distances exactly determined. It may be better to designate key or "core" loci and to store and retrieve computerized information on each locus relative to the nearest core locus. Dr. Eicher also outlines techniques that will lead toward assigning genes to particular chromosomal bands. Her presentation of specific cases helps greatly in understanding application of techniques.

Why should all of us be interested in formal genetics? Dr. Eicher's final point has to do with interspecies comparisons of linkage maps, which show considerable conservation of chromosomal segments, maintained intact through the long separate evolutionary paths stemming from a common ancestor to mouse, man, and rabbit. Her table demonstrating very extensive (45 autosomal loci on 17 chromosomes) homology, particularly for loci responsible for enzyme structure and for the arrangement of these loci in similar order in specific chromosome segments, is really exciting.

Why has the organization of chromosome segments, as well as particular genetic loci, been conserved? To answer that question, and to comprehend mammalian evolution, she says we must understand genomic organization thoroughly. I would like to add a physiological note here. Understanding all the intricacies of regulation of mammalian gene expression requires knowing which genes must stay close together through evolution and which *can* be distributed anywhere in the genome. Relationships between interacting genes may be very different in these alternative circumstances.

Perhaps this final point from Dr. Eicher's paper will serve as a link between her study at the gene-chromosome level and Dr. Leder's analysis of fine structure and regulation *within* the genes coding for the α-globin and β-globin chains of mouse adult hemoglobin. These elegant studies, using highly developed recombinant DNA technology, involve cloning of discrete genetic segments, coupled with rapid determination of nucleotide sequences. In 1977, these techniques disclosed that the coding sequence for mouse adult β-minor globin was interrupted at two points by sequences (one relatively short, one long) of nontranslated nucleotides. Electron microscope visualization of this cloned β-minor globin gene annealed to its mRNA clearly showed two regions of close pairing between DNA and RNA, interrupted by the longer of the two loops of unpaired DNA, which even Leder has called "Kilroy's nose." Dr. Leder's present paper tells a fascinating story of further findings about intervening sequences and makes very exciting suggestions about their implications for understanding the organization and evolution of the genes responsible for structures of α-globin and β-globin chains. Both α-globin and β-globin genes have transcribed-but-not-translated sequences inserted at the same points in their sequences, suggesting that *having* such interruptions in hemoglobin DNA sequences is an important feature that evolved more than 500 million years ago. The size of the smaller interruption, inserted between amino acids $\alpha 30$ and $\alpha 31$ or $\beta 31$ and $\beta 32$, is similar for both α- and β-DNA sequences, but sizes and constitutions of the longer insertions have diverged greatly during evolution. The second insertion is much longer in β-DNA sequences than in α-sequences. At least between β-major and β-minor genes, the divergence may have arisen largely by a complicated series of insertions and deletions. The elegant work of Phil and Ada Leder and their highly active research group is producing very rapid progress. Their work provides all of us with loads of very real information and important leads towards understanding both evolution and regulation. The organization of hemoglobin genes, at least, is considerably more complex than we had ever imagined.

Dr. Margaret Green's very neat summary of this session gives an experienced "formal" geneticist's perspective on gene and chromosome organization. She makes it clear that research at both the "gross" and the ultrafine levels must and will continue, and that findings at one level will have implications for the other.

Foundation for the Future: Formal Genetics of the Mouse

Eva M. Eicher

INTRODUCTION

While thinking about what to present for the 50th Symposium, I was struck with the importance of this occasion. My first impulse was to put together a really exciting research seminar showing masses of data points eloquently displayed on graphs and charts. Then a different type of desire overtook the first impulse; I would be representing the geneticists of the Jackson Laboratory — those of the past, whose contributions gave us our foundation, and those of the present, whose contributions are laying the foundation for the future. What I decided to present is a brief history of formal mouse genetics. Into the history I will weave contributions from my laboratory, both past and present, plus what I hope will be accomplishments in the near future. Along the way I will try to weave in contributions from others.

HISTORY OF THE MOUSE LINKAGE MAP

I want to begin the history of formal mouse genetics by looking at the 1979 linkage map compiled by Davisson and Roderick [1] (see Fig. 1). This map represents an accumulation of accomplishments spanning 64 years. As we look closely at the map, we note that 423 gene symbols are organized into numbered lists. Each list, displayed as a black vertical line, represents one of the mouse chromosomes. The varying lengths of the lines represent the relative lengths of the chromosomes at mitotic metaphase. Notice that loci have been assigned to all of the chromosomes, such that even the Y-chromosome contains a locus. A few gene symbols are written at the bottom of several chromosomes, indicating that each of these genes has been assigned to its particular chromosome by methods other than recombination and therefore cannot be specifically located within the linkage group. Because so much information has accumulated in the small area of Chromosome 17 occupied by the *H-2* locus, this region is "blown up" at the bottom to facilitate reading. With time, other chromosomal regions will have to be displayed in this manner. The black knob shown at the top of each gene list (or

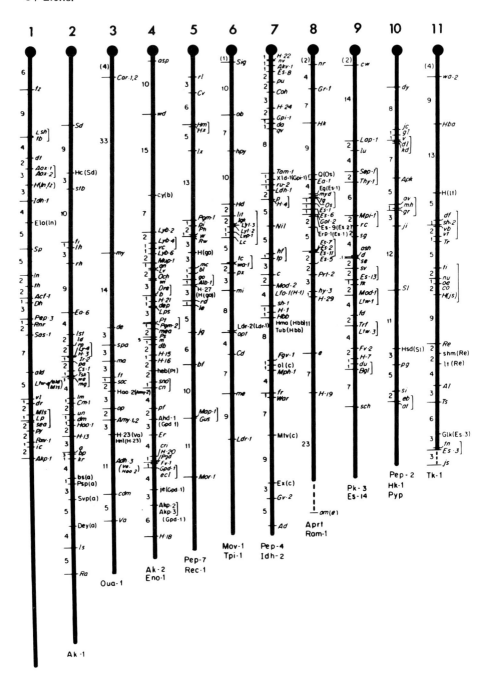

Fig. 1. 1979 Mouse linkage map, courtesy of M. Davisson and T. Roderick [1].

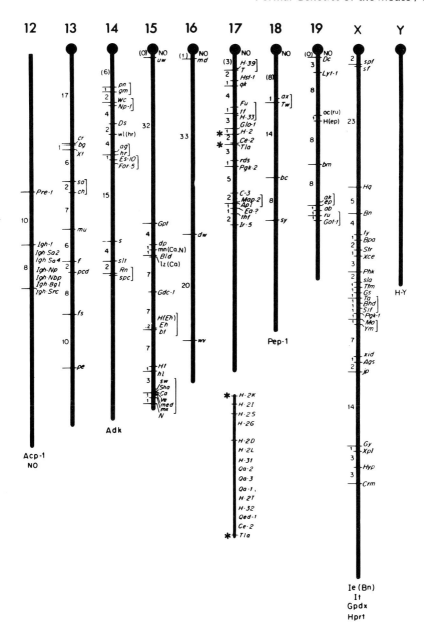

Fig. 1. Continued.

chromosome) marks the centromere position. Finally, numbers appearing to the left of each chromosome represent known distances between the genes, as measured by recombination.

Looking at this most recent mouse linkage map, however, does not convey a sense of the rate at which genetic information about this species has accumulated. To more easily portray this rate of increase, I have arranged this information in graph form (see Fig. 2). It is startling to see that the rate of gene mapping has doubled every ten years since the beginning, in 1915. If we extrapolate to 1985, we can predict that 600 mouse genes will be mapped. When one considers that there are currently over 300 identified genetic loci awaiting linkage assignment, the estimate of 600 mapped genes by 1985 may be conservative.

With this brief glimpse of the present map and a feeling as to how fast we traveled to get to this point, let us look back to the beginning, 64 years ago.

The first mouse gene mapping paper was published by J. B. S. Haldane, A. D. Sprunt, and N. M. Haldane in the Journal of Genetics in 1915 [2]. The authors called attention to the linkage of the albino (c) and pink-eyed dilution (p, called

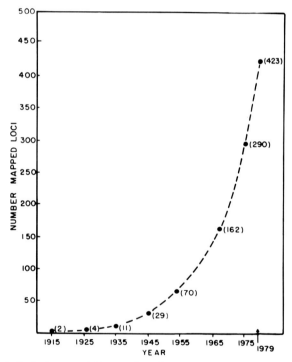

Fig. 2. Graph displaying cumulative number of mouse genes mapped as a function of time.

e by them) loci. Having noticed evidence of genetic linkage, "reduplication" as they put it, between these two loci in segregation data presented in a paper by Darbishire [3], they set up experiments to test their hypothesis of linkage. Because pure breeding lines (inbred strains) of mice were almost unknown in 1915, they were somewhat hampered in their approach since they did not know whether an albino mouse also carried the recessive *p* allele or whether a pink-eyed dilution mouse carried the recessive *c* allele. Nevertheless, they launched into their experiments, used a total of 362 mice, progeny-tested a number of F_2's, and discovered the first linkage group. Hereafter, the linkage groups would be assigned consecutive Roman numerals.

We next move to the symposium honoring the 25th anniversary of the Roscoe B. Jackson Memorial Laboratory. At that symposium Margaret M. Dickie presented a paper entitled "The Expanding Knowledge of the Genome of the Mouse" in which she surveyed the state of the mouse linkage map as of 1954 [4]. Three of her maps are presented in Figure 3. As is evident, by 1935, 13 loci were mapped; by 1945, 26 loci were mapped; and at the time of the 25th symposium, 70 loci were mapped. Those scientists familiar with present knowledge of the mouse genome will probably note that several erroneous linkages were listed in Dickie's 1945 and 1954 maps. These errors will be discussed later.

Probably the most artistically drawn mouse linkage map, and certainly the one that conveyed the most information about the mouse genome, appeared in the September 1945 issue of the Journal of Heredity, the authors being the Staff of the Roscoe B. Jackson Memorial Laboratory [5]. This wonderful map is shown in Figure 4. The photographs or drawings of mutant mice, or in one case a cross section through the retina of a mutant's eye, vividly displayed the phenotype produced by each mutation. No one needed to write a lengthy description to convey each phenotype. There is something sad in no longer being able to convey so clearly to others so much information so efficiently.

The last map using Roman numerals for designating linkage groups was produced by Margaret C. Green in 1971 [6] (see Fig. 5). Without a doubt, Green did more for organizing the mouse linkage map than anyone else. Every year she carefully sifted through newly published data together with recent contributions made in Mouse News Letter, an informal semiannual bulletin, and organized all of this information together with the previously known data into what was called "Marg's Map." Her 1971 map shows 19 linkage groups. Notice that nine of these gene groupings have centromere designations. These centromere locations were determined either by using segregation in mice heterozygous for a Robertsonian translocation, or by what I will call the brute-force method (scientifically called adjacent—2 segregation) using mice heterozygous for a reciprocal translocation. It is a credit to the early centromere placers that all of their assignments have withstood the test of time. I'll say more about centromere mapping later.

12 / Eicher

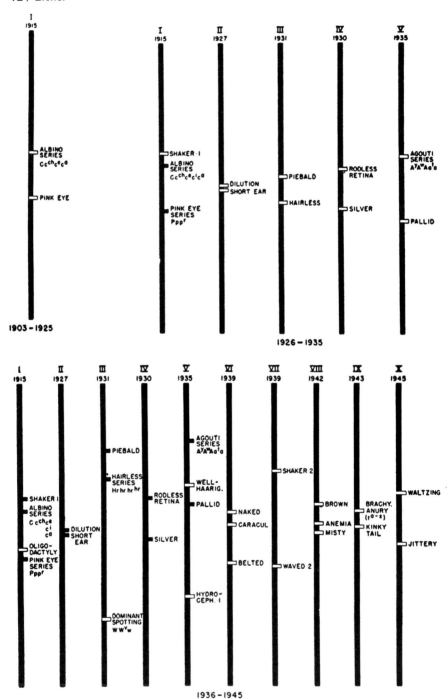

Fig. 3. Pictorial view of the history of mouse linkage prepared by Margaret Dickie for the Roscoe B. Jackson Memorial Laboratory 25th Anniversary Symposium [4].

Formal Genetics of the Mouse / 13

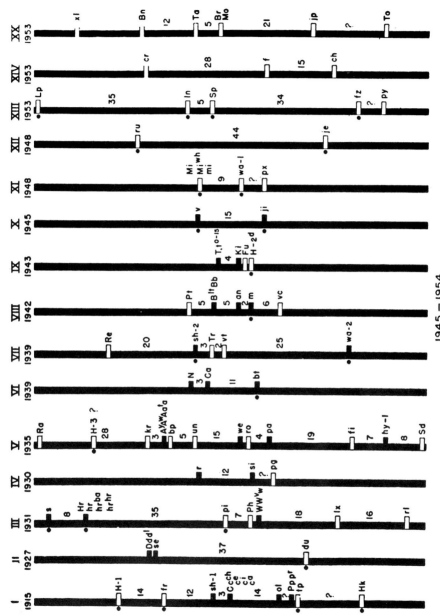

Fig. 3. Continued.

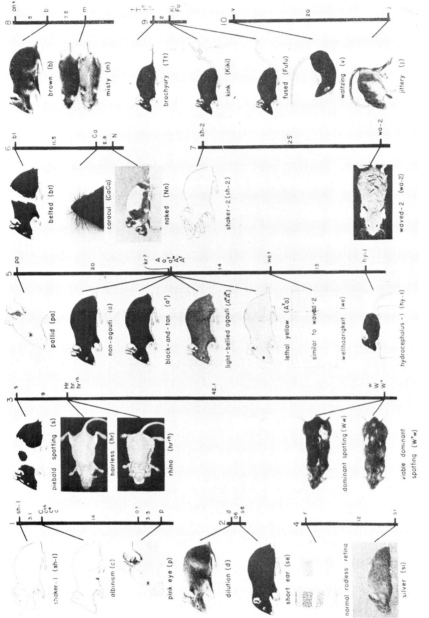

Fig. 4. Linkage map of the mouse prepared in 1945 by the Staff of the Roscoe B. Jackson Memorial Laboratory [5].

If we compare Green's 1971 map with the 25th symposium map of Dickie, we note that the *W* locus, shown to be on Linkage Group III, has, together with other loci, formed a separate linkage group, LG XVII [7]. Also the gene *je* (jerker), shown linked to *ru* (ruby), was later shown to be a part of LG VIII [8]. In addition to these minor corrections, major errors were displayed on this 1971 map. For example, in 1975, N. L. A. Cacheiro and L. B. Russell found, by accident, that LG IV and X were the same [9]. I think the reason no one had previously discovered this error had to do with the linkage testing stocks commonly used. One stock contained the *v* (waltzer) locus. Another contained the *Sl* (steel) locus. Once you found linkage of an unmapped gene to one of these loci, you didn't ask whether your gene also resided in another linkage group. Thus, a group of genes evolved around *v* and another group evolved around *Sl*. Some work still needs to be done in terms of fitting these two groups of loci together.

Around 1970, a revolution occurred in genetics that permanently affected how the mouse map was displayed: Caspersson and colleagues [10] discovered that quinacrine mustard bound to chromosomes in a specific and repeatable manner such that metaphase chromosomes were identifiable under the microscope. Quickly it was reported by four separate groups, Nesbitt and Francke in California, Miller and co-workers in New York, Schnedl in Germany, and Buckland and colleagues in England, that each mouse chromosome, when stained with quinacrine or treated with trypsin or specific salt solutions, banded in a specific pattern [11–14]. The next step in the puzzle was to identify which chromosome carried which linkage group.

The first assignment of a mouse linkage group to a specific chromosome was achieved in the late 1960s in the laboratory of Ernst Caspari at the University of Rochester [15]. I had received from Mary Lyon, in Harwell, England, two mouse translocations, a reciprocal translocation then designated *T(1;12)145H*, and a Robertsonian translocation called *T(2;12)163H*. (At that time, the Arabic numbers used inside the parentheses were the same as the Roman numbers used normally to designate the linkage groups.) Lyon had discovered that these two translocations shared a common linkage group, LG XII [16]. Earlier, Charles Ford had disclosed that specific chromosomes had clear staining regions (secondary constrictions) located near their centromeres [17]. What I noted was that one of these chromosomes with a secondary constriction, Chr 19, was involved in both translocations. Since LG XII was also involved in both, it followed that Chr 19 carried LG XII. Interestingly, in looking back, I remember being very happy at my simple discovery without appreciating the importance of the observation. To be sure, this was a step forward in assigning mouse linkage groups to chromosomes. Unfortunately, because only a few mouse chromosomes displayed secondary constrictions (later to be proven to be the site of ribosomal RNA synthesis [18]), this technique would have allowed only a few more linkage groups to be assigned to specific chromosomes.

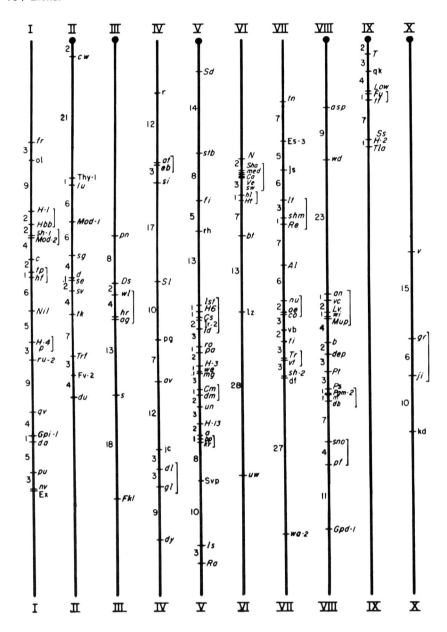

Fig. 5. Linkage map of mouse prepared in 1971 by Margaret Green [6]. This was the last map to display only linkage group numbers, as banding of chromosomes, thus assignment of linkage groups to specific chromosomes, had begun.

Formal Genetics of the Mouse / 17

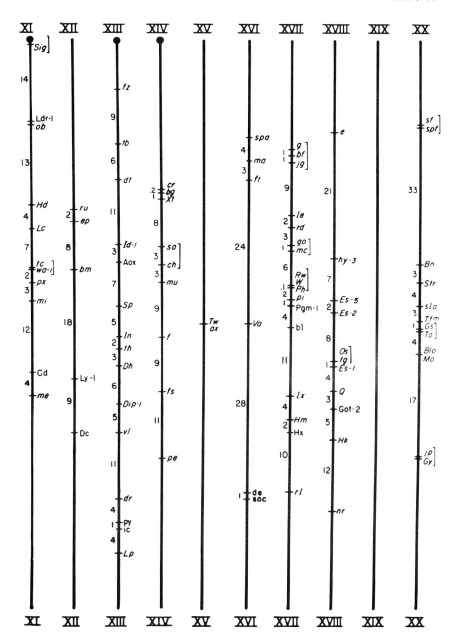

Fig. 5. Continued.

I do want to say that in the late 1940s and early 1950s two individuals were trying to construct male meiotic pachytene maps – Allen Griffen [19] and B. Slizynski [20]. It is possible that their efforts eventually would have led to the accurate assignment of mouse linkage groups to specific pachytene chromosomes, although probably much later than 1971. I am especially sorry that Allen Griffen did not live to see this dream of his come true.

In Table I, I have provided a listing of who found each linkage group and who assigned these linkage groups to specific chromosomes. One important aspect of assigning the linkage groups to specific chromosomes cannot be appreciated by studying the table. Two mouse geneticists, Mary Lyon and Tony (A. G.) Searle, had produced a large number of inherited translocations and had assumed the responsibility for seeing that these, together with the Carter, Lyon, and Phillips series of translocations [21], were not lost. In addition, they had diligently identified the linkage groups involved in most of these translocations. Remember, just perpetuating these translocation stocks before blood culture techniques were invented was no easy task. Finding the linkage group assignment for these translocations required a love for mouse genetics coupled with a belief in its future. On top of this, these two geneticists freely shared not only all of their unpublished information but also their mice. And with no strings attached! In my opinion, they are the heroine and hero of this era of mouse genetics because without their hard work and the sharing of their knowledge and valuable genetic stocks, mouse genetics could not have advanced to where it is today.

The logic used to assign mouse linkage groups to specific chromosomes was the same as I described for assigning LG XII to Chr 19. One looked at the banded chromosomes from individuals heterozygous for translocations sharing linkage groups. For example, one studied the chromosomes from an individual carrying the *T(2;9)138Ca* translocation and from an individual carrying *T(9;13)190Ca* (Arabic numbers designate linkage groups). Note that both translocations involve LG IX. In the case of the *T138Ca/+* individual, Chr 9 and 17 were involved. In the case of the *T190Ca/+* individual, Chr 1 and 17 were involved. Since both translocations involved Chr 17 and LG IX, it followed that LG IX was carried on Chr 17. It also followed that LG II was carried on Chr 9, because that was the other chromosome involved in the *T138Ca* translocation, and that LG XIII was carried on Chr 1 because that was the other chromosome involved in *T190Ca*. One could check these logical deductions by studying chromosomes from individuals carrying translocations involving LG II or XIII. By so doing, the linkage groups were assigned to specific chromosomes (see early review by Miller and Miller [22]).

One of the confusing aspects of the first mouse banding studies was that each investigative group had its own numbering system for designating the chromosomes. For example, the chromosome called number 5 by one group was called number 3 by one other group and 6 by yet a third group. These numbers, together

TABLE I. Assignment of Genes to Linkage Groups and Linkage Groups or Genes to Chromosomes

Reference (LG)	Genes	Linkage Group	Chromosome	Reference (Chr)
Haldane et al, 1915 [2]	p-c	I	7	Francke and Nesbitt, 1971 [76]; Kouri et al, 1971 [84]
Gates, 1927 [77]; Snell, 1928 [96]	d-se	II	9	Miller et al, 1971 [13]; Nesbitt and Francke, 1971 [91]
Snell, 1931 [97]	hr-s	III[a]	14	Eicher and Green, 1972 [72]
Keeler, 1930 [82]	r-si	} IV	10	Cacheiro and Russell, 1975 [9]
Falconer and King, 1953 [75]	si-pg			
Roberts and Quisenberg, 1935 [94]	pa-a	V	2	Miller et al, 1971 [89]
Cooper, 1939 [66]	Ca-N	VI	15	Eicher and Green, 1972 [72]
Snell and Law, 1939 [99]	wa-2-sh-2	VII	11	Cattanach and Moseley, 1973 [64]
Hertwig, 1942 [81]	an-b	VIII	4	Miller et al, 1971 [90]
Reed, 1937 [93]	T-Fu	IX	17	Miller et al, 1971 [13]
Snell, 1945 [98]	v-ji	X	10	Miller et al, 1971 [90]
Bunker and Snell, 1948 [63]	wa-1-Miwh	XI	6	Miller et al, 1971 [13]
Deol and Lane, 1966 [69]	Dc-ep	} XII	19	Eicher, 1971 [15]; Lyon and Glenister, 1971 [88];
Lane and Green, 1967 [85]	ep-ru			Miller et al, 1971 [13]; Nesbitt and Francke, 1971 [91]
Dickie and Woolley, 1950 [71]	fz-vn	XIII	1	Miller et al, 1971 [13]
King, 1956 [83]	cr-f	} XIV	13	Cattanach et al, 1972 [65]
Phillips, 1956 [92]	ch-f			
Lyon, 1958 [87]	Tw-ax	XV	18	Searle and Beechey, 1973 [30]
Curry, 1959 [67]	Va-de	XVI	3	Davisson et al, 1976 [68]; Eicher et al, 1976 [73]
Dickie and Woolley, 1946 [70]	pi-W^v	XVII[a]	5	Eicher and Lane, 1980 [33]
Green and Sidman, 1962 [79]	tg-Os	} XVIII	8	Cattanach et al, 1972 [65]
Green et al, 1963 [80]	Hk-Os			Miller et al, 1971 [89]
Falconer, 1953 [74]	Ta-Mo	XX	X	Miller et al, 1971 [13]; Schnedl, 1971 [14]; Francke and Nesbitt, 1971 [12]
Lane and Sweet, 1974 [28, 86]	dw-wv	—[b]	16	Eicher and Beamer, 1980 [29]; Roderick et al, 1976 [26]
Taylor et al, 1979 [34]	Igh-1-Pre-1	—[b]	12	Eicher et al, 1979 [35]

Note: Some references for assignment of genes to linkage groups were obtained from Robinson [95] and Green [78].
[a]Originally the genes on LG III were thought to be linked to the W locus. Lane in 1967 [7] showed that the W locus together with other loci belonged to a separate linkage group (LG XVII).
[b]These gene groupings were never given linkage group numbers.

with the linkage group numbers, created confusion as one always had to remember which numbering system was being used. Clearly, something had to be done.

In 1972 the Committee for Standardized Nomenclature of the Mouse proposed a chromosome numbering system that was welcomed and accepted by everyone [23]. The last part of the story was put together when Nesbitt and Francke [24] proposed a nomenclature system for banded mouse chromosomes that was both usable and, of even greater importance, open-ended so it would not become obsolete. Their nomenclature system is shown in Figure 6. Finally, because mouse geneticists could communicate about specific chromosomes and specific chromosomal regions, the level and rate of communication increased. The difficult part for most of us was to memorize which linkage groups belonged on which chromosome. Some of us are still struggling.

Now we move to a 1975 map compiled by Margaret Green [25] (Fig. 7). As expected, the Roman numerals have disappeared from the top of the map to appear out of order at the bottom. Replacing them are Arabic numbers, in numerical order, representing the banded mitotic chromosomes. From looking at the map one would conclude that genes had been assigned to all mouse chromosomes, except Chr 16, and that the centromere location had been determined for all gene groups, except Chr 18 (and of course, Chr 16). A few surprises, however, were yet to come. Let me illustrate what progress was made between 1975 and now, together with the "surprises," by returning to the 1979 map (Fig. 1).

One advancement was the assignment of the gene *md* (mahoganoid) to Chr 16 by Roderick et al [26]. Following this *dw* (dwarf), a gene discovered 50 years ago by George D. Snell [27], and *wv* (weaver) were shown by Lane and Sweet [28, 86], to be linked to each other and subsequently were mapped to Chr 16 [29]. Chr 18, long awaiting genes, was hypothesized by Searle and Beechey [30] to carry the *Tw* (twirler) and *ax* (ataxia) genes. Their hypothesis has now been shown to be correct [31].

The linkage map "surprise" involved Chr 3 and Chr 12. For years mouse geneticists had wondered: Where were the many genes that belonged on the long Chr 3? Finally, the answer came when Priscilla W. Lane and I showed that the groups of genes previously thought to be on Chr 12 were, in fact, not there [32] but on Chr 3 [33]. Now we all worried about where the genes on Chr 12 were hiding. Solving this mystery unexpectedly provided a solution to still another puzzle in formal mouse genetics. For years various scientists "had a go" at trying to find the linkage of the immunoglobin heavy chain locus (*Igh*). The first breakthrough came when Taylor and colleagues [34] found linkage between the *Pre-1* (prealbumin-1) locus and *Igh*. When it became known that the genes thought to be on Chr 12 were instead on Chr 3, it was immediately apparent that Chr 12 had never been tested for the residence of *Igh* or *Pre-1*. Ben Taylor, Roy Riblet, and I joined forces to test this possibility. We found that both loci were linked to a reciprocal translocation involving Chr 12, called *T(5;12)31H*, and thus were able to show

Formal Genetics of the Mouse / 21

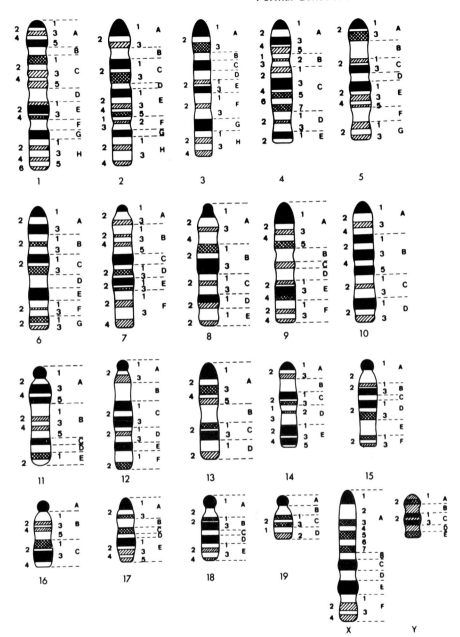

Fig. 6. Pictorial representation of banding nomenclature system proposed in 1973 by Muriel Nesbitt and Uta Francke [24]. The chromosomes are divided into major regions as designated by a capital letter. These regions are further subdivided into regions (a major band) as designated by a number. Thus, 19D2 refers to the last band of Chromosome 19.

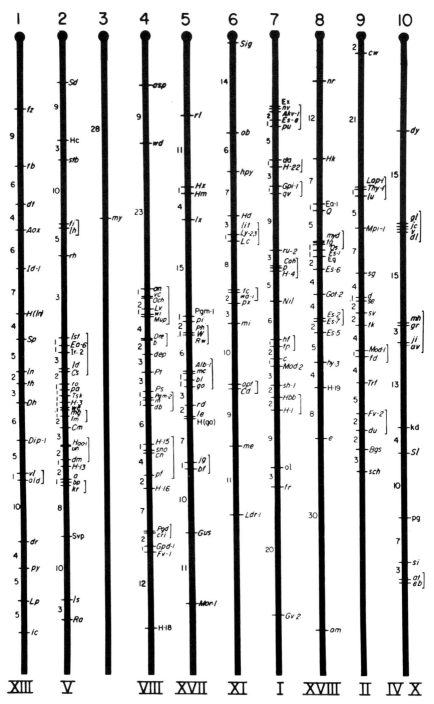

Fig. 7. 1975 mouse linkage map compiled by Margaret Green [25]. Note that all the genes shown on Chromosome 12 were later found to be on Chromosome 3.

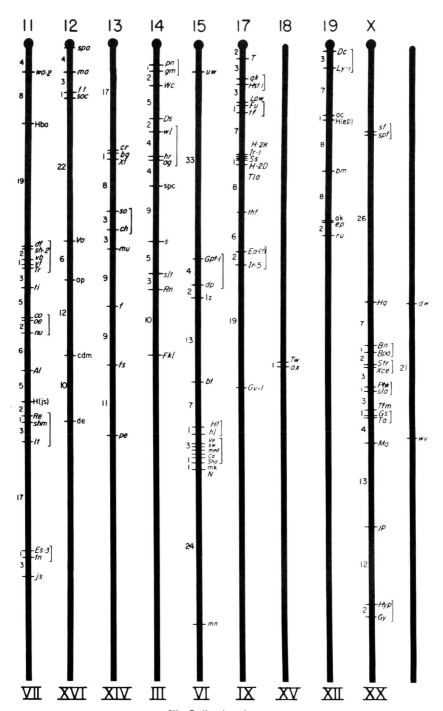

Fig. 7. Continued.

that both *Pre-1* and *Igh* resided on the distal half of Chr 12 [35]. At last, the notorious *Igh* complex had found a home. And finally, all mouse chromosomes contained at least one gene. I hope there are no more surprises.

SPECIAL LINKAGE APPROACHES INTRODUCED DURING THE LATE 1960s AND EARLY 1970s

Before I begin to speculate as to where formal mouse genetics will go in the next few years, I would like to acknowledge four successful approaches used in the 1970s to find and map mouse loci. These are the techniques of DNA hybridization in situ, isozyme electrophoresis, somatic cell hybridization, and recombinant inbred strains.

Around 1960–1970, a very important technical breakthrough came to mouse cytogenetics — hybridization of DNA to chromosomes (DNA hybridization in situ). Pardue and Gall [36] and Jones [37] demonstrated that the A-T-rich satellite DNA of the mouse was situated near the centromeres of all chromosomes except the Y chromosome. Following this, similar techniques were used to show that the secondary constrictions located near the centromeres of Chr 12, 15, 16, 18, and 19 were the sites of ribosomal DNA (rDNA) [18, 38]. What differs among chromosomes, and even between homologous chromosomes of mice from different strains, is the number of sites per chromosome [39]. Thus the technique of DNA hybridization in situ allowed RNA-producing genes or specific types of DNA to be localized onto a chromosome for the first time. Other such types of mouse genes await mapping by this technique.

Probably the most productive technique used in the 1970s for finding new loci and aiding in the mapping of old loci was isozyme electrophoresis [see review by Hutton, 40]. The people who come to mind when thinking about the pioneering search for mouse isozyme variants are Tom Roderick, Ray Popp, Ken Paigen, John Hutton, Frank Ruddle, Robert Selander, Tom Shows, and Verne Chapman. Later, other names joined this endeavor: Peter Nash, Jo Peters, Elizabeth Nichols, Rosemary Elliott, T. Watanabe, Roger Holmes, Jim Womack, and myself, to name but a few. I want to pay special credit to Verne Chapman for his discovery that wild subspecies of Mus, such as Mus mus castaneus and Mus mus molossinus, could provide, at many isozyme loci, the genetic variation missing in laboratory-inbred strains [41]. In fact, when one thinks of the castaneus or molossinus genome as a linkage testing stock, the reason some of us have been so successful at finding linkages can be more easily understood. For example, since 1972 we have either mapped to a chromosome, or more exactly, positioned on the map, more than 50 genes. Most of this was accomplished because of the availability of the isozyme differences between castaneus and standard mouse strains.

The above comments, of course, relate to electrophoretic variants found between individuals and used in standard genetic recombination mapping. The technique of isozyme electrophoresis is equally powerful when used to map genes using differences between species, as is done in somatic cell hybridization mapping.

I would like to acknowledge some of the people who have contributed to our knowledge about the mouse genome, using somatic cell genetic techniques (see references listed in Table II). They are Frank Ruddle, Peter Lally, Uta Francke, Christine Kozak, John Minna, and Carl Croce. This exciting approach to mapping the mouse genome is only beginning.

Finally, I want to mention a technique for mapping mouse genes that was introduced in the 1970s by Donald Bailey [42] and so beautifully perfected by Ben Taylor [43]. This is the use of recombinant inbred (RI) strains. As is true with many techniques, RI strains were first produced to solve a specific problem — the finding and mapping of non-*H-2* histocompatibility genes. Taylor saw this as a very powerful tool for quickly and efficiently finding and mapping genes that were almost impossible to deal with using conventional backcross or F_2 systems: genes that affected developmental parameters, such as enzyme levels in embryonic tissues; metabolism parameters, such as drug resistance; or susceptibility to specific types of cancers. I think the world has just started to appreciate the genetic power of the Recombinant Inbred Strain system for unraveling previously "complex" genetic systems.

THREE FUTURE PROBLEMS

Now let us turn our attention to the future of formal mouse genetics. Because of my limited vision, I can only address the immediate future and will concentrate on what I hope my laboratory will contribute.

Among a number of problems to be addressed by mouse geneticists, solutions to three seem most urgent. One is the problem of how to integrate more efficiently and accurately new linkage data and old linkage data into a more usable map. I call this the "information integration-retrieval" problem. The second problem concerns determining the centromere-to-nearest-gene distance for each chromosome. This can be called "anchoring of the linkage map onto the chromosome" problem. The third problem concerns fitting the linkage map to the banded chromosome, or, as I will call it, the "gene-to-band relationship" problem. I will now discuss more fully each of these problems and tell you the approach we are taking toward their solutions.

Information integration-retrieval. The first problem involves the need for the mouse geneticist to have available an accurate, easily updatable linkage map. As previously mentioned, this problem was addressed by Margaret Green, during the past decade. She assembled linkage information references, screened the data, and each spring integrated the data into an updated linkage map. The newly drawn map appeared in the Jackson Laboratory Annual Report and Mouse News Letter, and copies of it were distributed to all interested scientists.

The drawing of a fairly accurate mouse linkage map was not too difficult in the 1960s and early 1970s because a limited number of genes were available for the few "mappers" to shuffle. As mouse genetics became more popular, however, the rate of discovery of new linkage information increased. One misunderstanding developed about the map and is still with us. Margaret thought of her endeavor as a "working map." Unfortunately, many regarded it as the "gospel." This misunderstanding was further complicated because a set of references was not distributed with the map. Consequently, it was not clear which gene orders had been assumed by utilizing two-point cross information and which had been determined by three-point cross information. Unfortunately, many map users chose to "quote" the map rather than search for and review the linkage data relevant to their needs. The outcome has been that the "mappers" often are not credited for their work and assumed gene order errors have persisted.

When Margaret Green retired in 1976, the job of compiling the map was accepted by Jim Womack. Recently, this task was passed to Muriel Davisson and Tom Roderick. Not only are they confronted with the task of determining which of the previous orders are correct and which need verification, but also there is a flood of new gene order information. I believe the time has come for mouse geneticists to accept nonhuman help in compiling the linkage map. I recommend the computer for this task.

Janan Eppig, a graduate student in my laboratory, and I are putting together a computer program that will be able to assemble all of the past, present, and future published mouse linkage information and produce at any time an up-to-date linkage map. A general outline of the program is as follows.

The program is organized into three parts. The first part involves entrance into the computer of raw linkage data obtained from an individual reference. Included for each set of data is the mating used to produce the heterozygous parent(s) and the type of cross used to obtain the segregating progeny (backcross, intercross, repulsion, coupling, etc). Information regarding gene penetrance and genotype viability is also included. The recombination-calculation part of the program then computes the recombination distance for all of the segregating loci, two at a time. Finally, the genetic distances for two genes are computed using all data presented in the reference. Thus we can think of part one of the linkage map program as providing for each reference an analysis of each cross and a summary of all presented data. The second part of the program integrates the data from all references involving two genes and produces an average distance. The third part of the program uses the distances obtained from part two to place the genes in order with genetic distances between them. That is, it produces the linkage map.

A number of practical uses are available from this computer mapping program. For example, one can enter newly acquired raw data into part one of the program and have the computer integrate them into part two (all previously published data) so as to obtain an updated distance for two genes. Another practical use is to have the computer list all references pertaining to a given gene or two genes (a convenient way to search for references involving a given locus, chromosome, or chromosomal region). Of course, you can ask the computer to display the crosses made and data obtained from a given reference. A very important aspect of the whole program is that it can, for any group of loci, signify which gene orders were determined by three-point crosses and which by two-point orders. This is critical because an incorrect gene order may be deduced if one uses only data from two-point crosses. Finally, you can ask the computer to show how the map looked in different years, making an historical approach possible. Janan and I see our endeavor only as a beginning. Certainly, someone else will think of things that have not occurred to us. A solution to the handling of past and present mouse linkage data must come in the next few years.

Anchoring the linkage map onto the chromosome. The second problem facing the mouse geneticist is what I call the anchor problem. This problem consists of determining the genetic distance from the centromere of a linkage group to a closely linked gene. This is part of a larger problem of how to fit the linkage map onto the banded chromosome. The first step in solving the anchoring problem is to determine on which end of the linkage group to place the centromere. It must be remembered that laboratory mouse chromosomes are acrocentric; that is, their centromeres are at the end of the chromosome.

Quite early in the history of formal mouse genetics, George Snell became interested in determining the centromere ends of the linkage groups. The date was 1946. The method he used was a most complicated one, analysis by adjacent-2 segregation. Of historical importance is the fact that he accurately assigned the centromere of Linkage Group V (now Chr 2) to its *Sd* (Danforth short tail) end and the centromere for Linkage Group VIII (now Chr 5) to its *wd* (waddler) end [44].

A number of methods have been used to assign the centromere end to each mouse linkage group. Only three of these methods, however, provide a centromere-to-gene distance: segregation of a Robertsonian translocation, segregation of centromere-related heterochromatin, and analysis of ovarian teratomas. Robertsonian translocations have been the most frequent tool used to obtain centromere-to-gene distances. I am suspicious of the results, however, because Robertsonian translocations probably affect crossing over near the centromere since they are notorious for affecting crossing over in other chromosomal regions [45]. The use of quantitative differences in the amount of centromere-related heterochromatin to compute accurate centromere-to-gene distances is open to question because there are no available controls. Objections to both methods can be overcome using ovarian teratoma mapping.

In 1969 Linder and collaborators showed that human ovarian teratomas, also known as dermoid cysts, are derived from germ cells that have completed the first but not the second meiotic division (see Fig. 8) [46–48]. As expected, these teratomas are diploid. Because of the biological fact that homologous centromeres segregate at the first meiotic division, these teratomas are homozygous for all loci located near each centromere, unless a single crossing over event occurs between the centromere and the locus. Only in this case will they be heterozygous for the locus (see Fig. 8). Thus the origin of ovarian teratomas makes them the preferred tools for estimating a centromere-to-gene distance. One simply analyzes a number of teratomas derived from females that are heterozygous for the locus in question, and determines whether each has a heterozygous or homozygous genotype. To compute the genetic distance from the centromere to the locus, the formula Y = AB/2 is used, where Y is the percentage recombination and AB is the frequency of heterozygous tumors [49]. The reason we must divide the frequency of heterozygous teratomas by two is that one of the two chromosomes in a heterozygous tumor is a parental chromosome and one is a recombinant chromosome (see Fig. 8). The only assumptions we must make are that Y is less than 33 and that complete interference takes place. How do we use this beautiful system, developed in man, to get at the problem of centromere-to-gene distance in mice?

Normally ovarian teratomas are very rare in mice. Of significance to our problem of getting the linkage groups of the mouse anchored onto the map was the finding by Leroy Stevens and Don Varnum of the Jackson Laboratory that ovarian teratomas are common in females of the LT/Sv strain (a recombinant inbred strain derived from the C58 and BALB/c strains) [50]. In order to better understand the inheritance of this form of tumor, Stevens constructed a number of recombinant inbred strains using LT/Sv and C57BL/6J as the progenitor strains. He found that not only did females of one of these recombinant strains, LTXBJ, have a high frequency of ovarian teratomas, but so did 50% of the (LT × LTXBJ)F_1 females. Of interest to us is the fact that LT/Sv carries the $Gpi-1^a$ allele at the glucose phosphate isomerase-1 locus and the LTXBJ strain carries the $Gpi-1^b$ allele derived from C57BL/6J. Thus (LT × LTXBJ)F_1 females are heterozygous for the $Gpi-1$ locus carried on Chr 7. In collaboration with John Eppig, Leslie Kozak, and Leroy Stevens at the Jackson Laboratory, we analyzed 23 teratomas from (LT × LTXBJ)F_1 females. Ten were homozygous $Gpi-1^a$, 11 homozygous $Gpi-1^b$, and two heterozygous $Gpi-1^a/Gpi-1^b$. Since it was known that, like human ovarian teratomas, mouse ovarian teratomas are diploid, our results were easiest to interpret if we hypothesized that these tumors, as in man, arise from germ cells that have undergone the first meiotic division but have failed to complete the second division [51]. The two heterozygous tumors would have resulted from germ cells in which recombination occurred between the centromere and the $Gpi-1$ locus (see further discussion in Eicher [52]). To date Janan Eppig and I have analyzed 276 ovarian teratomas derived from $Gpi-1^a/Gpi-1^b$ females. Sixty-two of these were GPI-1AB. Thus for mouse Chr 7 the centromere-to-$Gpi-1$ gene distance is 11 map units.

The approach we are taking to determine the centromere-to-gene distances for the rest of the mouse chromosomes is to place, by repeated backcrossing, specific isozyme alleles from other strains and lines onto the LT/Sv inbred background. We have chosen isozyme loci because the isozyme constitution of tumors can be readily determined. Our experience is that, after 3–4 generations of backcrossing, females of the incipient LT/Sv congenic lines develop ovarian teratomas. These lines will enable us to determine the centromere-to-gene distances for a number of chromosomes.

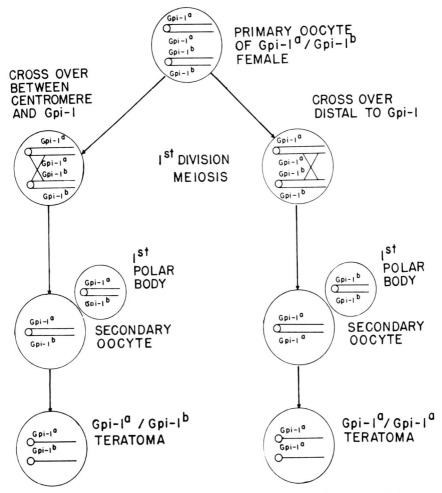

Fig. 8. Origin and genetic constitution of ovarian teratomas. The primary oocyte is from a $Gpi\text{-}1^a/Gpi\text{-}1^b$ female. After the first division of meiosis is completed, the resulting secondary oocyte carries two $Gpi\text{-}1^a$ alleles (two chromatids attached to centromere with $Gpi\text{-}1^a$ allele on both) *if* no recombination has taken place between the centromere and the $Gpi\text{-}1$ locus and both $Gpi\text{-}1^b$-containing chromatids went to the polar body, *or* the secondary oocyte contains both a $Gpi\text{-}1^a$ and a $Gpi\text{-}1^b$ allele (as does the polar body) *if* recombination occurred between centromere and $Gpi\text{-}1$ locus. The second polar body is either not formed or is reabsorbed into the egg resulting in a diploid teratoma.

Previously I criticized the use of Robertsonian translocations and centromeric-related heterochromatin for computing centromere-to-gene distances. Will teratoma mapping also introduce distance biases? As yet we do not know. Of importance, however, is the fact that we can introduce controls into our experiments. For example, teratomas can be used to determine gene-to-gene distances by simply making teratoma-producing females heterozygous for two linked loci. We can thus compute, in a three-point cross, the centromere-to-gene and gene-to-gene distances. Our control will be to compare the gene-to-gene distance obtained from teratoma data to that obtained by using similar females in a standard backcross. If the gene-to-gene distances are the same, it will be safe to conclude that teratoma-derived centromere-to-gene distances are correct.

A final note seems worthwhile in relation to ovarian teratomas. Since the chromosome constitution of ovarian teratomas is really that of 20 half-tetrads, half-tetrad analysis, thought possible only in Neurospora and Drosophila, is now possible in mice and man [52].

Band-to-gene relationship. The last problem I shall address involves understanding how the mouse genome is organized relative to the banded chromosomes. Two specific questions are: How are Mendelian genes spaced along the chromosome? Does genetic recombination take place randomly along each chromosome? The simplest approach to answering these and other questions is duplication-deficiency mapping. Because the chromosomal aberration most likely to be used for duplication-deficiency mapping is the reciprocal translocation, I shall introduce this type of translocation before discussing the mapping method.

A reciprocal translocation is what its name implies: The mutual exchange of material between chromosomes. Although reciprocal translocations can involve more than two chromosomes, the one we will discuss involves just two chromosomes.

We can think of a translocation as having permanently bisected a group of physically related genes. This cut, or breakpoint as it is called, can be determined either as a position on a banded chromosome or as a position within a linkage group. If we determine both the cytological and genetic breakpoints for one of the chromosomes of a specific translocation, you can see that we have defined which part of the linkage group resides between the cytological breakpoint and the centromere (the proximal part) and which part resides between the cytological breakpoint and the telomere (the distal part).

The easiest breakpoint to establish is the cytological one. One compares G-banded chromosomes, juxtaposing the two translocated and the two normal chromosomes, in a translocation carrier and determines where the "breaks" occur in the normal chromosomes to generate the two translocation products. The next step is to determine the genetic breakpoint.

In conventional crosses involving the mating of a mouse heterozygous for a reciprocal translocation to a normal mouse, the reciprocal translocation behaves as a dominant gene. Thus one can treat the translocation breakpoint as a dominant gene and locate it relative to other genes along the linkage group. If you now determine the cytologic and genetic breakpoints of a series of translocations involving a specific chromosome, you see that the associated linkage group can be divided into many regions whose boundaries are the genetic breakpoints of the translocations and that each of these gene groups is assigned to specific chromosomal regions whose boundaries are the cytological breakpoints. Recently Searle, Beechey, Eicher, Nesbitt, and Washburn [53] used this technique to partition Chr 2 into nine specific regions, as seen in Figure 9.

Theoretically, this approach could be used to partition a banded chromosome into smaller and smaller gene groupings. Unfortunately, there is a basic flaw in this approach. Reciprocal translocations are notorious for inhibiting recombination near their breakpoints. The consequence is that recombination can become so rare near a breakpoint that either a) the locus of interest no longer crosses over with the breakpoint, or b) when you find an informative individual you may not be able to prove that it was classified correctly (a disaster if the backcross young have to be killed to determine their phenotype). Thus, the genetic location of the *T11H* breakpoint is probably correctly located between *fi* (fidget) and *pa* (pallid). Whether *T11H* is closer to *pa* than *fi*, however, remains unknown since in the presence of *T11H* the entire region from *fi* to *pa,* normally occupying 26 map units, may be reduced to only a few map units. This example clearly illustrates that the standard method of determining the genetic breakpoint of a translocation by recombination allows us to "zero in" on the region where the breakpoint resides but does not allow us to determine the genetic position.

Does this mean that reciprocal translocations cannot be used to solve the problem of gene-chromosome organization? No. There is one beautiful fact about reciprocal translocations that, until recently, has been largely ignored. Individuals heterozygous for a reciprocal translocation not only produce genetically balanced gametes that give rise to live-born offspring, but they also produce genetically unbalanced gametes that result in chromosomally unbalanced offspring. It is this latter aspect of reciprocal translocations that provides us with the solution to our problem of fitting the linkage map to the banded chromosome. Let me illustrate using a specific reciprocal translocation *T(7;19)145H,* hereafter *T145H.*

A banded karyotype prepared from an individual heterozygous for *T145H* is shown in Figure 10. I have designated the two normal (nontranslocation) chromosomes as 7 and 19, and the two reciprocal translocation-containing chromosomes as 7^{19} and 19^7, where the base number represents centromere origin and the superscript number represents telomere origin. If chromosomes 7 and 19 or 7^{19} and 19^7 from a *T145H*/+ individual are incorporated into a gamete, the resulting fetus would be either normal (7, 7, 19, 19) or a translocation carrier (7, 7^{19}, 19^7, 19), respectively. If, however, a gamete receives Chr 7 and 7^{19}, or 19 and 19^7, or 7 and 19^7, or 7^{19} and 19, the fetus will be chromosomally unbalanced. These examples of chromosome distribution are products of what we call two-by-two segregation (two of the four involved chromosomes move to each pole during division). In many translocations, three-by-one segregation is also common. Usually 50% or more of the gametes from individuals heterozygous for reciprocal translocations are chromosomally unbalanced, resulting in chromosomal imbalance in 50% of the young. Because most of these individuals die in utero, the litter size from translocation heterozygotes is reduced by half.

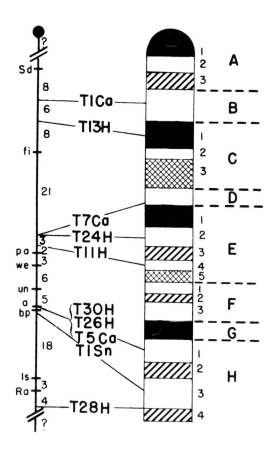

Fig. 9. Cytogenetic map of Chromosome 2 [53]. The genetic breakpoints displayed on the linkage map for 8 of the 10 different translocations listed were determined by recombination. The breakpoint of *T26H* is assumed to be at the *a* (non-agouti) locus and the breakpoint for *T1Sn* relative to *bp* (brachypodism) was determined by duplication-deficiency mapping. Chromosomal breakpoints were determined from G-banded metaphases. More details can be found in original paper.

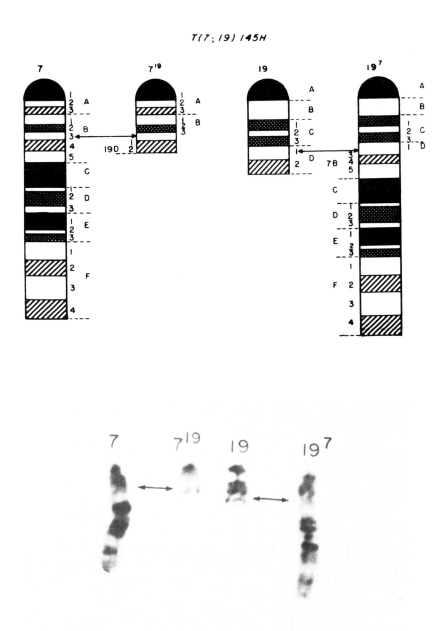

Fig. 10. Idiogram of *T145H* translocation together with G-banded karyotype from *T145H*/+ individual.

Let us look at a cross that was used to determine the genetic breakpoint relative to the Chr 7 loci *Gpi-1* (glucose phosphate isomerase-1) and *p* (pink-eyed dilution) (see Fig. 11) [54]. Matings were constructed such that $T145H/+$ females carried the Gpi-1^b and the recessive p^{2J} alleles on their normal Chr 7 and the Gpi-1^a and p^+ alleles on the translocated chromosomes. The females were mated to homozygous Gpi-1^b and p^{2J} nontranslocation carrier males. Near-term fetuses from this mating were examined for their *Gpi-1* and *p* genotypes and chromosomal constitution. Although more than one type of chromosomal unbalance were recovered, we will concentrate on those individuals that had 41 chromosomes, including two nor-

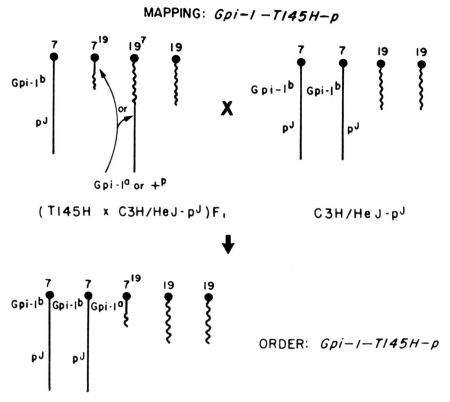

Fig. 11. Duplication-deficiency mapping of position of *Gpi-1* and *p* loci relative to *T145H* breakpoint. Note that individuals with 41 chromsomes (7, 7, 7^{19}, 19, 19) were Gpi-$1^b/Gpi$-$1^b/Gpi$-1^a (see Fig. 12) and had nonpigmented eyes. Thus the *T145H* breakpoint is located distal to the *Gpi-1* locus and proximal to the *p* locus.

mal 7s, two normal 19s, and the 7^{19} translocation product. The eyes of these fetuses were nonpigmented, indicating they had *not* inherited the p^+ allele from their mother. Their GPI-1 phenotype reflected neither a *Gpi-1ª/Gpi-1ᵇ* nor *Gpi-1ᵇ/Gpi-1ᵇ* genotype, but rather what could only be interpreted as resulting from a *Gpi-1ª/Gpi-1ᵇ/Gpi-1ᵇ* genotype (see Fig. 12). As is evident, this *Gpi-1* genotype was only possible if they had inherited *both* a *Gpi-1ª* and *Gpi-1ᵇ* allele from their mother. Since they inherited from their mother a normal 7 and the translocation product 7^{19} and both the *Gpi-1ª* and *Gpi-1ᵇ* alleles, we can conclude that the *Gpi-1* locus is located on 7^{19}, specifically between the centromere and breakpoint of *T145H*. Since the fetuses also inherited a p^{2J} allele and not the p^+ allele from their mother, as evidenced by eye color, we can conclude that the *p* locus is located on Chr 19^7 (the translocation product they did not inherit). Thus we have separated the linkage group of Chr 7 into two parts relative to a cytological band on the chromosome, represented by the *T145H* breakpoint. The beauty of the method we just used to accomplish this result, called duplication-deficiency mapping, is that it allowed us to place genes either proximal or distal to the *T145H* translocation breakpoint with-

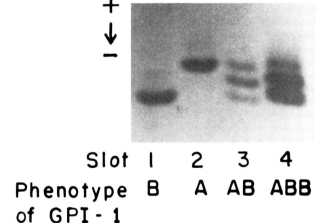

Fig. 12. Photograph of cellulose acetate electrophoresis gel stained for GPI. Samples in slots 1–4 are liver lysates from 1) a *Gpi-1ᵇ/Gpi-1ᵇ* individual, 2) a *Gpi-1ª/Gpi-1ª* individual, 3) a *Gpi-1ª/Gpi-1ᵇ* individual, and 4) a chromosomally unbalanced offspring (7, 7, 7^{19}, 19, 19) from *T145H*/+ female. This gel phenotype is compatible only with a *Gpi-1ª/Gpi-1ᵇ/Gpi-1ᵇ* genotype.

out relying on recombination. It is now easy to imagine that we could use a number of translocations scattered along Chr 7 to divide the linkage map into smaller and smaller regions relative to specific banded chromosomal regions.

Figure 13 depicts Chr 7 as it looks relative to the regional assignment of genes [54]. Notice the distance of 12 map units from the centromere to the *Gpi-1* locus, obtained using teratoma mapping. The position of *T145H* is shown together with positions involving another translocation, *T(7;X)1Ct*, and a deletion at the *c* (albino) locus, c^{25H}. Chr 7 has been divided into five gene groups each assigned to a specific banded region.

The duplication-deficiency method is being used in my laboratory to further divide Chr 7 as well as to partition other chromosomes. Shortly we should be able

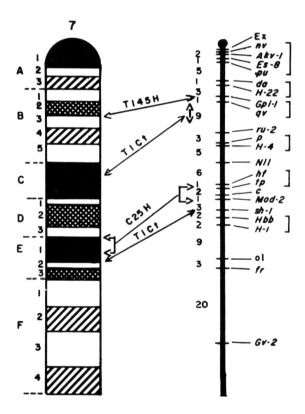

Fig. 13. Cytogenetic map of Chromosome 7. Distance of centromere to *Gpi-1* locus has been determined by teratoma mapping. Genetic location of *T145H* and *T1Ct* translocations were determined by duplication-deficiency mapping [54] and that for the c^{25H} deletion by deletion mapping [see 54].

to answer questions regarding the relationship of mitotic chromosome length to genetic recombination, as well as to know which genes are located in which banded region.

Before we leave the problem of gene-to-band assignment, I would like to point out one drawback to using duplication-deficiency mapping: You must be able to identify the lifespan of chromosomally unbalanced individuals and select genes expressed within the viable period. For example, if a gene expresses itself eight weeks after birth, it will not be usable if the individual to be genotyped dies before this time. In spite of this shortcoming, I am confident that with ingenuity, many loci can be used in duplication-deficiency mapping.

Many people would argue that there is an additional problem with using duplication-deficiency mapping: Because mice with gross chromosome imbalances die in utero, or shortly after birth, regional mapping will only be feasible for those chromosomal regions located near either the centromere or the telomere because these are the regions most often involved in minor chromosomal imbalances that are viable in live born young. I believe this dogma is not true. For example, we have discovered that trisomy for Chr 19, a condition previously considered to be lethal [55], is in fact viable [56]. These mice grow to adulthood and are strong enough to bite their handlers. And we have uncovered other equally striking "exceptions to the rule" [57].

FURTHER COMMENTS RELATED TO THE FUTURE

I would now like to discuss what I see as the future of formal mouse genetics. I predict that the amount of time spent obtaining some types of information will decrease, whereas that used to obtain other types of information will increase. Let me be more specific. If you look at Chr 2 (Fig. 1) you will note that the number of genes closely linked to the *a* (non-agouti) locus is quite large. No one is going to have the time or probably the money to pay for the mouse space needed to determine by the recombination method the exact order of 20–50 genes within this region. I do not mean to imply that it won't be important to *know* the order of these genes in the *a*-region of Chr 2; rather, I mean to convey that we must turn to mapping methods other than recombination for ordering such gene groupings.

I envision that, in future depictions of the mouse genetic linkage map, each linkage group will be shown with a number of genes scattered along its length at intervals of 5–10 map units. Each of these genes, which I will designate as "core genes," will have been carefully mapped relative to all other core genes and probably relative to a specific chromosomal band. Most of these genes will be fully penetrant and codominant, thus maximizing their usage (most isozyme loci fulfill these criteria), and will be easily scorable by everyone in the world. I have taken the present linkage map and circled some loci I believe will be core genes (see Fig. 14). Note that not all are codominant loci, but most are. Our well-known friends *a*, *T*, *W*, *mi*, *Sl*, *Sd*, *Re*, etc., will still be used widely. As new loci are found, they will be

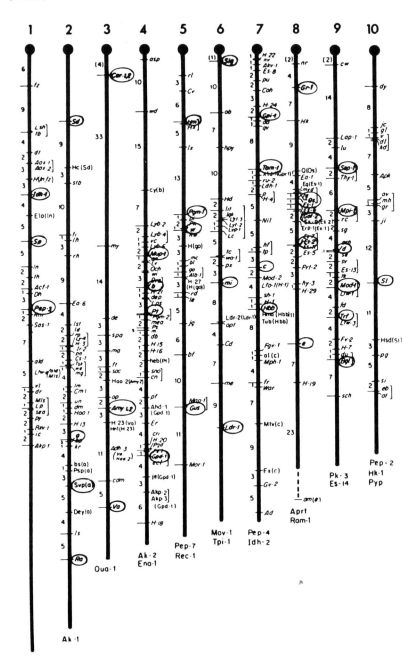

Fig. 14. 1979 linkage map [1] with core genes circled. As other core loci are discovered, they will be added. Eventually, core mapping loci should be located along each chromosome at 5–10 map unit intervals.

Formal Genetics of the Mouse / 39

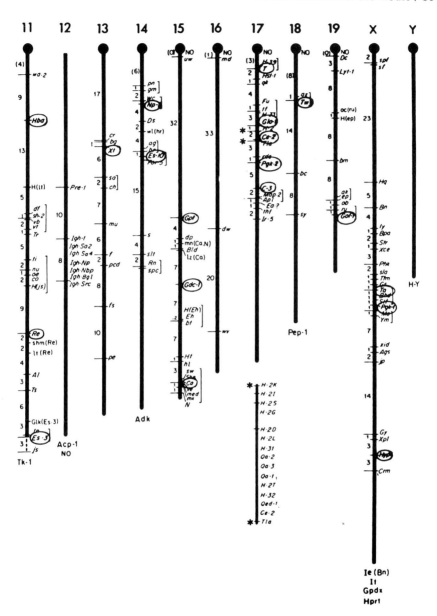

Fig. 14. Continued.

assigned quickly to a chromosome, utilizing crosses to one or two very efficient linkage stocks, such as castaneus or C57BL/6J. Then each locus will be located within the chromosome relative to the positions of core genes.

The next step, for those who want to determine more precisely a gene's location, will be accomplished by using a series of overlapping deletions known to involve the chromosomal region of interest. Thus the 20–50 (or more) genes located near the *a* locus will be referred to as being located within a specific chromosomal segment defined by complementation mapping using deletions. Furthermore, these regions will be defined as to their DNA content from recombinant DNA research. The only limitations I foresee in this type of marriage between formal and molecular genetics is our willingness to generate small chromosomal deletions, and the mouse genome's willingness to allow individuals carrying these deletions to be viable and fertile.

WHY MAP MOUSE GENES?

My final comments are directed toward the question "Why know the chromosomal assignment and order of the mouse genes?" I believe there are many reasons for wanting to know how the mouse genome is organized. One reason is that it is like the mountain waiting to be climbed. Another reason, more practical, is that by knowing more about mice we can learn more about ourselves. I must confess that it is easy to adopt either of these extreme positions. The position I favor, however, relates to the question of how evolution took the first life and mutated its genome, diploidized it over and over, and shuffled it around to get such diverse forms of life as appear today. As is the case with all overview questions, that's a difficult question to answer. One can, however, try to solve small bits of the puzzle and hope that, as we put together the pieces, the whole picture will emerge. For example, knowledge about the mouse genome can aid in obtaining information about the mammalian part of the puzzle.

One question that can be answered is: How conservative was evolution when it rearranged the mammalian genome to produce such diverse idiograms as are found among today's species? Two schools of thought exist on this subject. The first school, which I call the optimistic school, believes that, although chromosomal rearrangements occurred during mammalian evolution, these rearrangements were such that large pieces of DNA have remained together [58]. If this school is correct, man and mouse will be found to have extensive autosomal linkage homologies. The second school, which I call the pessimistic school, believes that so much rearrangement has occurred during mammalian speciation that autosomal linkage homologies between diverse species will be rare or the length of gene-homology units will be short [59]. My personal belief is that this hypothesis will be proven wrong. In fact, I think some blocks of genes will be found to be highly conserved because they share and need a common physically linked regulatory system [60]. Furthermore, if the pessimistic school is correct, the evolutionary puzzle may be impossible to solve and all scientists want to believe that biological questions are answerable.

I have prepared a table (Table II) to illustrate some of the information available for comparison of autosomal linkages of the two most genetically understood mammalian species, Mus musculus and Homo sapiens. (I have deliberately left out any discussion of the X-chromosome homology question.) One very striking homology concerns mouse Chr 4. Of the five mouse genes listed whose human equivalents are fairly certain, all are on the short arm of human Chr 1. Of these five mouse loci, two have genetic variants available for study *(Pgm-2* and *Pgd)*. Their genetic distance on Chr 5 (see again Fig. 1) occupies about 27 recombinational units. That's certainly not a short piece of chromosome!

One cannot escape the conclusion that this part of the genome ancestral to mouse and man has remained physically intact during the evolution of these two species. It doesn't take much imagination to predict that all of the mouse loci included within this region of Chr 4 are also on human Chr 1p. If this hypothesis is true, the advantage for the mouse and human geneticist is fantastic, because all the mouse "diseases" located within this region must have homologues located in the equivalent piece of human genome. For this genetic region at least, the human geneticist will be able to select an inherited disease in the mouse to serve as a good model to study a similar and specific human disease. Conversely, the mouse geneticist can better select the pertinent primate model of a specific murine disease.

Another practical aspect offered by this type of comparative mapping is the power to eliminate guessing and more logically approach mapping per se. One example will make the point.

An inherited electrophoretic variant of pyruvate kinase-3 was found in the mouse [61]. The homologous human gene, PK-3, was known to be located on Chr 15q near the mannose phosphate isomerase (MPI) locus [62]. If large chromosomal regions in man and mouse are homologous, the mouse *Pk-3* locus had a greater chance of being located on Chr 9 near *Mpi-1* than anywhere else in the genome. Matings were established that would segregate for *Pk-3* and several loci on Chr 9, including *Mod-1* (homology on human Chr 6q) and *Trf* (homology on human Chr 1). As predicted, the mouse *Pk-3* locus was found to be near *Mpi-1* [Peters, Nash, Eicher, and Bulfield, 142]. Thus a block of genes including *Mpi-1* and *Pk-3* has been kept together during the evolution of mouse and man.

The reverse, of course, is also true: Knowledge about the mouse genome will aid the human geneticist. For example, the fact that mouse *Pk-3* is closely linked to the dilute (*d*) locus may help identify the human equivalent of *d* because the human equivalent should be closely linked to *Pk-3* and *Mpi-1*.

Already the practical aspects of comparing mouse and human gene orders are so obvious that the frequency of consulting the genetic map of one species by those wanting information for other species is increasing. The final outcome of comparative mapping will be our ability to see how evolution generated all of these "various variations on a mammalian genome" by cutting up the genome of one

TABLE II. Homologous Chromosome Segments of Mouse and Man as Detected by Autosomal Synteny or Linkage Relationships

Mouse			Man[a]	
Gene	Chr	References	Gene	Chr
Pep-3[b]	1	[101]	PEPC	1
Idh-1[b]	1	[111]	IDH1	2q
Ak-1	2	[108]	AK1	9q
Amy-1	3	[32, 33, 107]	AMY1	1p
Amy-2	3	[32, 33, 107]	AMY2	1p
Pgm-2	4	[101]	PGM1	1p
Pgd	4	[100]	PGD	1p
Eno-1	4	[103, 119]	ENO1	1p
Ak-2	4	[119]	AK2	1p
Pep-7	5	[116]	PEPS	4
Pgm-1	5	[111]	PGM2	4
Alb	5	[129]	ALB	4
Gus	5	[120, 131]	GUSB	7
Mor-1	5	[140]	MDH2	7
Tpi-1	6	[122, 137]	TPI1	12
Gpi-1	7	[110]	GPI	19
Pep-4	7	[117]	PEPD	19
Ldh-1	7	[117, 130, 138]	LDHA	11p
Hbb	7	[132]	HBB	11
Idh-2	7	[117]	IDH2	15q
Aprt	8	[112]	APRT	16q
Gr-1	8	[128]	GSR	8p
Got-2	8	[104]	GOT2	6p
Mpi-1	9	[127]	MPI	15q
Pk-3	9	[118, 142]	PKM2	15q
Mod-1[b]	9	[111]	ME1	6q
Trf	9	[135]	TF	1
Bgs	9	[134]	GLB1	3
Hk-1	10	[115]	HK1	10
Pyp	10	[115]	PP	10
Pep-2[b]	10	[108, 122]	PEPB	12q
Hba[b]	11	[133]	HBA	16
Glk	11	[113, 125, 126]	GALK	17q
Tk-1	11	[113, 125]	TK1	17q
Acp-1	12	[108]	ACP1	2p
Adk	14	[121]	ADK	10
Np-1	14	[139]	NP	14q
Es-10	14	[139]	ESD	13q
Sod-1	16	[105]	SOD1	21q
Glo-1	17	[106, 123, 137]	GLO	6p
H-2	17	[109]	HLA	6p
Ss	17	[102, 114, 136]	C4	6p
Pep-1[b]	18	[108, 124]	PEPA	18q
Got-1	19	[107]	GOT1	10q

[a]References for assigning human loci to chromosomes can be found in McKusick, 1979 [141].
[b]Gene symbol Pep-3 was previously designated Dip-1, Idh-1 was Id-1, Pep-1 was Dip-2, Pep-2 was Trip-1, Mod-1 was Mdh-1, Hbb was Hb, and Hba was Sol.

and rearranging it to produce the genome of another and the mammalian piece of the evolutionary puzzle will be better understood.

SUMMARY

I think the future of formal mouse genetics looks bright, especially with the availability of recombinant DNA techniques. The challenge for all mouse geneticists is to share with graduate students and postdocs the beauty of mouse genetics. If we infect the young with our dreams, these dreams will become reality, for it is our students who will invent the new methods and provide the future.

DEDICATION AND ACKNOWLEDGMENTS

I would like to dedicate this paper to several Jackson Laboratory mouse geneticists who helped me at some stage in my career. First of all, I want to mention Earl Green. It was Earl who allowed me to be independent by hiring me and helping me polish my NIH grant applications. Earl believed in the power of the inbred strain and mutant gene; he spent hours with me discussing these wonderful genetic tools. Next I'd like to acknowledge all the help given to me by Margaret Green. Her belief in the importance of knowing the genetic location of a gene has strongly influenced me. I think Marg, more than anyone else, taught me the importance of accuracy and "doing it right the first time." I'd also like to thank Marg for helping me polish a number of papers. Another person who has been most helpful is Elizabeth Russell. It is from Tibby that I learned about the importance of having mutant genes on a defined genetic background. Tibby's favorite genome is C57BL/6J, and we all owe much to her for all the black-six congenic lines she made and shared with everyone. Next I'd like to acknowledge George Snell. George occupies a special place in my career because he made the first mouse translocation — Snell's translocation — and mapped the first mouse centromere. I was always inspired when I listened to George present his well-organized seminars. Finally, I would like to acknowledge all of the help given to me by Tom Roderick when I first came to the Jackson Laboratory.

At this time I would like to acknowledge the help given to me by the late Allen Griffen, who allowed me to work in his laboratory the summer of 1966, when I was a summer student with Marg Green. Griff listened to all my ideas and encouraged me to believe that there was a future in mouse cytogenetics. I remember I was very sad to say good-bye.

I want to express my appreciation to Linda Washburn and Elizabeth Russell for their valuable comments related to the manuscript, and to Janan Eppig for her help in preparing Table I. I am also indebted to Muriel Davisson and Tom Roderick for allowing me to use their unpublished 1979 linkage map and to Marg Green for the use of her 1971 and 1975 linkage maps.

The bibliographic search for this paper was completed in July 1980. I apologize to anyone whose work I slighted.

At this time I would like to thank the following people for the numerous contributions they have made to my research program: Janan Eppig, Ruth Farnsworth, Sally Fox, Sallie Reynolds Leighton, Janice Southard, and Linda Washburn. Without their supportive efforts, many of the findings reported herein would not have been possible.

The research reported in this manuscript was supported by NIH grants GM 20919, AM 17947, NS 09378, and RR 01183.

REFERENCES

1. Davisson M, Roderick TH: Linkage map of the mouse. Mouse News Letter 61:19, 1979.
2. Haldane JBS, Sprunt AD, Haldane NM: Reduplication in mice. J Genet 5:133, 1915.
3. Darbishire AD: On the result of crossing Japanese waltzing with albino mice. Biometrika 3:1, 1904.
4. Dickie MM: The expanding knowledge of the genome of the mouse. J Natl Cancer Inst 15:679, 1954.
5. Staff of the Roscoe B. Jackson Memorial Laboratory: Contributions in mouse genetics. J Hered 36:257, 1945.
6. Green MC: Linkage map of the mouse. In Whitten WK (ed): "42nd Annual Report." Bar Harbor, Maine: The Jackson Laboratory, 1971, p 56.
7. Lane PW: Linkage groups III and XVII in the mouse and the position of the light-ear locus. J Hered 58:21, 1967.
8. Eicher EM, Reynolds SC, Southard JS: Mouse News Letter 56:42, 1977.
9. Cacheiro NLA, Russell LB: Evidence that Linkage Group IV as well as Linkage Group X of the mouse are in chromosome 10. Genet Res 25:193, 1975.
10. Caspersson T, Zech L, Johansson C: Differentiation binding of alkylating fluorochromes in human chromosomes. Exp Cell Res 49:219, 1968.
11. Buckland RA, Evans HJ, Sumner AT: Identifying mouse chromosomes with the ASG technique. Exp Cell Res 69:231, 1971.
12. Francke U, Nesbitt M: Identification of the mouse chromosomes by quinacrine mustard staining. Cytogenetics 10:356, 1971.
13. Miller OJ, Miller DA, Kouri RE, Allderdice PW, Dev VG, Grewal MS, Hutton JJ: Identification of the mouse karyotype by quinacrine fluorescence, and tentative assignment of seven linkage groups. Proc Natl Acad Sci USA 68:1530, 1971.
14. Schnedl W: The karyotype of the mouse. Chromosoma 35:111, 1971.
15. Eicher EM: The identification of the chromosome bearing linkage group XII in the mouse. Genetics 69:267, 1971.
16. Lyon MF: Mouse News Letter 40:26, 1969.
17. Ford CE: The murine Y chromosome as a marker. Transplantation 4:333, 1966.
18. Henderson AS, Eicher EM, Yu MT, Atwood KC: The chromosomal location of ribosomal DNA in the mouse. Chromosoma 49:155, 1974.
19. Griffen A: A late pachytene chromosome map of the male mouse. J Morphol 96:123, 1960.
20. Slizynski BM: A preliminary pachytene chromosome map of the house mouse. J Genet 49:242, 1949.

21. Carter TC, Lyon MF, Phillips RJS: Gene-tagged chromosome translocations in eleven stocks of mice. J Genet 53:154, 1955.
22. Miller DA, Miller OJ: Chromosome mapping in the mouse. Science 178:949, 1972.
23. Committee on Standardized Genetic Nomenclature for Mice: Standard karyotype of the mouse, *Mus musculus*. J Hered 63:69, 1972.
24. Nesbitt MN, Francke U: A system of nomenclature for band patterns of mouse chromosomes. Chromosoma 41:145, 1973.
25. Green MC: Linkage map of the mouse. Mouse News Letter 53:10, 1975.
26. Roderick TH, Davisson M, Lane PW: Mouse News Letter 55:18, 1976.
27. Snell GW: Dwarf, a new Mendelian recessive character of the house mouse. Proc Natl Acad Sci USA 15:733, 1929.
28. Lane PW, Sweet HO: Mouse News Letter 50:44, 1974.
29. Eicher EM, Beamer WG: New mouse *dw* allele: Genetic location and effects on lifespan and growth hormone levels. J Hered 71:187, 1980.
30. Searle AG, Beechey CV: Probable assignment of LG XV to chromosome 18. Mouse News Letter 48:32, 1973.
31. Eicher EM: Chr 18. Mouse News Letter 58:50, 1978.
32. Lane PW, Eicher EM: Gene order in linkage group XVI of the house mouse. J Hered 70:239, 1979.
33. Eicher EM, Lane PW: Assignment of mouse LG XVI to Chromosome 3. J Hered 71:315, 1980.
34. Taylor BA, Bailey DW, Cherry M, Riblet R, Weigert M: Genes for immunoglobin heavy chain and serum prealbumin protein are linked in the mouse. Nature 256:644, 1975.
35. Eicher EM, Taylor BA, Riblet R: Assignment of *Igh* and *Pre-1* to Chromosome 12. Mouse News Letter 61:42, 1979.
36. Pardue ML, Gall JG: Chromosome localization of mouse satellite DNA. Science 168:1356, 1970.
37. Jones KW: Chromosome and nuclear location of mouse satellite DNA in individual cells. Nature 225:912, 1970.
38. Elsevier SM, Ruddle FH: Location of genes coding for 18S and 28S ribosomal RNA within the genome of *Mus musculus*. Chromosoma 52:219, 1975.
39. Henderson AS, Eicher EM, Yu MT, Atwood KC: Variation in ribosomal RNA gene number in mouse chromosomes. Cytogenet Cell Genet 17:307, 1976.
40. Hutton JJ: Biochemical polymorphisms — Detection, distribution, chromosomal location, and applications. In Morse HC III (ed): "Origin of Inbred Mice." New York: Academic Press, 1978, p 235.
41. Chapman VM: Biochemical polymorphisms in wild mice. In Morse HC III (ed): "Origin of Inbred Mice." New York: Academic Press, 1978, p 555.
42. Bailey DW: Recombinant-inbred strains. Transplantation 11:325, 1971.
43. Taylor BA: Recombinant inbred strains: Use in gene mapping. In Morse HC III (ed): "Origins of Inbred Mice." New York: Academic Press, 1978, p 423.
44. Snell GD: An analysis of translocations in the mouse. Genetics 31:157, 1946.
45. Cattanach BM, Moseley H: Nondisjunction and reduced fertility caused by the tobacco mouse metacentric chromosomes. Cytogenet Cell Genet 12:264, 1973.
46. Linder D: Gene loss in human teratomas. Proc Natl Acad Sci USA 63:699, 1969.
47. Linder D, McCaw BK, Hecht F: Parthenogenic origin of benign ovarian teratomas. N Engl J Med 292:63, 1975.
48. Linder D, Power J: Further evidence for post-meiotic origin of teratomas in the human female. Ann Hu Genet 34:21, 1970.
49. Ott J, Hecht F, Linder D, Lovrien EW, McCaw BK: Human centromere mapping using

teratoma data. In Baltimore Conference (1975): Third International Workshop on Human Gene Mapping. "Birth Defects: Original Article Series," vol 12, no. 7. New York: The National Foundation, 1976, p 396.
50. Stevens LC, Varnum DS: The development of teratomas from parthenogenetically activated ovarian mouse eggs. Dev Biol 37:369, 1974.
51. Eppig JJ, Kozak LP, Eicher EM, Stevens LC: Ovarian teratomas in mice are derived from oocytes that have completed the first meiotic division. Nature 269:517, 1977.
52. Eicher EM: Murine ovarian teratomas and parthenotes as cytogenetic tools. Cytogenet Cell Genet 20:232, 1978.
53. Searle AG, Beechey CV, Eicher EM, Nesbitt MN, Washburn LL: Colinearity in the mouse genome: A study of chromosome 2. Cytogenet Cell Genet 23:255, 1979.
54. Eicher EM, Washburn LL: Assignment of genes to regions of mouse chromosomes. Proc Natl Acad Sci USA 75:946, 1978.
55. White BJ, Tjio J-H, Van de Water LC, Crandall C: Trisomy for the smallest autosome of the mouse and identification of the T1Wh translocation chromosome. Cytogenetics 11:363, 1972.
56. Eicher EM, Washburn LL, Reynolds S: Mouse News Letter 56:42, 1977.
57. Eicher EM, Washburn LL: Mouse News Letter 56:43, 1977.
58. Ohno S: Frozen linkage groups and frozen accidents. Nature 244:259, 1973.
59. Ruddle FH, Creagan RP: Parasexual approaches to the genetics of man. Ann Rev Genet 9:407, 1975.
60. Eicher EM, Cherry M, Flaherty L: Autosomal phosphoglycerate kinase linked to mouse major histocompatibility complex. Mol Gen Genet 158:225, 1978.
61. Nash HR, Peters J, Bulfield G: Mouse News Letter 59:31, 1978.
62. van Heyningen V, Bobrow M, Bodmer WF, Gardiner SE, Povey S, Hopkinson DA: Chromosomal assignment of some human enzyme loci: Mitochondrial malate dehydrogenase to 7, mannosephosphate isomerase and pyruvate kinase to 15 and probably esterase D to 13. Ann Hum Genet 38:295, 1975.
63. Bunker H, Snell GD: Linkage of white and waved-1. J Hered 39:28, 1948.
64. Cattanach BM, Moseley HJ: Assignment of LG VII to Chromosome 11. Mouse News Letter 48:31, 1973.
65. Cattanach BM, Williams CE, Bailey H: Identification of the linkage groups carried by the metacentric chromosomes of the tobacco mouse (*Mus poschiavinus*). Cytogenetics 11:412, 1972.
66. Cooper CB: Linkage between naked and caracul in the house mouse. J Hered 30:212, 1939.
67. Curry GA: Genetic and developmental studies on droopy eared mice. J Embryol Exp Morphol 7:39, 1959.
68. Davisson MT, Eicher EM, Green MC: Genes on chromosome 3 of the mouse. J Hered 67:155, 1976.
69. Deol MS, Lane PW: A new gene affecting the morphogenesis of the vestibular part of the inner ear in the mouse. J Embryol Exp Morphol 6:543, 1966.
70. Dickie MM, Woolley GW: Linkage studies with the pirouette gene in the mouse. J Hered 37:335, 1946.
71. Dickie MM, Woolley GW: Fuzzy mice. J Hered 41:193, 1950.
72. Eicher EM, Green MC: The *T6* translocation in the mouse: Its use in trisomy mapping, centromere localization, and cytological identification of Linkage Group III. Genetics 71:621, 1972.
73. Eicher EM, Stern RH, Womack JE, Davisson MT, Roderick TH, Reynolds SC: Evolution of mammalian carbonic anhydrase loci by tandem duplication: Close linkage of *Car-1*

and *Car-2* to the centromere region of Chromosome 3 of the mouse. Biochem Genet 14:651, 1976.
74. Falconer DS: Total sex-linkage in the house mouse. Zeits Indukt Abstamm Vererbungslehre 85:210, 1953.
75. Falconer DS, King JWB: Mouse News Letter 9:15, 1953.
76. Francke U, Nesbitt M: Cattanach's translocation: Cytological characterization by quinacrine mustard staining. Proc Natl Acad Sci USA 68:2918, 1971.
77. Gates WH: Linkage of short ears and density in the house mouse. Proc Natl Acad Sci USA 13:575, 1927.
78. Green MC: Short history of the linkage map of the mouse. Mouse News Letter 50:8, 1974.
79. Green MC, Sidman RL: Tottering, a neuromuscular mutation in the mouse and its linkage with oligosyndactylism. J Hered 53:233, 1962.
80. Green MC, Snell GD, Lane PW: Linkage group XVIII of the mouse. J Hered 54:245, 1963.
81. Hertwig P: Neue Mutationen und Koppelungsgruppen bei der Hausmaus. Zeits Indukt Abstamm Vererbungslehre 80:220, 1942.
82. Keeler CE: Hereditary blindness in the house mouse with special reference to the linkage relationship. Harvard Med Sch, Howe Lab Ophthalmol Bull 3:11, 1930.
83. King JWB: Linkage group XIV of the house mouse. Nature 178:1126, 1956.
84. Kouri RE, Miller DA, Miller OJ, Dev VG, Grewal MS, Hutton JJ: Identification by quinacrine fluorescence of the chromosome carrying mouse linkage group I in the δ Cattanach translocation. Genetics 69:129, 1971.
85. Lane PW, Green EL: Pale ear and light ear in the house mouse. J Hered 58:17, 1967.
86. Lane PW, Sweet H: Mouse News Letter 51:23, 1974.
87. Lyon MF: Twirler: A mutant affecting the inner ear of the house mouse. J Embryol Exp Morphol 6:105, 1958.
88. Lyon MF, Glenister P: Cytological identification of LG XII. Mouse News Letter 45:24, 1971.
89. Miller DA, Kouri RE, Dev VG, Grewal MS, Hutton JJ, Miller OJ: Assignment of four linkage groups to chromosomes in *Mus musculus* and a cytogenetic method for locating their centromeric ends. Proc Natl Acad Sci USA 68:2699, 1971.
90. Miller OJ, Miller DA, Kouri RE, Dev VG, Grewal MS, Hutton JJ: Assignment of linkage groups VIII and X to chromosomes in *Mus musculus* and identification of the centromeric end of linkage group I. Cytogenetics 10:452, 1971.
91. Nesbitt M, Francke U: Linkage groups II and XII of the mouse: Cytological localization by fluorochrome staining. Science 174:60, 1971.
92. Phillips RJS: The linkage of congenital hydrocephalus in the house mouse. J Hered 47:302, 1956.
93. Reed SC: The inheritance and expression of fused, a new mutation in the house mouse. Genetics 22:1, 1937.
94. Roberts E, Quisenberg JH: Linkage of the genes for non-yellow and pink-eyed-2 in the house mouse. Am Nat 69:181, 1935.
95. Robinson R: "Gene Mapping in Laboratory Mammals," part B. New York: Plenum Press, 1972.
96. Snell GD: A crossover between the genes for short-ear and density in the house mouse. Proc Natl Acad Sci USA 14:926, 1928.
97. Snell GD: Inheritance in the house mouse; the linkage relations of short-ear, hairless and naked. Genetics 16:42, 1931.
98. Snell GD: Linkage of jittery and waltzing in the mouse. J Hered 36:279, 1945.

99. Snell GD, Law LW: Linkage between shaker-2 and wavy-2 in the house mouse. J Hered 30:447, 1939.
100. Chapman VM: 6-Phosphogluconate dehydrogenase (PGD) genetics in the mouse: Linkage with metabolically related enzyme loci. Biochem Genet 13:849, 1975.
101. Chapman VM, Ruddle FH, Roderick TH: Linkage of isozyme loci in the mouse: Phosphoglucomutase-2 (*Pgm-2*), mitochondrial NADP malate dehydrogenase (MOD-2) and dipeptidase (Dip-1). Biochem Genet 5:101, 1971.
102. Curman B, Östberg L, Sandberg L, Malmheden-Eriksson I, Stalenheim G, Rask L, Peterson PA: H-2 Linked Ss protein is C4 component of complement. Nature 258: 243, 1975.
103. D'Ancona GG, Croce CM: Assignment of the gene for enolase to mouse chromosome 4 using somatic cell hybrids. Cytogenet Cell Genet 19:1, 1977.
104. DeLorenzo RJ, Ruddle FH: Glutamate oxalate transaminase (GOT) genetics in *Mus musculus:* Linkage polymorphism, and phenotypes of the *Got-2* and *Got-1* loci. Biochem Genet 4:259, 1970.
105. Eicher EM, Gerald PS: Unpublished.
106. Eicher EM, Reynolds S: Mouse News Letter 58:50, 1978.
107. Eicher EM, Reynolds S, Southard JS: Mouse News Letter 56:42, 1977.
108. Francke U, Lalley P, Moss W, Ivy J, Minna J: Gene mapping in *Mus musculus* by interspecific cell hybridization: Assignment of the genes for tripeptidase-1 to Chromosome 10, dipeptidase-2 to Chromosome 18, acid phosphatase-1 to Chromosome 12, and adenylate kinase-1 to Chromosome 2. Cytogenet Cell Genet 19:57, 1977.
109. Gorer PA, Lyman S, Snell GD: Studies on the genetic and antigenic basis of tumor transplantation. Linkage between a histocompatibility gene and 'fused' in mice. Proc R Soc Lond B 135:499, 1948.
110. Hutton JJ: Linkage analysis using biochemical variants in mice. I. Linkage of the hemoglobin beta chain and glucosephosphate isomerase loci. Biochem Genet 3:507, 1969.
111. Hutton JJ, Roderick TH: Linkage analysis using biochemical variants in mice. III. Linkage relationships of eleven biochemical markers. Biochem Genet 4:339, 1970.
112. Kozak CA, Nichols EA, Ruddle FH: Gene linkage analysis in the mouse by somatic cell hybridization: Assignment of adenine phosphoribosyltransferase to Chromosome 8 and α-galactosidase to the X-chromosome. Somat Cell Genet 1:371, 1975.
113. Kozak CA, Ruddle FH: Assignment of the genes for thymidine kinase and galactokinase to *Mus musculus* Chromosome 11 and the preferential segregation of this chromosome in Chinese hamster/mouse somatic cell hybrids. Somat Cell Genet 3:121, 1977.
114. Lachmann PJ, Grennan D, Martin A, Démant P: Identification of Ss protein as murine C4. Nature 258:242, 1975.
115. Lalley PA, Francke U, Minna JD: Assignment of the genes coding for pyrophosphatase and hexokinase-1 to mouse Chromosome 10: Implication for comparative gene mapping in man and mouse. Cytogenet Cell Genet 22:570, 1978.
116. Lalley PA, Francke U, Minna JD: Comparative gene mapping: The linkage relationships of the homologous genes for phosphoglucomutase and peptidase S are conserved in man and mouse. Cytogenet Cell Genet 22:573, 1978.
117. Lalley PA, Francke U, Minna JD: Comparative gene mapping in man and mouse: Assignment of the genes for lactate dehydrogenase-A, peptidase-D, and isocitrate dehydrogenase-2 to mouse Chromosome 7. Cytogenet Cell Genet 22:577, 1978.
118. Lalley PA, Francke U, Minna JD: The genes coding for pyruvate kinase (M2) and mannophosphate isomerase are linked in man and mouse. Cytogenet Cell Genet 22:581, 1978.
119. Lalley PA, Francke U, Minna JD: Homologous genes coding for enolase, phosphogluconate dehydrogenase, phosphoglucomutase, and adenylate kinase are syntenic on mouse

chromosome 4 and human chromosome 1p. Proc Natl Acad Sci USA 75:2382, 1978.
120. Lalley PA, Shows TB: Lysosomal and microsomal glucuronidase genetic variant alters electrophoretic mobility of both hydrolases. Science 185:442, 1974.
121. Leinwand L, Fournier REK, Nichols EA, Ruddle FH: Assignment of the gene for adenosine kinase to chromosome 14 in *Mus musculus* by somatic cell hybridization. Cytogenet Cell Genet 21:77, 1978.
122. Leinwand LA, Kozak C, Ruddle FH: Assignment of the genes for triose phosphate isomerase to chromosome 6 and tripeptidase-1 to chromosome 10 in *Mus musculus* by somatic cell hybridization. Somat Cell Genet 4:233, 1977.
123. Leinwand L, Nichols E, Ruddle FH: Assignment of the gene for glyoxylase I to mouse chromosome 17 by somatic cell genetics. Biochem Genet 16:659, 1978.
124. Leinwand LA, Ruddle FH: Assignment of the gene for dipeptidase 2 to *Mus musculus* chromosome 18 by somatic cell hybridization. Biochem Genet 16:477, 1978.
125. McBreen P, Orkwiszewski KG, Chern CJ, Mellman WJ, Croce CM: Synteny of the genes for thymidine kinase and galactokinase in the mouse and their assignment to mouse chromosome 11. Cytogenet Cell Genet 19:7, 1977.
126. Mishkin JD, Taylor BA, Mellman WJ: *Glk*: A locus controlling galactokinase activity in the mouse. Biochem Genet 14:635, 1976.
127. Nichols EA, Chapman VM, Ruddle FH: Polymorphisms and linkage for mannose phosphate isomerase in *Mus musculus*. Biochem Genet 8:47, 1973.
128. Nichols EA, Ruddle FH: Polymorphism and linkage of glutathione reductase in *Mus musculus*. Biochem Genet 13:323, 1975.
129. Nichols EA, Ruddle FH, Petras ML: Linkage of the locus for serum albumin in the house mouse, *Mus musculus*. Biochem Genet 13:551, 1975.
130. O'Brien D, Linnenbach A, Croce CM: Assignment of the gene for lactic dehydrogenase A to mouse Chromosome 7 using mouse-human hybrids. Cytogenet Cell Genet 21:72, 1978.
131. Paigen K, Noell WK: Two linked genes showing a similar timing of expression in mice. Nature 190:148, 1961.
132. Popp RA, St. Amand W: Studies on the mouse hemoglobin locus. I. Identification of hemoglobin types and linkage of hemoglobin with albinism. J Hered 51:141, 1960.
133. Russell ES, McFarland EC: Genetics of mouse hemoglobins. Ann NY Acad Sci 241:25, 1974.
134. Seyedyazdani I, Lundin LG: Association between β-galactosidase activities and coat color in mice. J Hered 64:295, 1973.
135. Shreffler DC: Linkage of the mouse transferrin locus. J Hered 54:127, 1963.
136. Shreffler DC, Owen RD: A serologically detected variant in mouse serum: Inheritance and association with the histocompatibility-2 locus. Genetics 48:9, 1963.
137. Minna JD, Bruns GAP, Krinsky AH, Lalley PA, Francke U, Gerald PS: Assignment of a *Mus musculus* gene for triosephosphate isomerase to Chromosome 6 and for glyoxalase-I to Chromosome 17. Somat Cell Genet 4:241, 1978.
138. Soares ER: Mouse News Letter 59:11, 1978.
139. Womack JE, Davisson M, Eicher EM, Kendall D: Mapping of nucleoside phosphorylase (NP-1) and esterase 10 (*Es*-10) on mouse Chromosome 14. Biochem Genet 15:347, 1977.
140. Womack JE, Hawes NL, Soares ER, Roderick TH: Mitochondrial malate dehydrogenase (*Mor-1*) in the mouse: Linkage to chromosome 5 markers. Biochem Genet 13:519, 1975.
141. McKusick VA: "Mendelian Inheritance in Man," fifth Ed. Baltimore: The Johns Hopkins University Press, 1978.
142. Peters J, Nash HR, Eicher EM, Bulfield G: Polymorphism of mouse kidney pyruvate kinase is determined by a gene, *Pk-3,* on Chromosome 9. Biochem Genet (in press),1981.

The Organization and Evolution of Cloned Globin Genes

Philip Leder, David A. Konkel, Yutaka Nishioka, Aya Leder, Dean H. Hamer, and Marian Kaehler

INTRODUCTION

The genes responsible for the production of mouse hemoglobin are subject to at least two interesting and different forms of regulation. One form operates in *trans* (at a distance) and controls the expression of alpha and beta genes located on different chromosomes. The other operates in *cis*, regulating the expression of individual alpha or beta genes that are located close to one another on the same chromosome. These modes of regulation are only two of the especially interesting and useful aspects of this system. In addition, globins from a variety of different species have been studied in considerable structural detail, and evolutionary comparisons have been made between them. In many cases, specific amino acid alterations have been correlated with alterations in the physiologic and physical properties of the hemoglobin molecule and the selective pressures that have maintained the globin structure during evolution have been assessed.

In what follows, we would like to describe studies we have carried out during the past four years, which were directed toward understanding both the control and the evolution of these genes. This work, which has allowed us to detail the structure of several alpha- and beta-globin genes, has been made possible by the development of the new and powerful techniques of recombinant DNA technology [1]. These techniques — which permit the cloning of discrete genetic segments — coupled with rapid techniques for the determination of nucleotide sequences [2], have allowed us to identify specific regions in each gene that are likely to be concerned with its ordered expression. They also have allowed us to trace the mutational events that led to the evolution of discrete globin genes, and to identify two molecular mechanisms that are responsible for this divergence. And finally, these techniques have allowed us to take these well-characterized genes and to return them to a cell in which their expression can be assessed.

THE TECHNOLOGY

While much of what we have learned recently about the globin system, both in the mouse and in other species, has depended upon the development of recombinant DNA technology [1] and rapid nucleotide-sequencing techniques [2], it is beyond the scope of this article to discuss them in detail. It is important, however, to emphasize that recombinant DNA technology provides a means of selecting a particular segment of the genome from among a million or so similar segments. This segment can then be amplified and made amenable to detailed structural studies as well as to genetic manipulation and assay.

Most of our cloning has been done utilizing the bacteriophage lambda (λ). Using a phage originally constructed by Ron Davis and his co-workers [3], we have introduced a number of mutations into the λgtWES · λB system that make it appropriate for studies in vertebrate organisms [4]. This phage contains two sites for the restriction endonuclease, Eco R1 (Fig. 1). These sites divide it into three fragments, the middle one of which is expendable, in that it contains no genes necessary for phage growth. The two end fragments, containing all the genes necessary for phage propagation, do not by themselves provide a sufficient length of DNA for phage assembly. Therefore, if the central fragment of DNA is eliminated (either by purification or by enzymatic degradation), viable phage production will depend upon the insertion of a foreign fragment of DNA. This property provides a positive selection for phage that have incorporated a fragment of foreign DNA, and a negative selection against parental-type recombinants. Thus, the vector can be used in conjunction with several extremely efficient phage packaging systems [5] or with systems that allow assembly of complete phage particles to permit the generation of anywhere from 10^5 to 10^6 recombinants per microgram of inserted DNA.

Several strategies also have been developed for finding the desired gene among the large population of recombinants. These strategies can involve useful techniques for the purification of a given fragment [6] or the generation of "libraries" of recombinants that contain, on the average, every sequence present in the genome of an organism [7, 8]. Both of these strategies rely on the use of a powerful technique developed by Benton and Davis [9] that involves printing, or blotting, onto a nitrocellulose filter anywhere from 2,000 to 10,000 phage plaques on a single petri dish. The imprint of these plaques can then be annealed to a radioactively-labeled probe (such as the globin cDNA probe) and detected by autoradiography and subsequent image alignment. This technique readily allows the screening of a million phage plaques and makes any fragment for which there is an appropriate probe readily accessible.

Our original cloning efforts involved the purification of specific globin gene-containing fragments of DNA [10, 11]. These have the advantage of allowing one to focus on a given gene from a complex genetic system. The library technology, on the other hand, has the advantage of providing all fragments in a single prepara-

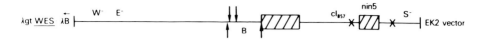

Fig. 1. Diagrammatic representation of the phage lambda vector λWES · λB. The virus is approximately 40,000 bases in length; all its essential genetic information is encoded on the left and right Eco R1 fragments; the central fragments can be deleted and replaced with a 1-kb to 17-kb fragment of foreign DNA. Upward arrows indicate sites cleaved by the restriction endonuclease, Eco R1. Downward arrows indicate SaqI sites. Hatched boxes stand for deleted segments of lambda DNA. Letters stand for specific mutations in the phage [14]. X's stand for Eco R1 sites.

tion and permits the construction of overlapping genetic maps from random cloned fragments that physically overlap one another.

The remarkable and, at the time [10, 12], unexpected structures of the globin genes were revealed by yet another powerful technique, this time involving electron microscopy. The R-loop technique, developed by Ray White and David Hogness [13], allows the direct visualization of any cloned gene for which an RNA transcript is available. The technique involves the melting out of the double-stranded DNA fragment under conditons in which the two DNA strands just begin to separate. Under these conditions, RNA will form a more stable hybrid with its DNA complement and — in this case — by introducing globin mRNA in the presence of the partially melted-out globin gene, the RNA-DNA duplex will displace the second strand of DNA, that is, the strand identical to the globin message, and form a loop. This loop structure is readily visualized in the electron microscope, and the initially cloned globin gene visualized in this way is shown in Figure 2.

GENERAL ORGANIZATION OF GLOBIN GENES

The earlier notion that genetic information is completely colinear would have led to the expectation of a smooth R loop, but that was not the structure observed in these first electron micrographs (Fig. 2). Instead, the now-classical "Kilroy" structure appeared where a large double-stranded loop of DNA is formed between two segments of the gene that anneal to the globin message. Obviously, the globin mRNA sequence is interrupted in the genome, and the structure formed by R-loop analysis resembles the eyes and nose of the World War II-vintage Kilroy cartoon. Actually, analysis of this structure revealed a second, smaller, interruption in the gene (arrow in Fig. 2). These two interruptions have been found in each of the two mouse beta [13, 14] and two mouse alpha genes [15, 17] that we have analyzed.

This general form of organization is indicated diagrammatically in Figure 3. Using this type of electron microscopic analysis and the sequencing noted below, we found that both the adult beta major and minor genes of the mouse are interrupted by two intervening sequences that occur in exactly the same position within both coding sequences. Since these genes have been apart from one another over a period of

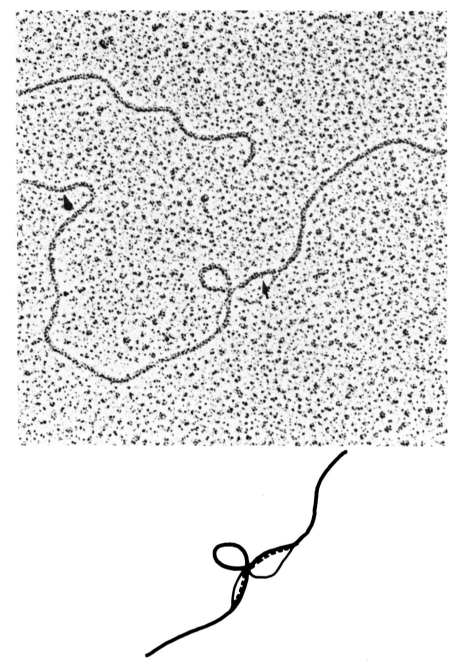

Fig. 2. An electron micrographic visualization of the cloned beta globin minor gene annealed to globin mRNA. The diagrammatic representation illustrates the annealing of globin messenger RNA to the coded globin gene. The heavy line represents double-stranded DNA or RNA-DNA hybrid. The dashed line represents the globin mRNA. An arrow points to the smaller intervening sequence in the electron microgram. The double-stranded structure visualized in the middle of the gene is the large intervening sequence. The data are from Tiemeier et al [6].

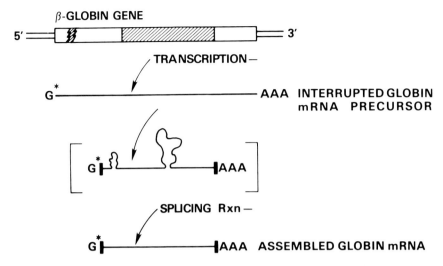

Fig. 3. Diagrammatic representation of a representative globin gene, indicating splicing pathway leading from primary transcript to mature mRNA. The uppermost figure is the gene, with hatched areas representing intervening sequences. The thin lines are RNA, with G* and AAA representing the 5'-cap and 3'-polyA, respectively.

approximately 50 million years, we can begin to assume that they play some essential function in the expression of these genes (see below).

But this evolutionary analysis can be extended by an order of magnitude. The alpha and beta genes arose from a gene duplication event that occurred early in vertebrate evolution in a primitive, ancestral globin sequence. As the alpha and beta genes are present in all vertebrate organisms, we may presume that they have existed and evolved as separate genetic elements for the entire period of vertebrate evolution, some 500 million years. The organization of the alpha genes turned out to be identical to that of the beta genes [16]. That is, the alpha gene sequence also is interrupted at locations that are precisely homologous to those of the beta genes, dividing the alpha gene into three separate coding blocks. Thus, these interruptions have been preserved for over 500 million years, further suggesting that they play some essential function in gene expression.

The next question that arose regarding this remarkable gene arrangement focused upon how the coding segments could be joined to one another in perfect frame, so as to form the continuous coding sequence found in globin mRNA. This question was quickly answered by taking advantage of the fact that the globin mRNA was known to have been synthesized via an approximately 1,500 nucleotide-long precursor [18, 19]. This precursor, which is thought to be the initial transcript of the globin genes, could be shown to contain transcripts that include both coding and intervening sequences [20]. Using the large precursor isolated by Curtiss and Weissman [18] and in collaboration with them [20], we were able to show that the R loop formed between the globin gene and the precursor RNA was a smooth, interrupted structure of the expected length, clearly containing intervening sequences. We therefore concluded that the contiguous sequence finally represented in the mRNA was formed by splicing out and covalently rejoining internal segments of the mRNA precursor. Such a mechanism is shown diagrammatically in Figure 3. An enzyme that has such an activity and operates to splice yeast tRNA precursors has already been discovered by Ableson and his colleagues [21].

In our earliest studies, we were able to focus on a region of preserved homology that surrounded the two beta-globin genes [22]. This was done by forming heteroduplexes between beta major and beta minor fragments and visualizing these structures in the electron microscope. These structures revealed a homology that was preserved only in a short segment of each fragment. This segment included the coding regions, the small intervening sequence, and several hundred bases surrounding the transcribed area of the gene. Such studies allowed us to focus on this region for subsequent sequencing. Although we may assume that the original duplication event that formed the beta major and beta minor genes involved a large segment of DNA, the fact that this homology has been preserved in only a small region surrounding the gene was puzzling and led us to suggest that this loss of homology may play some role in the stabilization of duplicate genes [22]. Since these genes are located close to one another on the chromosome, large segments of homology between them presumably would provide large targets for homologous recombination. Such unequal crossing-over might lead to gene amplification or loss. Thus, if there were a selective advantage in having duplicate genes (and duplicate beta and alpha genes appeared to occur in all vertebrate organisms), then loss of homology in the surrounding regions would serve to stabilize them by reducing the target for recombination. This general relationship between the beta-globin genes, first seen in the electron microscope [6], has been confirmed by the direct sequence analyses shown below (Figs. 5 and 6).

THE COMPLETE SEQUENCES OF THE ALPHA AND BETA-GLOBIN GENES AND THE MECHANISM OF THEIR EVOLUTION

Given the availability of rapid nucleotide-sequencing techniques [2], all the impressions gained from the direct visualization of these genes could be and were confirmed by their direct sequence analysis. The complete sequences of particular alpha, beta major, and beta minor genes are given in Figures 4, 5 and 6, respectively. The sequence data confirm that the cloned genes indeed correspond to the adult alpha and beta genes. In some instances, they resolve ambiguities that arise from the amino acid sequencing data. In each case, they confirm that the genes are interrupted by intervening sequences of DNA at precisely the same location. Moreover, they indicate that intervening sequence homology need not be preserved either in primary structure or in length. While this is certainly true of the second intervening sequence of both beta globin genes, interestingly, the smaller, first intervening sequences have been preserved very closely between the two beta genes. This may be the

```
              10        20        30        40        50        60        70        80        90
   0    GTAAGCAGGTTGTGGTTGAGAAAGGAAAGTGTGAAACAGGGACCCAGAGGGAGAGGTGGGGGATGGCGCTGCTCAGTTTGGTTTGAGGGACTTGCTTCT
 100    CTGACCAAGGTAGGAGGATACTAACTTCTTCCCAAACTGCCATCACTGGAGACATAGTAAGGGGTAAGAAAGTGTGTCCGGGCAACTGATAAGGATTCCC
 200    TGCACCTAGGGGAAGCACAACCCAGCCCCAGAATCTCAGGGGCCCTAACAAGTTTTACTGGGTAGAGCAAGCACAAACCAGCCAATGAGTAACTGCTCCA
                                                  tataag                  cAp
 300    AGGGCGTGTCCACCCTGCCTGGAGGACAGCCCTTGGAGGGCATATAAGTGCTACTTGCTGCAGGTCCAAGACACTTCTGATTCTGACAGACTCAGGAAGA
 400    AACCATGGTGCTCTCTGGGGAAGACAAAAGCAACATCAAGGCTGCCTGGGGGAAGATTGGTGGCCATGGTGCTGAATATGGAGCTGAAGCCCTGGAAAGG
              INIValLeuSerGlyGluAspLysSerAsnIleLysAlaAlaTrpGlyLysIleGlyGlyHisGlyAlaGluTyrGlyAlaGluAlaLeuGluArg
 500    TGAGAACAGGACCTTGATCTGTAAGGATCACAGGATCCAATATGGACCTGGCACTCGCTCAGTGGGCAGCTTCTAACTATGCTTTTCTGTGACCTCAACT
        - 31
 600    TCTCTTCTCTCCTTCTCCCAGGATGTTTGCTAGCTTCCCCACCACCAAGACCTACTTTCCTCACTTTGATGTAAGCCACGGCTCTGCCCAGGTCAAGGGT
                           32-MetPheAlaSerPheProThrThrLysThrTyrPheProHisPheAspValSerHisGlySerAlaGlnValLysGly
 700    CACGGCAAGAAGGTCGCCGATGCGCTGGCCAGTGCTGCAGGCCACCTCGATGACCTGCCCGGTGCCTTGTCTGCTCTGAGCGACCTGCATGCCCACAAGC
        HisGlyLysLysValAlaAspAlaLeuAlaSerAlaAlaGlyHisLeuAspAspLeuProGlyAlaLeuSerAlaLeuSerAspLeuHisAlaHisLysL
 800    TGCGTGTGGATCCCGTCAACTTCAAGGTATGCGCTGGGACCTGGCAGGCGGCATCTGGGACCCCTAGGAAGGGCTTGGGGGTCCTCGTGCCCAAGGCAGG
        euArgValAspProValAsnPheLys- 99
 900    GAACATAGTGGTCCCAGGAAGGGGAGCAGAGGCATCAGGGTGTCCACTTTGTCTCCGCAGCTCCTGAGCCACTGCCTGCTGGTGACCTTGGCTAGCCACC
                                                               100-LeuLeuSerHisCysLeuLeuValThrLeuAlaSerHisH
1000    ACCCTGCCGATTTCACCCCCGCGGTACATGCCTCTCTGGACAAATTCCTTGCCTCTGTGAGCACCGTGCTGACCTCCAAGTACCGTTAAGCTGCCTTCTG
        isProAlaAspPheThrProAlaValHisAlaSerLeuAspLysPheLeuAlaSerValSerThrValLeuThrSerLysTyrArgTER-142
                                                                     aataaa                |pA...
1100    CGGGGCTTGCCTTCTGGCCATGCCCTTCTTCTCTCCCTTGCACCTGTACCTCTTGGTCTTTGAATAAAGCCTGAGTAGGAAGAAGCCTGCATGCCTGGTT
1200    CTCTGCGTCTGCAAAGGTGTCATGTTTAGTGTGGGGATGCCGCAGCTCATTTGCCATGGGGCAGTAAAGACAAGGTTCAGAGCAAAAAGCATAATTGGAT
1300    GCCTACACACACACACATATGTCTTCTGAGTCTGGGCAGCAGTCCCTCCCAAGCCCTCCACTGACAGCCATGTGTCTTCTCCTCGAGCCAAAGAAGCCAA
1400    AGATCGTCTTTGGAGGGTCCTTATCACAGGACCTCTGAGGG
```

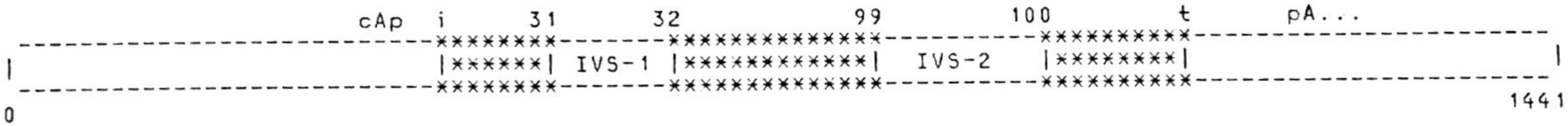

Fig. 4. The nucleotide sequence of a mouse alpha globin gene. Coding sequences are indicated by the amino acid abbreviations below the sequence. INI represents an initiation codon; TER represents a termination codon. The numbers within the sequence refer to the amino acid coding position. The lower case superscripts indicate a putative promoter-site homology sequence, TATAAG, the capping site, cAp, the polyA addition site hexanucleotide, AATAAA, and the putative polyA addition site, pA. . . . A diagrammatic representation of the gene is shown below the sequence. The data are from Nishioka and Leder [16].

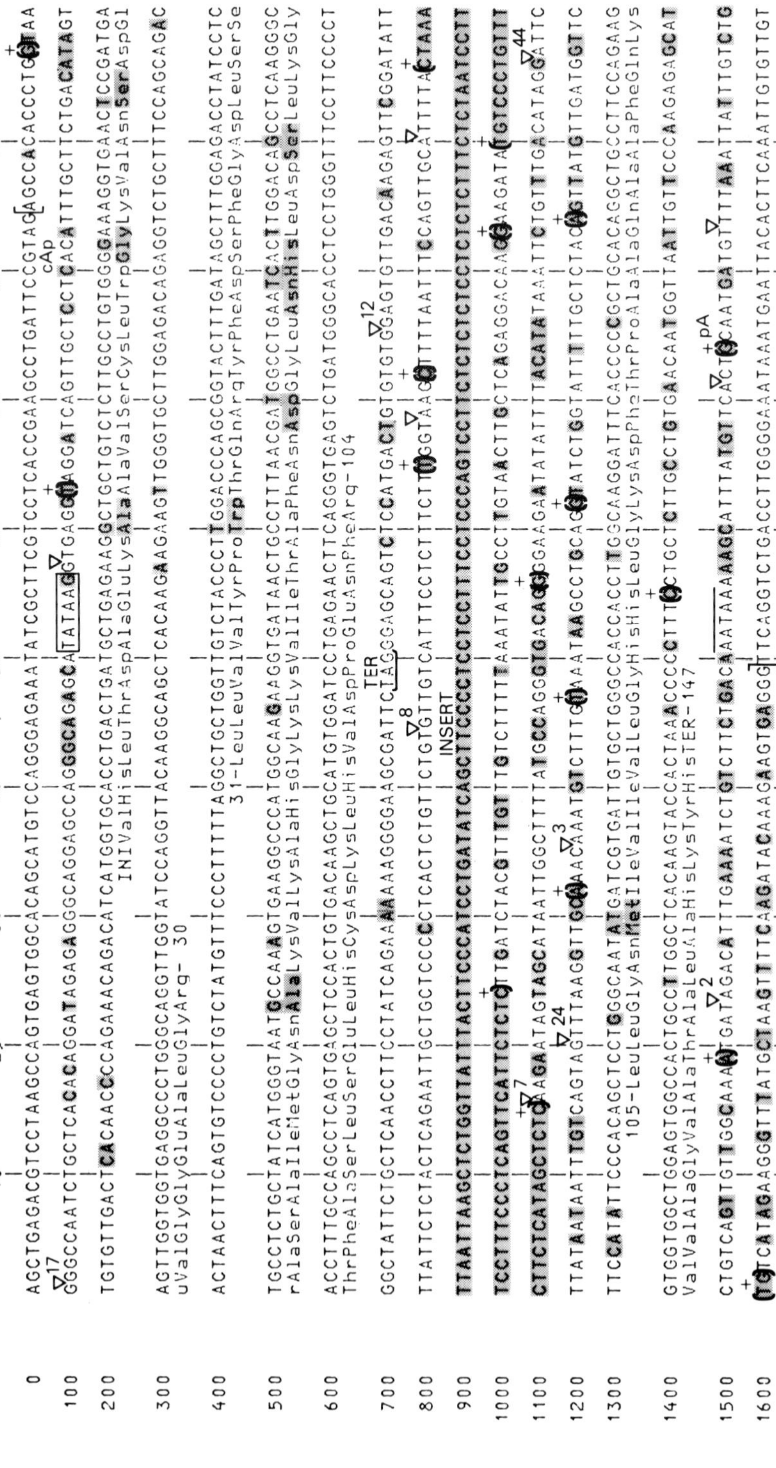

result of the fact that the insertion-deletion mechanisms (see below) that bring about their divergence are less likely to operate within a short sequence where they are more likely to overlap coding or splicing regions.

A comparison of the structure of the alpha and beta genes is also quite instructive. These sequences conserve considerable (about 55%) sequence homology in their coding regions. This homology — with certain interesting exceptions — is lost immediately in the highly divergent flanking and intervening sequences. Indeed, there is a great discrepancy of length between the larger, approximately 650-base-long second intervening sequences of the beta globin genes and the smaller, approximately 130-base-long intervening sequence of the alpha genes.

While the alpha and beta genes have diverged to an extent (except in the coding region) that makes it virtually impossible to reconstruct the events that led to their evolution, the two beta globin genes provide valuable clues as to this mechanism. We already have noted that their sequences diverge within a few hundred nucleotides of their transcribed regions and within their large intervening sequences. The coding sequences themselves and the small intervening sequence are quite tightly homologous. Such differences that occur involve point mutations and the homology of these regions is greater than 95%. In contrast, the region within the intervening sequence that has diverged appears to have achieved this divergence by an insertion-deletion mechanism.

We have used a computer program derived by Dr. Jacob V. Maizel [14] to reconstruct these events in the diagram shown in Figure 7. Here it can be seen that the homology alignments in this region involve about seven insertion-deletion events. Point mutations also have occurred in this region, but the major differences arise through the insertion-deletion mechanism. Obviously, the coding sequence tolerated no such changes. It is impossible to imagine that insertion-deletion events do not occur within the coding region. It is more likely that the entire genome is subject to the same chemical events that occur within the large intervening sequence. On the other hand, selection would be likely to operate against permitting insertions and deletions within the coding sequence. This is so, for at least two reasons. First, such insertion-deletions generally would involve major amino acid-sequence alterations that might not be consistent with the function of the gene product. And second, two-thirds of all random insertions or deletions would alter the phase of the genetic code and create a nonsense sequence. Both outcomes presumably would

---

Fig. 5. The nucleotide sequence of a beta-major gene. The amino acid sequence is displayed below gene sequences (strand corresponding to mRNA). Symbols are as represented in Figure 4, with the following additions: the 5′-putative promoter region sequence TATAAG is boxed; the first in-phase termination codon in IVS-2 is underlined and overscored "TER"; bold-faced bases and amino acids differ from the corresponding position in the β maj sequence (Fig. 4); a ▽ indicates the position of bases deleted in β min relative to β maj; if there are n bases deleted, it is represented by "▽n"; insertions relative to β maj are enclosed by parentheses and overscored with a +; the overall region of detectable homology is bracketed. The data are from Konkel et al [13], as modified [14].

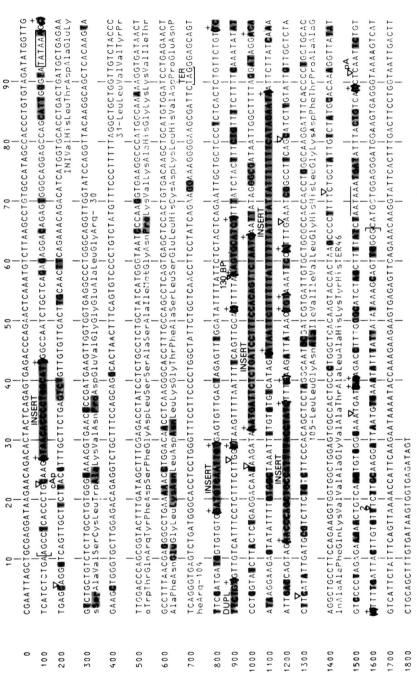

Fig. 6. The nucleotide sequence of the mouse beta minor gene. See legend for Figures 4 and 5. The data are from Konkel et al [14].

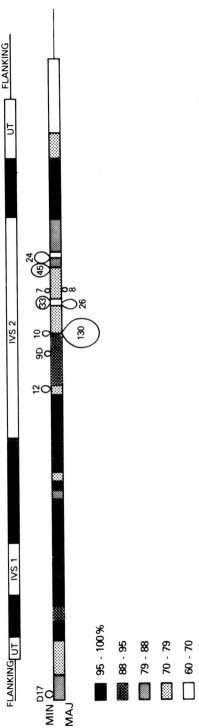

Fig. 7. Homology map of the beta major and beta minor genes. Increased homology is indicated as an increase in density of stipple. Large insertions are shown as loops, with the number indicating the number of bases not represented in the other sequence.

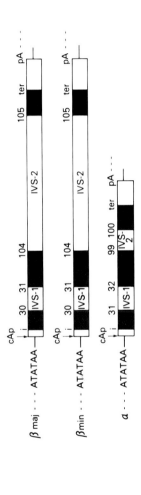

Fig. 8. Comparison of the mouse alpha and beta globin genes. The location and sequence of signals putatively involved in transcriptional initiation RNA processing, RNA splicing, and polyA termination in the globin genes. The diagram represents the general structure of the alpha and beta globin genes. The filled regions represent coding sequences; the open regions represent intervening and flanking sequences, as indicated. The sequences indicated above are preserved regions corresponding (from 5′–3′) to a putative promoter and capping site, splicing signal, and polyA addition site.

The GT/AG Rule of Intervening Sequence Excision

Beta Major

    IVS-1   GGCAG(GT........AG)GCTGCT

    IVS-2   TCAGG(GT........AG)CTCCTG

Beta Minor

    IVS-1   GGCAG(GT........AG)GCTGCT

    IVS-2   TCAGG(GT........AG)CTCCTA

Alpha

    IVS-1   GAAAG(GT........AG)GATGT

    IVS-2   TCAAG(GT........AG)CTCCTG

Fig. 9. A comparison of the regions surrounding the intervening sequences of alpha and beta globin genes. The intervening sequences are indicated as being spliced out between the dinucleotides, GT and AG.

provide selective disadvantages to the organism. Point mutations, on the other hand, involving the alteration of only a single amino acid, might be more easily tolerated. Thus, we imagine that sequences not subject to strong pressures undergo evolutionary drift thorugh both point mutation and deletion-insertion mechanisms. These same mechanisms affect coding or other critical sequences, but, at least in the coding regions, they are selected against by virtue of their drastic consequences. Indeed, these results suggest far greater mutational activity than could have been predicted from a study of the amino acid sequence alone.

PRESERVED REGULATORY REGIONS

We already have noted that there are exceptions to the loss in sequence homology between the alpha and beta genes. These exceptions, in addition to preserved coding sequences, are short, preserved sequences that are positioned so that they might encode signals for transcriptional initiation, polyadenylation, and RNA splicing. Such signals are indicated diagrammatically in Figure 8 and in precise detail in Figures 4, 5, and 6. The gene positions of the capped nucleotides of the alpha and beta mRNAs can be determined from a comparison with the sequences

Orientation Maps of SV40–α-Globin Recombinants

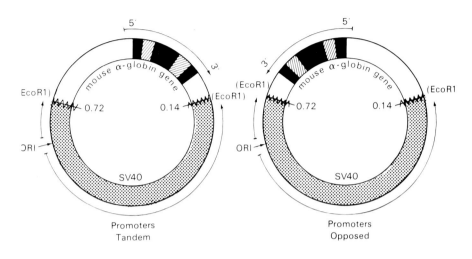

Fig. 10. A diagrammatic representation of the cloned alpha genes in SV40. The gene has been cloned in opposite orientation in each hybrid virus, so that the sense strand of the alpha gene will be under control of the late SV40 promoter in the "tandem" orientation and under control of the antisense in the "opposed" orientation. Tandem and opposed refer to the presumed orientation of the late SV40 and putative globin gene promoters. Numbers inside the circle represent positions on the SV40 map. Stippled areas are SV40 sequences; clear, filled, and hatched areas represent alpha globin flanking, coding, and intervening sequences respectively. The "sense" orientation of the globin gene insert is given by the $5'-3'$ arrow. "ORI" indicates the SV40 origin of replication. The globin gene has been inserted using synthetic Eco R1 linker sequences.

of relevant RNAs [23]. Some evidence suggests that these sequences correspond to the initial nucleotides incorporated into the mRNA [24]. While evidence that these sites constitute the promoter region is not firm, they remain the most promising candidates for these functions. In each of the three globin genes, located exactly 25 nucleotides to the 5' side of these capping sites, a sequence TATAA/G may be found. Very similar sequences are preserved in the same region on the 5' side of the histone genes [25] and in certain adenovirus mRNA promoter regions [26, 27]. Interestingly, these sequences are located almost exactly three turns of a double helix from one another, in terms of the three-dimensional structure of DNA. They would, therefore, be accessible through the major groove of the DNA double helix on the same surface of the DNA structure. Such sites bear an obvious analogy to the Pribnow boxes [28] or bacterial gene-promoter sequences that have been located on the 5' side of known bacterial promoter sequences.

Additional regions of homology may be noted between the borders of the intervening and coding sequences of all three genes. These are illustrated in Figure 9. In each case, an intervening sequence may be spliced out between the dinucleotides, GT and AG. Additional sequence homologies are present, but these quickly break down, particularly within the compared intervening sequences of the alpha and beta globin genes (see Figs. 4, 5, and 6). And finally, as had been known from studies of the RNA [29], the ubiquitous pentanucleotide AATAA can be located approximately 80 nucleotides from the presumed polyA addition site of the beta major and alpha genes. While these correlation analyses suggest that these sequences might be involved in transcription, splicing, or polyA addition functions, direct experimental tests must be applied.

DEVELOPMENT OF A BIOLOGIC TEST SYSTEM: PROOF OF UNIVERSALITY OF RNA-SPLICING MECHANISM AND A PHYSIOLOGIC ROLE FOR INTERVENING SEQUENCES

Obviously, one would hope to be able to introduce these genes into biologic systems in which their expression could be assessed. We have been using the small, well-defined monkey virus, SV40, for this purpose [30]. Appropriate vectors that will accommodate slightly more than 2,000 base pairs of foreign genetic information can be derived from the virus. Making use of appropriate restriction sites, and the alpha globin genes have been introduced into this virus [31, 32]. An example of such a cloned gene within the viral vector, (the gene for alpha globin) is shown in Figure 10.

Using a variety of inserted fragments, we were able to show that the presence of at least one intervening sequence, either derived from the virus or from the globin gene, was necessary for the production of a stable transcript in SV40-hybrid-infected cells [33]. These results strongly suggest that the intervening sequence and/or the splicing mechanism play an essential role in the expression of these genes. Moreover, by using the vectors containing either the cloned beta globin or alpha globin genes to infect African green monkey kidney cells, it was shown that such infected cells produce relatively large amounts of authentic alpha and beta globin [31, 32]. These experiments indicate that the splicing mechanism operates nonspecifically across species boundaries and is not specific for a particular gene product.

In our earlier experiments it was shown that the appropriate splicing signal as well as the appropriate polyA addition sites was utilized in the infected cell [30]. These experiments were done with the beta gene in an orientation such that it would utilize the SV40 late-region promoter site. Naturally, we were anxious to determine whether or not the globin gene fragment carried its own promoter and whether such a promoter could function in SV40-infected cells.

In order to test this possibility, we have inserted the smaller alpha globin gene into two orientations, with respect to the late SV40 promoter (Fig. 10). In one orientation, the alpha globin promoter is located in tandem with and 3' to the late SV40 promoter. In the other orientation, the late SV40 promoter is opposed to the SV40 promoter, and the latter directs the transcription of an antisense copy of the alpha globin gene, while the putative alpha globin promoter should direct the synthesis of the globin sense mRNA. Cells were infected with SV40 hybrids carrying the alpha globin gene in both orientations. In both cases, alpha globin was produced.

Such results suggest that a region on the 5' side of alpha globin gene does contain an active promoter and, furthermore, that such an active mouse-globin promoter functions efficiently in an African green monkey kidney cell. While we still have reservations about the possibility that alpha globin expression depends upon the SV40 early region promoter, such results encourage us to believe that the SV40 system constitutes a useful biologic assay for the function of many gene sequences derived from a variety of sources.

SUMMARY

The globin genes represent a complex set of sequences that are expressed in a coordinate fashion during the development of red blood cells. While this complex family of genes may consist of as many as ten to fourteen members [34], three of these genes have now been cloned and their entire nucleotide sequence determined. As was initially observed in the case of beta globin major gene, all are encoded in three distinct coding blocks separated by two intervening sequences of DNA. Their intervening sequences of DNA are preserved, with respect to location, but are widely divergent, with respect to size and sequence. The divided information in each gene is edited and spliced together at the level of its initial RNA transcript which is complementary to the entire gene sequence inclduing its intervening sequences. Structural correlation analyses have allowed us to identify sites in all three genes that might be responsible for the initiation of transcription, RNA splicing, and polyA addition. The function of these sites has been tested by cloning these genes in an animal virus vector, SV40. Such animal virus hybrids have been used to infect tissue culture cells and have directed the synthesis of both alpha and beta mouse globin in cells of monkey origin. These studies indicate that such signals operate across species barriers and further indicate that the animal virus vector system will be useful in elucidating their function.

REFERENCES

1. Science Vol 196, No. 4286, 1977.
2. Maxam AM, Gilbert W: Proc Natl Acad Sci USA 74: 560–564, 1977.
3. Thomas M, Cameron JR, Davis RW: Proc Natl Acad Sci USA 71: 4579–4583, 1974.

4. Leder P, Tiemeier D, Enquist L: Science 196: 175–177, 1977.
5. Sternberg N, Tiemeier D, Enquist L: Gene 1: 255–280, 1977.
6. Tiemeier DC, Tilgham SM, Leder P: Gene 2: 173–191, 1977.
7. Maniatis T, Hardison RC, Lacy E, Lauer J, O'Connell C, Quon D, Sim GK, Efstratiadis A: Cell 15: 687–701, 1978.
8. Clarke L, Carbon J: Cell 9:91–99, 1976.
9. Benton WD, Davis RW: Science 196: 180–182, 1977.
10. Leder P, Honjo T, Seidman J, Swan D: Cold Spring Harbor Symp Quant Biol 41:855–862, 1977.
11. Tilghman SM, Tiemeier DC, Polsky F, Edgell MH, Seidman JG, Leder A, Enquist LW, Norman B, Leder P: Proc Natl Acad Sci USA 74: 4406–4410, 1977.
12. Tilghman SM, Tiemeier DC, Seidman JG, Peterlin BM, Sullivan M, Maizel JV, Leder P: Proc Natl Acad Sci USA 75:725–729, 1978.
13. Konkel DA, Tilghman SM, Leder P: Cell 15: 1125–1132, 1978.
14. Konkel DA, Maizel JV, Leder P: Cell 18: 865–873, 1979.
15. Leder A, Miller HI, Hamer DH, Seidman JG, Norman B, Sullivan M, Leder P: Proc Natl Acad Sci USA 75:6187–6191, 1978.
16. Nishioka Y, Leder P: Cell 18:875–882, 1979.
17. Nishioka Y, Leder A, Leder P: Proc Natl Acad Sci USA 77:2806–2809, 1980.
18. Curtis PJ, Weissman C: J Mol Biol 106:1061–1075, 1976.
19. Ross J: J Mol Biol 106:403–420, 1976.
20. Tilghman SM, Curtis PJ, Tiemeier DC, Leder P, Weissman C: Proc Natl Acad Sci USA 75:1309–1313, 1978.
21. Osden RC, Beckman JS, Abelson J, Kans HS, S'oll D, Schmidt O: Cell 17:399–406, 1979.
22. Tiemeier DC, Tilghman SM, Polsky FI, Seidman JG, Leder A, Edgell MH, Leder P: Cell 14:237–245, 1978.
23. Baralle FE, Brownlee GG: Nature 274: 84–87, 1978.
24. Curtis PJ, Mantei N, Weissman C: Cold Spring Harbor Symp Quant Biol 42: 971–984, 1977.
25. Goldberg ML, Hogness DS: Personal communication.
26. Ziff EB, Evans RM: Cell 15:1463–1475, 1978.
27. Akusjärvi G, Petterson U: J Mol Biol 134:143–158, 1979.
28. Pribnow D: J Mol Biol 99:419–443, 1975.
29. Proudfoot NJ, Brownlee GG: Nature 263:211–214, 1976.
30. Hamer DH, Leder P: Cell 17:737–747, 1979.
31. Hamer DH, Leder P: Nature 281: 35–40, 1979.
32. Kaehler M, Hamer DH, Leder P: Unpublished results.
33. Hamer DH, Leder P: Cell 18: 1299–1302, 1979.
34. Konkel DA, Leder A: Unpublished results.

Gene and Chromosome Organization: Chairman's Summary

Margaret C. Green

The talks we have heard in this session of the symposium have considered gene and chromosome organization on two very different levels. The first or more gross level, described by Dr. Eicher, is based on information obtained mostly from classic genetic crosses and related types of experiments. For the mouse, classic linkage crosses have been in use since 1915 and, together with recombinant strains and somatic cell hybrids, have been accumulating the information summarized in the 1979 linkage map, with its approximately 450 loci distributed on 20 chromosomes. At the second or fine structure level, as we have heard from Dr Leder, new techniques for studying the molecular structure of genes have been developed in the last few years, which allow individual genes, naturally present as pairs in the cell, to be produced in large enough quantities to be sequenced and compared with their messenger RNA (mRNA). These experiments have yielded the surprising information that the genes contain long intervening sequences that are transcribed into RNA in the nucleus and then spliced out of the RNA to make the mature mRNA found in the cytoplasm. Intervening sequences have been known for less than three years.

I suppose a summarizer should try to say something about the significance of the research described. In doing this, I have to speak as one who has spent most of her professional life concerned with the first level of gene and chromosome organization. It hardly seems necessary to talk about the significance of the findings about split genes and spliced mRNA. These findings have opened the door to many new areas of research and have raised many questions. Some of the questions are: How general is the occurrence of intervening sequences? How does splicing of mRNA take place? Do the intervening sequences have a regulatory function, or any function at all? How did intervening sequences evolve? Dr. Leder's paper begins to answer some of these fascinating questions. In addition, I wonder if it is at all likely that sequencing of intervening and flanking sequences will tell us anything about the nature of recombination in higher organisms, particularly whether the process is localized to a particular kind of sequence. Answers to these and other questions will surely take us a long way toward understanding how genes function.

What about the significance of the work reported by Dr. Eicher? Is the new methodology likely to put the old out of business? I think not. A linkage map provides information not discoverable by the methods of DNA sequencing but necessary, nevertheless, for understanding how chromosomes are organized. As Dr. Eicher has told us, from classical linkage crosses using visible as well as biochemical and antigenic loci and from recombinant inbred strains and congenic strains, we learn about the gross distribution of loci on chromosomes. This information is important because it can give us some clue as to how activity of genes with related functions is controlled.

We can see which kinds of loci appear to be scattered randomly and whose functional relationship, if any, therefore, does not depend on their position on the chromosomes. The loci of many enzymes that catalyze sequential steps in a chain of reactions appear to be randomly distributed. Another example, as Dr. Leder has pointed out, is the location of the genes for the α- and β-globin chains on different chromosomes.

We can see which kinds of loci are clustered and, therefore, which may be related; and we can speculate about whether the clustering is necessary for the proper functioning of the genes, or whether it is accidental and, perhaps, results from historical processes such as origin of the genes from a common ancestor. Clustered genes in the mouse include the *H-2,* or major histocompatibility complex and its associated immune response and complement-determining loci; the immunoglobulin heavy-chain *(Igh)* region, containing a large number of loci determining the constant and variable regions of the heavy chain; and the cluster of at least seven esterase loci within 10 recombination units on Chromosome 8. Other closely linked clusters comprise loci with apparently unrelated functions such as the group around the caracul (Ca) locus on Chromosome 15, which affect the nervous system, skin and hair, and blood formation. These clusters may result from a localization of chiasmata, or crossover points, in some other part of the chromosome, so that recombination rarely occurs within the cluster. It is important to remember that linkage maps measure distance in terms of recombination frequency between loci, which is assumed to be roughly proportional to physical distance, but will not be so in any cases where the crossover points tend to be localized.

Information about the chromosome assignment of genetic loci also comes from the techniques of somatic-cell genetics, which Dr. Eicher has described; and these techniques are useful for loci for which there are no genetic variants. Somatic-cell hybrids recently have been used by Swan et al [1] to demonstrate that both the constant and variable regions of the immunoglobulin kappa-chain complex are on Chromosome 6, although the experiment does not tell us how far apart they are on that chromosome. The location of the constant region of kappa was not known before, because there are no known genetic variants for this region.

It also is not possible to show by analysis of restriction endonuclease fragments that the two regions of DNA are on the same chromosome, if there is a sequence containing endonuclease sites between them; and this is the case for the kappa chain as well as for the lambda chain.

This is an example of the usefulness of information obtained by the methods of level one for work at level two. It should make considerable difference for theories about the rearrangement of immunoglobulin genes in immunoglobulin-producing cells to know whether the constant and variable regions are located near each other, far apart on the same chromsome, or on different chromosomes, in the progenitor cells. As another example, Dr. Leder pointed out that any hypothesis about the control of production of α- and β-globin chains must take into account the fact (known from classical genetic studies) that the two β-chain loci are very closely linked and are on a separate chromosome from the α-chain loci.

It seems extremely unlikely that no more important loci remain to be discovered and mapped. The techniques of isoelectric focusing and two-dimensional electrophoresis are identifying new protein variants at an increasing rate. Other methods — even use of endonuclease fragments and gene sequencing — will probably yield many new variants. These loci will all be more useful when their chromosome locations are known.

Finally, a small take-home message from this session is that science is full of surprises. Back in 1969–1970, when it seemed to mouse cytogeneticists that the genetic characteristics of individual mouse chromosomes would remain a mystery through our inability to tell one chromosome from another, the sudden and surprising discovery that chromosomes had banding patterns that made them individually identifiable opened up a whole new area of research. This was a relatively minor discovery, to be sure, but a pleasant surprise, nevertheless. Of more far-reaching significance, the sudden and surprising discovery within the last two or three years of intervening sequences and mRNA splicing is producing a small revolution in molecular genetics that promises to bring great advances in understanding how gene expression is controlled.

Who knows what may happen tomorrow? The immediate product of most genes that cause morphological and pigment abnormalities is not known. The agouti locus, for example, has at least 17 different alleles affecting the relative amounts of black and yellow pigment in the coat, and we have very little idea how the alleles do this. Perhaps someone will invent a mehtod for identifiying the DNA of genes in the absence of knowledge of their immediate product, say by their close linkage to an identifiable gene or by the use of very small deletions. By the 100th anniversary, we may know what such inscrutable genes as those at the agouti locus do.

REFERENCE

1. Swan D, D'Eustachio P, Leinwand L, Seidman J, Keithly D, Ruddle FH:Chromosomal assignment of the mouse *k* light chain genes. Proc Natl Acad Sci USA 76:2735, 1979.

SESSION II. ANALYSIS OF MAMMALIAN DIFFERENTIATION

JAMES D. EBERT, Chairman

Introduction to Session II

In 1929 (the starting date of our Century of Progress in Mammalian Genetics and Cancer), mammalian developmental biology was a distinctly limited field, restricted by the protective, attached, internal position of embryos and fetuses. Available approaches allowed only descriptions based on dissections and histological preparations, each of which ended the life of the developing individual. Potentialities are much more extensive now. Embryos and fetuses still grow implanted in that same protective environment, but methodology has been developed to manipulate mouse zygotes and embryos at preimplantation stages, to grow these treated embryos to the blastocyst stage in culture, and to arrange for implantation of these treated blastocyst-stage embryos into receptive surrogate mothers, where they continue development appropriately. The first and third papers in this session, by Virginia Papiaoannou and Karl Illmensee, report very recent and very marvelous manipulations yielding results that have greatly increased our understanding of the control of mouse differentiation and development. The second paper, by Roy Stevens, gives us a history of development and utilization of another research tool valuable for analysis of differentiation; that is, mouse teratomas and teratocarcinomas, in which many tissues can differentiate, but not as organized embryos. Analysis of abnormal development has been extremely productive in helping us to understand the normal process.

Almost all of the progress reported in these papers has been achieved in the past quarter century, and much of it within the past decade, but I feel that it is appropriate to remind ourselves that one of the most important steps, successful development of young following transfer of early embryos from one inbred strain into surrogate mothers of a different strain, was achieved by Dr. Elizabeth Fekete early in the history of the Jackson Laboratory (1942). Her chief interests were in late-life characteristics (incidence of mammary cancer) rather than in prenatal development, but her successful methods provided inspiration for the later investigations. While the central feature of analysis of mouse development is manipulation of preimplantation embryos, these efforts would come to naught if the treated embryos could not continue their differentiation. First, mouse embryos must be maintained successfully in vitro until they reach the blastocyst stage. Then they must be implanted into surrogate mothers. Developmental biologists should give a sincere vote of thanks to the reproductive physiologists for their attention to techniques that assure regular production of appropriately timed pseudopregnant recipient females to support implantation and normal development of those manipulated embryos.

In the first paper of this session, Dr. Virginia Papaioannou points out that experimental chimaeras, individuals that continue development following fusion of cells from two or more genetically different embryos, provide excellent research tools for elucidating paths of development. She describes two elegant methods for making chimaeric mice. The aggregation techniques of Tarkowski and of Mintz bring about cohesion of cells from four to eight cell embryos which divide in culture to form a single blastocyst. Gardner has developed a different method in which single cells are placed in the cavity of a blastocyst embryo and continue their development integrated with cells of the recipient. Wonderful results have been achieved by each of these methods.

I must not "predigest" Dr. Papaioannou's lovely story; you will get your "fate map" information much better directly from her. Nevertheless, I want to stress three points. 1) Predetermination is delayed and much less rigid in mammals than in Drosophila and many other invertebrates; at the four-cell cleavage stage each cell can contribute to all embryonic and extraembryonic tissues. 2) The setting aside of trophectoderm, whose giant cells invade maternal tissue at implantation, is a very important feature of differentiation. 3) Determination of separate fates seems to be a gradual process, and observation of Gardner-type chimaeras suggests strongly that not more than four cells from the inner cell mass of a four-day embryo contribute to the fetus proper.

There have been many "operation bootstraps" in the career of our second speaker, Dr. Leroy C. Stevens. More than the rest of us, he has created his own research material. One of his first discoveries at The Jackson Laboratory in the early 1950s was that 1% of male mice of strain 129 had spontaneous testicular teratomas. One percent is a pretty low number, but it is real and significant, a "find" for a student of development.

Teratomas usually arise in the gonads and are derived from germ cells. The tumors contain many kinds of tissues, derived from all three primary germ layers. Some, known as teratocarcinomas, contain stem (embryonal carcinoma) cells that are very much like early embryonic cells. Teratocarcinomas may be established as transplantable tumors. Both original and transplanted tumors are valuable for study of differentiation and development.

Through many tests, Stevens learned that certain named genes could increase incidence of teratomas, so he backcrossed these genes into the 129 inbred background, markedly raising teratoma incidence. At this point fate was kind to him! A spontaneous mutation, later given the symbol *ter*, occurred in his special "doctored" 129 strain, and increased teratoma incidence to 30%. Now he has really tractable research material.

Steven's work has not been limited to these spontaneous teratomas and teratocarcinomas. He has developed and maintained transplantable teratocarcinoma cell lines. Further, he has found that 12–13-day fetal gonadal ridges, when implanted inside the testicular capsule, develop into teratomas, particularly in some inbred strains. Still further, he has found that six-day embryonic ectoderm implanted into adult testis or kidney yields teratomas and teratocarcinomas.

His studies also include ovarian teratomas, which develop in LT/Sv females from parthenogenetically activated retained ovarian eggs.

Dr. Stevens will tell you of experimental uses of these fascinating "differentiating" tumors, but I feel it's up to me to let you know that throughout his career, Steve has been very generous in "sharing his wealth," freely passing out his special stocks and transplantable tumors to many, many other investigators.

The final speaker in this session, Dr. Karl Illmensee, turns our attention back to experimental manipulation of developing mouse embryos and carries analysis to the intracellular level. He tells us first about experimentally induced uniparental mice. To produce these "instant homozygous" individuals, Hoppe and Illmensee removed one of the haploid pronuclei from each zygote immediately after fertilization. The remaining pronucleus (paternal or maternal) was encouraged by cytochalasin B to diploidize, and the embryos continued cleavage. The resultant blastocysts developed normally in surrogate mothers. Isogenicity was confirmed by assessing a variety of biochemical and other genetic markers.

In other experiments, cells from one of Stevens' teratocarcinomas were incorporated into chimaeras, where they participated in normal development, expressing their own normal genetic repertoire and contributing to the germline.

Cultured mouse teratocarcinoma cells have been fused with human cells, and resulting hybrid cells (containing some human as well as many mouse genes) have been integrated into chimaeras that grew to adulthood. Human genes were expressed in some of the tissues of these chimaeric mice.

These examples are only a slight introduction to Illmensee's story, but I hope I make it clear that experimental manipulations of embryos have already led to great advances in understanding of differentiation, and that future developments will be very exciting.

Experimental Chimaeras and the Study of Differentiation

Virginia E. Papaioannou

INTRODUCTION

If embryos were merely preformed versions of adult animals there would be no need to study determination and differentiation, the processes by which cells become committed to fates and become different from one another. Embryology would then be confined to studying the unfolding of a preformed package. But even Aristotle formulated convincing arguments against this view, arguments drawn from logical considerations as well as from his own observations on embryos. He believed that all the parts of an adult exist potentially in an embryo, not actually, so a process of differentiation must occur to produce the final product. Embryonic development could then be seen as a changing in the quality of the parts of the embryo so that finally they become in actuality what they are potentially.

As modern and reasonable as these ideas seem to us today, they were not generally accepted until the late 18th century, the controversy between preformation and epigenesis having been sustained for centuries. The epigenetic view held that the embryo was created anew from a structureless egg that, during development, was subject to external forces. Paradoxically, the development of the microscope boosted preformation ideas for a time, when with its help, early investigators imagined they could see minute but fully formed creatures in each and every sperm. The only difficulty seemed to be how to explain Nature's apparent wastage of so many souls, since only a small number of sperm actually settled into the womb to unfold as individuals. In the early 19th century the mammalian egg was finally discovered. This finding had long been hampered by the limits of eyesight and of early lenses and also by the blinkered vision of scientists searching for a more recognizably adult animal form. With the use of technically better microscopes and the beginning of experimentation on embryos, preformationist ideas finally were laid to rest and the science of experimental embryology came into being to investigate processes involved in the development of a multicellular adult from a single undifferentiated cell.

Development has been described by Needham [1] as consisting of the closing of doors, of determination or the progressive restriction of the possible fates of cells. But how and when does this happen for any particular group of cells in an embryo? When are cells committed to specific developmental pathways and how do they become differentiated? The development of mammals, occurring as it does in the uterus of the mother, presents special difficulties for elucidating these processes. It is only very recently that innovative techniques for the production of experimental chimaeras in mammals have given us a way of "watching" the cells of a mammalian embryo in utero, not in actual fact, but by inference from the positions of the genetically different cell populations in the chimaera at different stages in development. In this way the early cell lineages in the mammalian embryo can be traced and the differentiation of cells into specific cell and tissue types can be studied.

METHODS OF PRODUCING CHIMAERAS

As in many other areas of research, features of the laboratory mouse make it the animal of choice for experimental embryology and especially for the production of chimaeras. The abundance of genetic markers, the resilience of the embryos to perturbations, and the ease with which preimplantation embryos can be transferred to a foster mother's uterus for development all contribute to the usefulness and suitability of the mouse embryo as a model for mammalian differentiation and for exploiting the unique features of experimental chimaeras.

There are two general methods of producing experimental chimaeras. The first, devised by Tarkowski [2] and by Mintz [3, 4], is by the aggregation of cells from different cleaving mouse embryos (Fig. 1) and the second, devised by Gardner [5], involves the injection of cells into embryos at the blastocyst stage (Fig. 2). Both methods have been elaborated in various ways to address a variety of embryological, genetic, immunological and other types of questions (see McLaren [6] for an extensive review of the many uses of chimaeras). Different mammalian species have been used [7–10], interspecific chimaeras made [11–13], and the numbers and stages of embryos have been varied [14–18]. The injection procedure has been modified for the construction of chimaeric blastocysts with an inner cell mass (ICM) from one embryo, trophectoderm from another [19; also Papaioannou, unpublished] (Fig. 3), and for the injection of cells into postimplantation embryos both in utero [20] and in vitro [21].

In all of these experiments, naturally occurring genetic variation has been used to distinguish the two or more different components in the chimaeras. In the analysis of adult chimaeras, pigmentation differences commonly have been used (Fig. 4) to detect coat and eye chimaerism, while a variety of biochemical differences and cytogenetic markers have been exploited to analyze internal organs and tissues. In the analysis of chimaeras at embryonic stages, the useful markers are more limited, and allozyme differences of the ubiquitous enzyme glucose phosphate isomerase have been used most often. With this biochemical marker

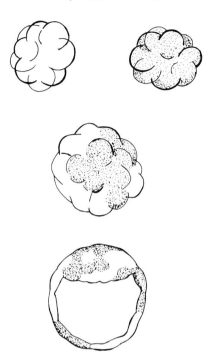

Fig. 1. Morula aggregation for the production of chimaeras. Genetically different morulae (top) flushed from the oviducts of pregnant mice are cultured in contact with one another after removal of the nonliving zonae pellucidae. A single, large morula results (center), which then forms a blastocyst (bottom) with the trophectoderm, the outer cell layer, and the inner cell mass composed of cells of the two genotypes. The blastocyst is transferred to the uterus of a foster mother for further development.

there is a severe limitation of resolution due to the necessity for tissue homogenization; nonetheless, it is the best general marker available at present for tracing the progeny of cells in a chimaera.

CHIMAERAS AND THE STUDY OF EARLY DIFFERENTIATION

The principle involved in tracing embryonic cell lineages with chimaeras is the same as in using vital dyes, but instead of marking an existing cell of the embryo with a dye, a genetically marked cell(s) from a different source is added to the embryo. The capacity of the early mammalian embryo to regulate for disturbances in cell number means that the marked cell or cells can be incorporated into the embryo at different stages, resulting in a single normal foetus with two distinct but coexisting cell populations. This regulation is an essential feature for our purposes, since the accuracy with which we can define the fate of a cell depends on the degree of disturbance that the cell suffers as a result of the transplantation

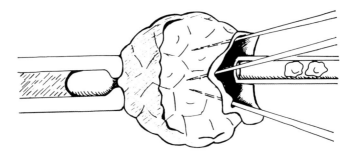

Fig. 2. Blastocyst injection method for producing chimaeras. A preimplantation blastocyst, flushed from the uterus of a pregnant mouse, is held in position with a suction pipette (left). A hole is made in the trophectoderm opposite the inner cell mass with three glass needles, and another pipette (containing genetically different cells from another embryo) is introduced into the blastocoelic cavity. The cells are released from the second pipette into the cavity or directly into the inner cell mass. The needles and pipette are withdrawn and the embryo collapses. After the blastocyst heals and reexpands, it can be transferred to the uterus of a foster mother for development.

procedure. If the cell is undisturbed by transplantation into its normal position in another embryo, or at least behaves as if it were undisturbed, then it is possible to determine its normal fate. On the other hand, the potential of a cell, that is, the full range of developmental performances of which it is capable, can also be investigated if a chimaera can be made by putting the cell either temporally or spatially into an abnormal position. It follows that the stability of cellular differentiation is being tested, making it possible to analyze the time course of commitment and differentiation in early embryogenesis.

In the field of experimental embryology, chimaeras made within a single species with normal embryonic tissue yield the most useful information on cell fate and lineage. However, in some situations, heterogeneous combinations of tissue such as parthenogenetic, mutant, or malignant with normal can be used to pose specific questions about normal differentiation. The survival of chimaeras produced from normal and parthenogenetic mouse embryos [22] provides scope for the study of the developmental problems that lead to the death of parthenotes alone. The various other types of chimaeras will be discussed in the following sections with respect to the information they provide on the earliest differentiative steps in mammalian development.

Chimaeras Incorporating Normal Genetic Variants

The cells of cleaving mouse embryos are very similar in appearance until the formation of the blastocyst. There appears to be no cytoplasmic regionalization of specific morphogenetic determinants; the destruction of one blastomere at the

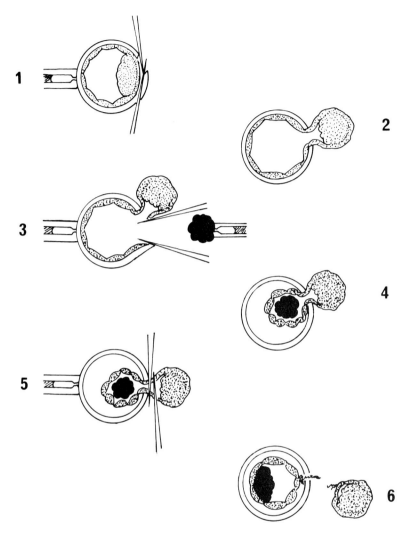

Fig. 3. Reconstitution method for producing chimaeras. Blastocysts are flushed from the uterus of pregnant mice and held in position with a suction pipette. 1. Glass needles are used to open a slit in the zona pellucida directly over the inner cell mass end of the blastocyst. 2. Fluid pressure in the blastocyst forces this end of the embryo out through the slit. 3. An inner cell mass isolated from a genetically different blastocyst (solid) is introduced into that part of the blastocyst remaining inside the zona, with the aid of needles and a suction pipette. 4. The embryo collapses and the zona pellucida holds the two inner cell masses in position. 5. Glass needles are used to cut off the host inner cell mass, along with its overlying trophectoderm. 6. A reconstituted blastocyst, with a host trophectoderm (stippled) and donor inner cell mass (solid) is left inside the zona and can be transferred to the uterus of a foster mother for development.

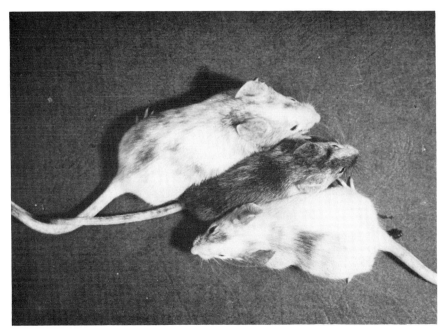

Fig. 4. Three adult chimaeras produced by aggregating morulae from a pigmented strain with morulae from an albino strain.

two-cell stage does not lead to the development of half a mouse, but, rather, to a complete individual [23]. At the four-cell stage it is more difficult to get isolated blastomeres to develop as complete individuals but this is probably due to the inability of the embryo to compensate for such a drastic reduction in cell number, rather than to a restriction in the developmental potential of the individual cells [24] or to the segregation of cytoplasmic factors. Experimental chimaeras provide a novel way of looking at this old embryological problem. The successful regulation of aggregates of even very large numbers of morulae argues against the idea of rigid predetermination of cells [14, 25]. The potential of each cell of a four-cell embryo can be examined in chimaeras by separating the four cells and restoring each to the correct cell number by aggregation with other genetically different blastomeres. This has been done by Kelly [26], who found that all four blastomeres contributed cell progeny to all parts of the later conceptus, foetus, membranes, and placenta, a strong argument against predetermination and one for the totipotency and equipotency of all cells at the four-cell stage. Experiments with eight-cell embryos indicated a similar totipotency of eight-cell stage blastomeres, but the technical difficulty of producing and recovering all eight chimaeras from a single eight-cell embryo is formidable and the results of the experiment were inconclusive [26]. However, if the indications are correct, and the cells of the eight-cell embryo are indeed equal in their capacity, then the lack of overt differentiation probably reflects a lack of covert differentiation at these early stages.

The formation of trophectoderm in the blastocyst stage embryo marks the first step of differentiation. It is thought that the transition from undifferentiated morula cell to differentiated blastocyst cell must, in the absence of cytoplasmic determinants, be an epigenetically controlled phenomenon. A likely controlling factor is the microenvironment of a cell produced by its position on the inside or outside of the morula [27]. Studies with chimaeras have corroborated this idea. A number of experiments using different cell markers (tritiated thymidine label, GPI, or silicon fluid drops), but all using embryo aggregation to position the marked cells either on the inside or outside of the embryos (Fig. 5), have indicated that cells in an outside position of the morula most often differentiate as trophectoderm, while cells in an interior position contribute mostly to the ICM [14, 28]. The specific microenvironmental features of cell position that operate on these cells to influence differentiation are not at all well understood, but experimental chimaeras have indicated the sound footing of the hypothesis.

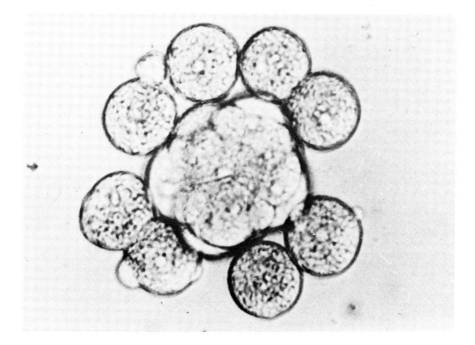

Fig. 5. The whole of one embryo outside another. A four-cell embryo was disaggregated and each blastomere was allowed to divide once. All these blastomeres were then arranged on the outside of another genetically different 8–16 - cell embryo. From Hillman et al[14], with permission from the Company of Biologists.

In mammals, the evolution of intrauterine development necessitates the rapid establishment of maternal-foetal relations for the nutrition of the foetus and thus, the first tissues to develop are sequestered for this purpose. At the blastocyst stage the embryo is not yet attached to the uterus, but implantation soon occurs and a number of further differentiative steps take place quite rapidly. There is differentiation within the trophectoderm. This is a single-cell-thick, uniform layer at the blastocyst stage, but two regions can be defined according to their position relative to the ICM: the polar trophectoderm overlying the ICM, and the mural trophectoderm situated away from the ICM and forming the blastocoel. The mural cells are the first to establish contact with maternal tissue in the implantation crypt and, at the time of implantation, they are undergoing terminal differentiation. They have ceased mitotic division and have begun to endoreduplicate their DNA on the way to becoming the primary giant cells.

There is continued cell division in the polar region, however, and it is here that an ectoplacental cone, and later the placenta form. A large number of secondary giant cells appear at the periphery of the ectoplacental cone and eventually surround the entire conceptus. From the static picture provided by fixed, sectioned material, we can make shrewd guesses about the fate and lineage of these various cells, but it has only been through the use of chimaeras that our interpretations could be critically tested.

ICMs can be dissected either mechanically [29] or by immunosurgery [30] and put into genetically different vesicles of mural trophectoderm made by cutting off the ICM end of a blastocyst (Fig. 3). The normal development of chimaeric foetuses following this procedure indicates that at least some of the mural cells of the preimplantation embryo are not yet committed to forming the primary giant cells and that, if put in contact with an ICM, they can be induced to continue proliferation [19]. Analysis of the normal chimaeras later in development also provides cell lineage information indicating that the trophectoderm does indeed form all the secondary giant cells and the ectoplacental cone. In addition to these tissues, it also appears that the trophectoderm grows down into the egg cylinder to form the entire extraembryonic ectoderm, later the chorionic ectoderm [19; also, V.E. Papaioannou, unpublished], whereas the foetus and all of the other extraembryonic membranes are derivatives of the ICM (Fig. 6).

At about the time of implantation at 4½ days postcoitum, there is visible differentiation within the ICM as a layer of primitive endoderm forms on its exposed surface and migrates around the inner surface of the trophectoderm (Fig. 6). The production of chimaeras by the injection into blastocysts of single cells or groups of cells from the primitive endoderm or the nonendoderm cells of the ICM — the primitive ectoderm — have established the cell lineages of these two tissues [31, 32]. The primitive endoderm cells do not appear to contribute at all to the foetus. Their progeny are found exclusively in the endoderm of the extraembryonic membranes of the embryo by midgestation. The primitive ectoderm cells of the

ICM, on the other hand, form the extraembryonic mesoderm as well as the entire foetus, including the definitive endoderm. Thus, what we have traditionally called ectoderm at the blastocyst stage is, in fact, a multipotential tissue that forms all three germ layers of the adult. The other early-differentiating tissues of the embryo are entirely concerned with the formation of nutritive and supportive extra-embryonic tissues.

Many studies such as these, which use normal genetic variation as cell markers in chimaeric mice, have allowed us to draw up a fate map of early mouse development (Fig. 6). The resulting cell lineage diagram (Fig. 7) indicates the stepwise nature of determination — the closing of doors — that characterizes development.

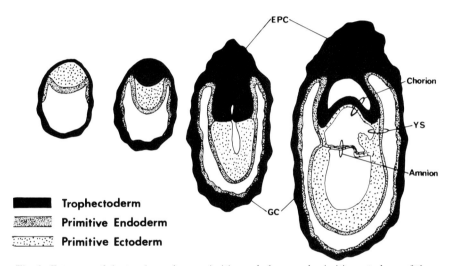

Fig. 6. Fate map of the trophectoderm, primitive endoderm, and primitive ectoderm of the late blastocyst projected onto successive stages through the seventh to eighth day of gestation. The primitive ectoderm-derived tissue (in the bottom half of the drawing on the right) will give rise to the entire foetus. EPC = ectoplacental cone; YS = yolk sac; GC = giant cells.

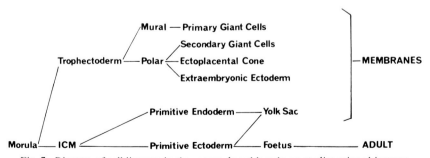

Fig. 7. Diagram of cell lineages in the mouse, based largely on studies using chimaeras.

Chimaeras Incorporating Developmental Mutants

Chimaeras incorporating mutants usually are made with a view to understanding the effect of the mutant gene, but implicit in any study of mutant effect is the hope that a better understanding of normal processes will result from understanding an aberration. A growing number of mutants are being examined in chimaeric combinations and a few of these can potentially provide information on early developmental events. Different mutants at the T/t locus that act at various pre-implantation and post-implantation stages, the agouti locus genes lethal yellow and lethal nonagouti, which act around the time of implantation, and the oligosyndactylism, blind, and velvet genes, which act shortly after implantation, are examples of genes whose lethal effects might be altered in chimaeras in a way that would provide information about their effects on developmental processes and differentiation. A few of these genes already have been the subject of studies with chimaeras.

Mintz [4] was the first to combine lethal with normal tissue. She used the t^{12} mutation which affects embryos at the late morula stage. Although the normal-mutant aggregates formed blastocysts in culture, the mutant cells were still recognizable by features characteristic of the mutant. Their physiology was thus not obviously altered by the proximity of normal cells. Similar results also have been obtained with chimaeras incorporating t^6, t^{w5}, and t^{w32}, three other early-acting alleles (33–35). In all these studies, the lack of alteration of the mutant phenotype in chimaeric combination with normal tissue meant that their usefulness for elucidating normal developmental processes was very limited. There was, however, never any indication that the mutant cells brought about the demise of the normal embryo and this at least indicates that the mutants are not capable of causing a disorganization by disrupting developmental signals between normal cells.

A later-acting allele, t^{w18}, was examined in chimaeric combinations with normal embryos by Spiegelman [33]. In t^{w18} homozygous mutants, the segregation of mesoderm cells from the primitive streak appears to be blocked, since few and very abnormal mesoderm cells appear. Unlike the normal stellate reticular associations of mesoderm cells, aggregated groups of rounded cells appear in the mutant, and the primitive streak becomes abnormally large; further development is prevented. Among the chimaeric embryos, a new class of abnormal embryos that was more advanced than the mutants was seen. These were assumed to be the mutant-normal chimaeras and a distinguishing feature was the presence of some mesoderm cells with a phenotype unlike either mutant or normal mesoderm cells. The tentative conclusion was that the mutant cells lack some signal component that usually prevents mesoderm segregation from the primitive streak, but that the mutant presumptive mesoderm cells in a chimaera are capable of responding to signals from the normal cells. The unusual phenotype of these cells, however, indicates a further block to normal mesoderm differentiation. Nonetheless the change in mutant phenotype in these experiments offers intriguing possibilities for studying the control of mesoderm segregation and migration.

It should perhaps be emphasized that the lack of specific cell-autonomous cell markers linked to particular lethal mutants makes unequivocal identification of individual cells in a chimaera impossible, thus, the recognition of mutant-normal chimaeras is subject to error. In some cases, cellular characteristics of the mutant can serve to identify the mutant cells, but in cases where rescue of the mutant is complete or partial, the attribution of genotypes must remain tentative.

A study of the lethal mutant, A^y, involving the production of chimaeras by ICM injection into blastocysts, gave evidence for a similar partial rescue of lethal tissue by normal cells [36]. Mutant embryos usually die at the time of implantation with death of the trophectoderm cells occurring first. When a mutant ICM is put into a normal blastocyst, however, the mutant tissue appears to be capable of extensive normal proliferation and differentiation in the resulting chimaera, as indicated by the high proportion of mutant cells in normal-appearing tissue at midgestation. The reciprocal chimaera made with a normal ICM in a mutant blastocyst, on the other hand, does not appear to be able to survive and dies at the time the mutant embryo alone normally would die. These experiments support the idea of the gene acting primarily in the trophectoderm. However, since the yellow mutant does not produce a recognizable cellular phenotype prior to cell death, the attribution of genotypes in these chimaeras is not possible, and the conclusions are based on statistical arguments.

Eventually, it may be possible to elucidate the role of cell or tissue interactions in development with the aid of mutants that are not merely cell lethals. Cell-autonomous markers related to or linked to mutants would facilitate this line of investigation greatly.

Chimaeras Incorporating Teratocarcinoma Cells

Another abnormal cell type that has been tested extensively for its potential to form chimaeras is the teratocarcinoma stem cell [37–46]. These cells are obtained from tumours of the embryo or gonad called teratocarcinomas and are extremely interesting cells in that they retain many features of undifferentiated early embryonic cells. Their developmental potential to differentiate into a variety of cell types either in a tumour or in culture reflects their origin from early embryonic cells or germ cells (see L.C. Stevens, this volume). The apparent normality of the differentiated cell progeny of teratocarcinomas, in spite of the malignant features of the stem cell, has led to their widespread use as models for differentiation [47]. The advantages of studying a large number of cells differentiating in culture as opposed to a few cells differentiating in an embryo in utero are considerable. The biochemical events of differentiation, for example, can be much more easily followed in vitro.

Chimaeras incorporating teratocarcinoma cells, like chimaeras incorporating mutant cells, are excellent systems in which to test the potential of the abnormal cells. The chimaeras have been used in studies of gene expression and malignancy, and their potential to transmit mutant genes selected in teratocarcinoma cell culture to the mouse germ line has not been overlooked [42, 46, 48]. In developmental studies, these chimaeras present tantalizing possibilities for the study of differentiation and control of proliferation although it is still not well understood how the embryo exerts the control over this malignant cell to direct its growth in a chimaera.

Perhaps the most valuable contribution teratocarcinoma cell chimaeras have made so far to the study of development, however, is a validation of the use of teratocarcinoma differentiation as a model for normal differentiation. Teratocarcinoma cells from either spontaneous or experimentally induced tumours, taken from culture or direct from the tumour are capable of participating in normal development in a chimaera if placed in a blastocyst stage embryo. Different tumour lines behave differently in chimaeras and some show retention of malignancy [45]. However, the normal chimaeras developing after introduction of cells from several different teratocarcinoma cell lines or tumour lines [37–46] indicate the basic similarity of teratocarcinoma and embryonic cells.

The teratocarcinoma cells do not always participate in development in a manner identical to normal embryonic cells and chimaeras with uneven tissue composition are often seen. The reasons for this are, as yet, not clearly understood, but may in some cases reflects a limitation in potential or a partially differentiated cell state. For example, very few chimaeras were produced with cell line C145b [44] but in every case the C145b cells were detected only in the yolk sac. Since cultured C145b cells easily differentiate into cells resembling endoderm, it is likely that the contribution to the chimaeric yolk sacs was also endoderm.

Taken together, these observations suggest that C145b may be an example of a line of cells with limited, endoderm-forming potential. Other lines might well provide material for studying other specific developmental pathways.

Chimaeras incorporating teratocarcinoma cells undoubtedly will continue in their role of validating the use of the differentiation of this malignant cell as a model for differentiative steps in development. The normal chimaeras also have encouraged the study of cell-cell interactions in normal cell-teratocarcinoma cell combinations [49] in the expectation that cellular interactions within an embryo will be similar.

The various sorts of chimaeras discussed have provided a new means of examining a variety of embryological problems, especially in the early developmental stages. They have provided interesting new material and raised exciting new questions for the future.

ACKNOWLEDGMENTS

I would like to thank John West and Richard Gardner for helpful advice on the manuscript, and Teresa Clements and Joan Brown for help in its preparation. I also thank the Medical Research Council for financial support.

REFERENCES

1. Needham J: "Order and Life." New Haven: Yale University Press, 1936.
2. Tarkowski AK: Mouse chimaeras developed from fused eggs. Nature 190:857, 1961.
3. Mintz B: Formation of genotypically mosaic mouse embryos (abstract). Am Zool 2:432, 1962.
4. Mintz B: Formation of genetically mosaic mouse embryos, and early development of 'lethal' (t^{12}/t^{12})-normal mosaics. J Exp Zool 157:273, 1964.
5. Gardner RL: Mouse chimaeras obtained by the injection of cells into the blastocyst. Nature 220:596, 1968.
6. McLaren A: "Mammalian Chimaeras." Cambridge: Cambridge University Press, 1976.
7. Gardner RL, Munro AJ: Successful construction of chimaeric rabbit. Nature 250:146, 1974.
8. Mayer JF, Fritz HI: The culture of preimplantation rat embryos and the production of allophenic rats. J Reprod Fertil 39:1, 1974.
9. Moustafa LA: Chimaeric rabbits from embryonic cell transplantation. Proc Soc Exp Biol Med 147:485, 1974.
10. Tucker EM, Moor RM, Rowson LEA: Tetraparental sheep chimaeras induced by blastomere transplantation. Changes in blood type with age. Immunology 26:613, 1974.
11. Gardner RL, Johnson MH: Investigation of early mammalian development using interspecific chimaeras between rat and mouse. Nature New Biology 246:86, 1973.
12. Mulnard MJ: Embryologie – Formation de blastocystes chimeriques par fusion d'embryons de rat et de souris au stade VIII. C R Acad Sci (Paris). 276:379, 1973.
13. Mystkowska ET: Development of mouse-bank vole interspecific chimaeric embryos. J Embryol Exp Morphol 33:731, 1975.
14. Hillman N, Sherman MI, Graham C: The effect of spatial arrangement on cell determination during mouse development. J Embryol Exp Morphol 28:263, 1972.
15. Markert CL, Petters RM: Manufactured hexaparental mice show that adults are derived from three embryonic cells. Science 202:56, 1978.
16. Mintz B: Allophenic mice of multi-embryo origin. In Daniel JC Jr (ed): "Methods in Mammalian Embryology." San Francisco: Freeman, 1971, p 186.
17. Moustafa LA, Brinster RL: Induced chimaerism by transplanting embryonic cells into mouse blastocysts. J Exp Zool 181:193, 1972.
18. Rossant J: Investigation of the determinative state of the mouse inner cell mass. I. Aggregation of isolated inner cell masses with morulae. J Embryol Exp Morphol 33:979, 1975.
19. Gardner RL, Papaioannou VE, Barton SC: Origin of the ectoplacental cone and secondary giant cells in mouse blastocysts reconstituted from isolated trophoblast and inner cell mass. J Embryol Exp Morphol 30:561, 1973.
20. Weissman I, Papaioannou VE, Gardner RL: Fetal hematopoietic origins of the adult hematolymphoid system. In Clarkson B, Marks PA, Till JE (eds): "Differentiation of

Normal and Neoplastic Hematopoietic Cells. Cold Spring Harbor Conferences on Cell Proliferation Vol 5 Book A." Cold Spring Harbor Laboratory, 1978, p 33.
21. Copp AJ, Beddington R: Production of chimaeric foetuses after grafting of tissues into postimplantation mouse embryos in vitro. Manuscript in preparation, 1979.
22. Stevens LC, Varnum DS, Eicher EM: Viable chimaeras produced from normal and parthenogenetic mouse embryos. Nature 269:515, 1977.
23. Tarkowski AK: Experiments on the development of isolated blastomeres of mouse eggs. Nature 184:1286, 1959.
24. Rossant J: Postimplantation development of blastomeres isolated from 4- and 8-cell mouse eggs. J Embryol Exp Morphol 36:283, 1976.
25. Mintz B: Experimental genetic mosaicism in the mouse. In Wolstenholme GEW, O'Connor M (eds): "Preimplantation Stages of Pregnancy." London: JA Churchill Ltd 1965, p 194.
26. Kelly SJ: Studies of the potency of early cleavage blastomeres of the mouse. In Balls M, Wild AE (eds): "The Early Development of Mammals. British Society for Developmental Biology Symposium 2." London: Cambridge University Press, 1975, p 97.
27. Tarkowski AK, Wroblewska J: Development of blastomeres of mouse eggs isolated at the 4- and 8-cell stage. J Embryol Exp Morphol 18: 155, 1967.
28. Gardner RL: Manipulations on the blastocyst. Adv Biosci 6:279, 1971.
30. Solter D, Knowles BB: Immunosurgery of mouse blastocyst. Proc Natl Acad Sci USA 72:5099, 1975.
31. Gardner RL, Papaioannou VE: Differentiation in the trophectoderm and inner cell mass. In Balls M, Wild AE (eds): "The Early Development of Mammals. British Society for Developmental Biology Symposium 2." London: Cambridge University Press, 1975, p 107.
32. Gardner RL, Rossant J: Investigation of the fate of 4.5 day post coitum mouse inner cell mass cells by blastocyst injection. J Embryol Exp Morphol 52:141, 1979.
33. Spiegelman M: Fine structure of cells in embryos chimeric for mutant genes at the T, t locus. In Russell LB (ed): "Genetic Mosaics and Chimeras in Mammals." New York: Plenum Press, 1978, p 59.
34. Wudl LR, Sherman MI: In vitro studies of mouse embryos bearing mutations at the T locus: t^{w5} and t^{12}. Cell 9:523, 1976.
35. Wudl LR, Sherman MI: In vitro studies of mouse embryos bearing mutations in the T complex: t^6. J Embryol Exp Morphol 48:127, 1978.
36. Papaioannou VE, Gardner RL: Investigation of the lethal yellow A^y/A^y embryo using mouse chimaeras. J Embryol Exp Morphol 52:153, 1979.
37. Brinster RL: The effect of cells transferred into the mouse blastocyst on subsequent development. J Exp Med 140:1049, 1974.
38. Cronmiller C, Mintz B: Karyotypic normalcy and quasi-normalcy of developmentally totipotent mouse teratocarcinoma cells. Dev Biol 67:465, 1978.
39. Dewey MJ, Martin DW, Martin GR, Mintz B: Mosaic mice with teratocarcinoma-derived mutant cells deficient in hypoxanthine phosphoribosyltransferase. Proc Natl Acad Sci USA 74:5564, 1977.
40. Illmensee K: Reversion of malignancy and normalized differentiation of teratocarcinoma cells in chimeric mice. In Russell LB (ed): "Genetic Mosaics and Chimeras in Mammals." New York: Plenum Press, 1978, p 3.
41. Illmensee K, Mintz B: Totipotency and normal differentiation of single teratocarcinoma cells cloned by injection into blastocysts. Proc Natl Acad Sci USA 73:549, 1976.

42. Mintz B, Illmensee K: Normal genetically mosaic mice produced from malignant teratocarcinoma cells. Proc Natl Acad Sci USA 72:3585, 1975.
43. Mintz B, Illmensee K, Gearhart JD: Developmental and experimental potentialities of mouse teratocarcinoma cells from embryoid body cores. In Sherman MI, Solter D (eds): Teratomas and Differentiation." London: Academic Press, 1975, p 59.
44. Papaioannou VE, Evans EP, Gardner RL, Graham CF: Growth and differentiation of an embryonal carcinoma cell line (C145b). J Embryol Exp Morphol, 54:277–295, 1979.
45. Papaioannou VE, Gardner RL, McBurney MW, Babinet C, Evans MJ: Participation of cultured teratocarcinoma cells in mouse embryogenesis. J Embryol Exp Morphol 44:93, 1978.
46. Papaioannou VE, McBurney MW, Gardner RL, Evans MJ: Fate of teratocarcinoma cells injected into early mouse embryos. Nature 258:70, 1975.
46. Martin GR: Teratocarcinomas as a model system for the study of embryogenesis and neoplasia. Cell 5:229, 1975.
48. Mintz B: Teratocarcinoma cells as vehicles for introducing mutant genes into mice. Differentiation 12:25, 1979.
49. Gaunt SJ, Papaioannou VE: Metabolic co-operation between embryonic and embryonal carcinoma cells of the mouse. J Embryol Exp Morphol 54:263, 1979.

Genetic Influences on Teratocarcinogenesis and Parthenogenesis

Leroy C. Stevens

INTRODUCTION

The role of genes in carcinogenesis has always been a prominent theme at the Jackson Laboratory. This report is an example of how a nongeneticist has used simple genetic tools to investigate the origin and development of testicular, ovarian, and embryo-derived teratomas.

Teratomas usually occur in the gonads. They are different from other kinds of tumors, in that they may be composed of many kinds of cells and tissues derived from all three primary germ layers. The stem cells, referred to as embryonal carcinoma cells, resemble early embryonic cells morphologically [1–3], antigenically [4–6], biochemically [7–9], and in developmental potential [10–12]. When teratomas contain embryonal carcinoma cells as well as differentiated cells, they are referred to as teratocarcinomas. When they are composed only of differentiated cells, they are benign. Embryonal carcinomas and teratocarcinomas may be malignant and may be established as transplantable tumors.

Transplantable teratocarcinomas have been established from ovarian [13], embryo-derived [14], spontaneous [15, 16] and experimentally-produced [17] testicular teratocarcinomas. When some transplantable teratocarcinomas are converted to the ascites form, thousands of embryoid bodies may be obtained, suspended in the ascitic fluid. They resemble early-egg cylinder stages of normal development [18]. They are composed of an outer layer of endodermal cells that surround a core of pluripotent or even totipotent undifferentiated embryonal cells.

TESTICULAR TERATOMAS

During the 25th year of The Jackson Laboratory, the first testicular teratomas of the mouse were described [19, 20]. About 1% of male inbred-strain-129 mice had spontaneous congenital teratomas, suggesting that this strain was genetically different from other strains in susceptibility to testicular teratocarcinogenesis. Seven sublines of strain 129 at The Jackson Laboratory were surveyed for their incidence of these tumors, and the frequency ranged from 0.26–1.7%, again indicating a genetic basis for susceptibility to teratocarcinogenesis.

Strain-129 mice were reciprocally crossed with animals of seven other inbred strains [21]. The offspring from these crosses were backcrossed to strain 129. From these crosses and backcrosses, more than 11,000 males were examined for tumors, and only one teratoma was observed. These results indicated that susceptibility to testicular teratocarcinogenesis in mice is determined by multiple genetic factors present in inbred-strain-129 mice.

Single, specific genes have been associated with the incidence of several types of tumors in mice [22–25] and some have been found to affect teratocarcinogenesis.

Influence of the Gene Yellow (A^y)

Heston and Vlahakis [26] found that the gene yellow (A^y) increased the incidence of pulmonary and mammary gland tumors in females, hepatomas in males, and induced skin tumors in both sexes. I introduced the gene yellow onto the Strain-129 genetic background by repeated backcrosses of A^y-carrying animals to Strain-129 mice to determine if this gene would increase susceptibility to teratocarcinogenesis [21]. The incidence of teratomas in yellow mice was only one-tenth of that in their nonyellow littermates. Apparently, the agouti or a very closely linked locus contributes to resistance to teratocarcinogenesis.

Influence of the Gene Steel (Sl)

Homozygous steel (Sl/Sl) mice lack pigment cells and are anemic. They also are sterile because their gonads contain few or no germ cells. This gene was introduced onto the strain 129 genetic background, and it was found that heterozygous (Sl/+) mice had twice as many teratomas as their normal (+/+) littermates [27].

Influence of the Gene Teratoma (ter)

The gene-dominant spotting (W) has pleotropic effects similar to those of steel. Homozygotes have few or no germ cells. I introduced W onto the strain-129 genetic background to see if it, like Sl, would increase susceptibility to testicular teratocarcinogenesis [28]. After eight backcross generations, the incidence of teratomas was the same in the mutants as in their normal littermates.

At the eighth generation backcrossing of W/+ mice to Strain 129, a W/+ female mated to a strain 129/Sv male produced 38 offspring and 8 of them had testicular teratomas. This was a high incidence, compared with 3% for the closely-related strain 129/Sv mice. A colony derived from that mating was established and designated 129/terSv (later changed to 129/Sv-ter). About 30% of the males of this colony had spontaneous testicular teratomas. The high incidence of teratomas in 129/Sv-ter compared with 129/Sv is probably due to a single gene mutation, ter.

Influence of the Gene Situs Inversus Viscerum *(iv)*

In addition to genetic influences on teratocarcinogenesis, there are environmental influences. Animals in second and later litters have twice as many teratomas as those in first litters [27]. Teratomas occur twice as frequently in the left testis as in the right.

The gene situs inversus viscerum *(iv)* causes left–right transposition of the thoracic and abdominal viscera and associated blood vessels including the origin of the spermatic artery [29]. To try to help explain why twice as many teratomas develop in the left testis as in the right, situs inversus viscerum was introduced onto the strain 129 genetic background. The penetrance of the gene was incomplete on the strain 129 background. About half of the homozygous *(iv/iv)* animals had normally oriented viscera, and the other half had reversed viscera. There were about twice as many teratomas in the left testis as in the right in the normal-appearing males, conversely, the animals with reversed viscera had twice as many teratomas in the right testes as in the left testis.

Normally, the right spermatic artery originates from the aorta more anteriorly than the left spermatic artery. Perhaps this difference in the branching of the spermatic artery plays a role in susceptibility to teratocarcinogenesis.

Strain Differences in Susceptibility to Experimental Teratogenesis

Testicular teratomas may be experimentally induced in some strains of mice by grafting genital ridges from 12- or 13-day fetuses to the testes of adults [30]. When 12-day male genital ridges of strain 129 mice were grafted to the testes of adults, about 80% of them developed into testes with teratomas. The incidence of teratomas in testes developing from grafted 13- or 14-day fetal genital ridges dropped sharply to about 8% – similar to the incidence of spontaneous teratomas.

Genital ridges from 12- and 13-day male fetuses of 29 inbred strains were grafted to the testes of adults [31]. The incidence of induced tumors was low in strains A/J (1/40), A/WySn (2/88), C57BL/10ScSn (1/62), C57BK/Ks (1/34), and SJL/J (1/48). The incidence was intermediate in strains C57BL/6By (2/20), LP/J (3/33), and B10.129(14M)/Sn (3/24). The incidence was high in strains 129/Sv, A/He, B10.LP (20/64), LG/Chi (24/51), and B10.129(13M)/Sn (14/45).

Considering that spontaneous testicular teratomas have been observed commonly only in strain 129, that only two have been observed in strain A/He, and that only one has been observed in strains DBA/2J and LG, it was surprising that these extremely rare tumors could be induced in several inbred strains.

Twelve-day genital ridges were grafted to adult testes from fetuses produced by crossing strains 129 or A/He with mice of 12 other inbred strains [17]. The incidence of induced tumors was low for crosses to BALB/cBy (2/37); intermediate for A/J (5/29), C3H/DiSn (2/17); and high for strains B.10LP (29/58), C57BL/6By (7/14), C57BL/10ScSn (14/25), DBA/2J (10/26), LG/Chi (5/6), LP (9/19), and St/Ks (6/21). No teratomas were obtained from grafts of genital ridges from F_1 hybrids between strains 129 and AL/Ks (0/35), BALB/cJ (0/30), and BALB/cWt (0/22). Apparently, susceptibility to experimental teratocarcinogenesis depends upon genetic background.

Genetic Evidence That Testicular Teratomas Are Derived From Germ Cells

All teratomas in 15- and 16-day fetuses were located within the seminiferous tubules [32]. This means that they must be derived from either primordial germ cells or from cells that would become Sertoli cells. This morphological evidence supports the theory that teratomas are derived from germ cells, since they are the most pluripotent cells of the body, and there is no evidence that cells that have become determined to form Sertoli cells have the developmental capacity to differentiate into other kinds of cells.

I have obtained genetic evidence that strongly supports the morphological observation that teratomas are derived from germ cells [33]. Animals that are homozygous for the gene steel *(Sl)* are white because they lack pigment cells, anemic because they have few red blood cells, and sterile because they have few, if any, primordial germ cells.

I developed a new inbred strain that is congenic with strain 129/Sv. I introduced the genes *Sl* and the wild-type alleles at the albino *(C)* and pink-eyed dilution *(P)* loci onto the strain 129/Sv genetic background by repeated backcrosses of animals with *Sl*, *C*, and *P* to strain 129/Sv. The wild-type alleles of *C* and *P* were introduced to enable us to identify the *Sl*/+ mice. Strain 129/Sv are albino *(c/c)*, or their coat color is much reduced by the pink-eyed dilution gene *(p/p)*. In *Sl C P* mice, the coat color is slightly dilute, and the tail tips are unpigmented. As previously mentioned, animals with the gene *Sl* had twice as many spontaneous teratomas as their *Sl*/+ and +/+ littermates. We were unable to determine the incidence of teratomas in homozygous *Sl/Sl* animals, because they die during fetal life.

I attempted to "rescue" the primordia of the testes of *Sl/Sl* mice by dissecting the genital ridge and transplanting them to the testes of adults. We were able to identify the genotype of the grafts as being *Sl/Sl* or *Sl*/+ and +/+ by grafting a piece of dorsal skin along with the genital ridge. If the skin that developed from the grafts had pigmented hairs, we knew they were *Sl*/+ or +/+. If the skin had white hairs, we knew that they were *Sl*/*Sl*. Seventy-five percent of the grafted *Sl*/+ and +/+ genital ridges developed into testes with teratomas. This was the same incidence as when strain 129/Sv genital ridges were similarly grafted. In contrast, the *Sl*/*Sl* genital ridges developed into testes, none of which had teratomas. A major difference between the two groups was that *Sl*/+ and +/+ genital ridges had primordial germ cells, whereas *Sl*/*Sl* genital ridges lacked them. This genetic evidence supports the morphological evidence that teratomas are derived from germ cells.

Karyotype

The karyotype and sex of primary testicular teratomas of 7- to 10-day-old strain 129 mice were studied and compared with normal 7- to 10-day-old testicular and embryonal cells [34]. These tumors are initiated at 12.5 days of gestation [35], and the tumors examined were estimated to be 14- to 17-days old. The cells of these early teratomas were indistinguishable from normal cells, on the basis of chromosome number and morphology indicating that gross chromosomal changes are not involved in teratocarcinogenesis in strain 129 mice. In 87 of 89 cells, a Y chromosome was detected.

EMBRYO-DERIVED TERATOMAS

There is another method of experimentally producing teratocarcinomas. When early embryos (two-cell to six-day) are grafted to the testis [14] or kidney [16], the cells of the grafts become disorganized, and about half form teratomas (tumors composed solely of differentiated tissues), while the other half form teratocarcinomas (with embryonal carcinoma cells).

The developmental capacities of embryonic and extraembryonic ectoderm and of endoderm of six-day embryos were studied by grafting them to the testes of adults [36]. Grafts of embryonic ectoderm gave rise to teratomas and teratocarcinomas composed of undifferentiated embryonal cells and derivatives of all three germ layers including respiratory and alimentary epithelium. Grafts of extraembryonic ectoderm gave rise to trophoblastic giant cells. Grafts of endoderm produced only distal endoderm with Reichert's membrane.

Damjanov and Solter [37] found that transplanted egg cylinders of C57BL rarely gave rise to teratocarcinomas, and they were unable to establish transplantable tumors from them. When we grafted six-day embryos of C57BL/6 embryos to the testes of adults, about half gave rise to teratocarcinomas; but we were also unable to establish transplantable tumors from them [17].

OVARIAN TERATOMAS

Only about seven cases of ovarian teratomas have been described in mice. In the inbred strain LT/Sv, however, they occur in about half of the females three months of age [13]. They originate from eggs that become parthenogenetically activated in the ovary. They undergo cleavage and form blastocysts that appear normal by light microscopy. Usually they become disorganized at the early egg-cylinder stage, but occasionally they form ectoderm, endoderm, primitive streaks, Reichert's membrane, and trophoblastic giant cells.

When strain LT/Sv oocytes are ovulated, about 10% of them cleave in the oviduct and form normal looking blastocysts that successfully implant in the uterus. Occasionally they develop to the primitive streak stage, but they are usually aborted at the early egg-cylinder stage for unknown reasons. We considered the possibility that parthenogenetic embryonic cells might be rescued as chimeras if the cells were aggregated with normal embryonic cells [38] in a manner similar to the experimental chimeras composed of normal and genetically lethal embryos to transmit a recessive X-linked lethal mutation [39]. Accordingly, we aggregated parthenogenetic eight-cell embryos of an albino strain and transferred them to uteri of pseudopregnant females. We obtained offspring that were white with pigmented patches. The pigment could only have been derived from the parthenogenetic embryonic cells. The parthenogenetic embryos also differed from the normal embryos at the glucose-phosphate-isomerase-1 locus. The chimeric offspring contained gene products of the locus contributed by the parthenogenetic embryonic cells. This experiment showed that even though parthenogenetic embryos cannot survive as organisms in utero, they are able to participate in normal development and to survive to adulthood as normal cells in chimeras.

One chimeric female was mated to an albino male. She produced several litters of albino offspring and one litter with two fully pigmented offspring. The pigmented offspring could have been derived only from eggs of parthenogenetic origin [40]. Thus, parthenogenetic cells may be totipotent; it remains a mystery why parthenogenetic embryos cannot survive in utero.

The stem cells (embryonal carcinoma cells) of ovarian teratomas are morphologically similar to those of testicular teratomas. They also are similar in developmental potential. Nearly all of them have neural tissue, respiratory and alimentary epithelia, muscle, and cartilage. There are some conspicuous differences, however, Nearly all testicular teratomas have notochord, whereas ovarian tumors do not. Trophoblastic giant cells are common in ovarian teratomas, but not in testicular teratomas.

Genetic Basis for Ovarian Teratocarcinogenesis

Inbred Strain LT mice originated from a cross between a C58 mouse (in which a mutation called light (B^{lt}) occurred [41]) and a strain BALB mouse. All descendants from this mating were produced by brother × sister matings. Although the gene B^{lt} was found not to be responsible for teratocarcinogenesis [17], apparently a combination of genes present in C58 and in BALB that became fixed during inbreeding determined the high susceptibility to ovarian teratocarcinogenesis.

Recombinant Inbred Strains

Strain LT mice were crossed with mice of seven other inbred strains [17]. The incidence of teratomas in the F_1 offspring was very low or zero for all but the cross with C57BL/6, which had about 4.5%. When the F_1 generation was backcrossed to LT, the incidence was about 25%, considerably higher than for backcrosses of other F_1 hybrids. It appeared that C57BL/6 mice differed from the other strains in their genetic susceptibility to ovarian teratocarcinogenesis.

Fifteen recombinant inbred strains were established, with LT/Sv and C57BL/6 as progenitors. The incidence was low or intermediate for 14 RI strains. The incidence in one strain, designated LTXBJ, was about 95% — significantly higher than for LT/Sv. The recombination of genes from LT/Sv and C57BL/6 led to an increased incidence of ovarian teratomas. When LTXBJ was crossed with LT, the offspring also had a higher incidences of teratomas than LT.

Origin of Teratomas From Oocytes That Have Completed the First Meiotic Division

Linder [42] found that teratomas in women who were heterozygous for alleles producing allozymes of glucose-6-phosphate dehydrogenase and phosphoglucomutase were homozygous for the alleles in most cases. He concluded that the teratomas arose from oocytes that had completed the first division of meiosis.

Eppig et al [43] examined teratomas electrophoretically in (LT/Sv × LTXBJ)F_1 females. These mice were heterozygous for the *Gpi-1* allele. Of 23 teratomas, 21 revealed a homozygous A- or B-allozyme banding pattern. The other two were heterozygous. They concluded that the teratomas in (LT/Sv × LTXBJ)F_1 originated from parthenogenetically cleaved oocytes that had completed the first meiotic division [43]. In the two cases where the heterozygous banding pattern was found, the most probable conclusion was that crossing-over had occurred between the centromere and the *Gpi-1* locus, so that the *Gpi-1^a* and *Gpi-1^b* alleles were located on two chromatids attached to the same centromere. Eicher [44] has used mouse ovarian teratomas as a new tool to estimate centromere-to-gene distances.

Strain-LT/Sv ovaries have a unique population of follicles that have fewer layers of granulosa cells surrounding the oocytes than enclose oocytes of similar sizes in other strains [45]. The parthenogenetic embryos that give rise to ovarian teratomas in Strain LT/Sv are usually found in these granulosa-cell deficient follicles. The frequency of ovarian teratocarcinogenesis correlated with the simultaneous occurrence of two atypical conditions. First, the capability of the matured ova to undergo spontaneous parthenogenetic activation and, second, the high frequency of granulosa-cell deficient follicles.

X-Chromosome Activity

The study of X-chromosomal activity in early mouse embryos is difficult, because of the small number of cells involved and because half of the embryos are male. Martin et al [46, 47] found that both X chromosomes are genetically active in clonal cultures of undifferentiated female mouse teratocarcinoma stem-cells derived from a spontaneous ovarian tumor of strain LT/Sv origin. As the cells differentiated in vitro, one of the X chromosomes became inactivated. Since both X chromosomes were activated in undifferentiated LT/Sv cells, it was inferred that they were similar to embryonic cells at a stage before X-chromosome inactivation.

Kahan [48] observed an extraordinary pattern of growth of a retransplantable ovarian teratocarcinoma. When cells derived from a clone were injected intraperitoneally, large, rapidly growing tumors were found that, in most females, were located exclusively in the ovaries and were bilateral in many cases. Males inoculated intraperitoneally with the same number of cells had no tumors. In females, the tumors, composed mostly of embryonal carcinoma cells, are located within the ovary and are probably derived from cells that have arrived there via the bloodstream. Kahan [48] made the fascinating suggestion that the specificity with which these cells arrive at and grow only in the ovary perhaps reflects the specificity shown by normal germ cells, from which the tumor is derived, in their migration to the genital ridge during embryonic development. To quote Kahan [48], "It would be a pleasing analogy to find that, just as their normal counterparts appear to do when they migrate into the embryonic ovary, so female teratocarcinoma stem cells "recycled" through the ovary also reactivate an X chromosome."

In contrast to Martin et al [46, 47], Kahan, using a different retransplantable ovarian teratocarcinoma, obtained evidence suggesting X inactivation [48]. It is possible that Martin's tumor was derived from a parthenogenetic embryo in an earlier stage of development (before X inactivation) than Kahan's.

PARTICIPATION OF EMBRYONAL CARCINOMA CELLS IN NORMAL DEVELOPMENT AND REVERSION FROM THE MALIGNANT STATE TO NORMAL

The differentiated cells of teratocarcinomas appear to be normal [20]. Even though they are derived from potentially malignant cells, when they differentiate they are no longer proliferative. Differentiated cells in teratomas are functional. They produce pigment, mucus, red blood cells. Skeletal and cardiac muscular movements are frequently seen in teratomas of freshly killed animals. All testicular, ovarian, and embryo-derived teratomas in their early stages of development contain undifferentiated embryonal cells that are potentially malignant. However, most teratomas become benign through differentiation of these cells into apparently normal tissues. I have established many retransplantable teratocarcinomas that have grown progressively for as many as 40 transplant generations, however, they finally stop growing because all of the embryonal cells differentiate and become benign.

Kleinsmith and Pierce [49] placed single embryonal carcinoma cells in small capillaries and implanted them under the skin of an adult. The cells proliferated and some produced clines that differentiated into several kinds of tissues — all derived from the single injected cell. This demonstrated that embryonal carcinoma cells can be pluripotent.

Brinster [50] injected embryonal carcinoma cells near the inner-cell mass of a normal blastocyst. The embryonal carcinoma cells were derived from a pigmented strain, and the host blastocyst from an albino strain. He transferred the injected blastocyst to the uterus of a pseudopregnant female. It developed normally to term; and he obtained a white mouse with pigmented patches. The patches could have been derived only from the injected embryonal carcinoma cells. This indicated that embryonal carcinoma cells could participate in normal development and differentiate, at least, into normal melanocytes.

Mintz and Illmensee [10] performed experiments similar to Brinster [50] but they used teratoma and host blastocyst combinations with many genetic markers that made it possible to identify the origin of many kinds of cells as deriving from either the teratoma or blastocyst. As donor cells, they used the core cells of embryoid bodies of a retransplantable teratocarcinoma that I had established about eight years earlier by grafting a 6-day embryo into the testis of an adult [14]. They obtained mice that were chimeras — partly derived from the injected embryonal carcinoma cell and partly from the normal host blastocyst. They were able to identify gene products derived from the embryonal carcinoma cells in all organs such as hemoglobin in red blood cells, immunoglobulins in plasma cells, liver proteins in liver cells, black melanin in epidermal cells, and electrophoretic variants of glucose-phosphate isomerase. Two male chimeras gave rise to sperm that fertilized eggs, which developed into normal adult mice.

Illmensee and Mintz [11] then grafted single embryonal carcinoma cells into blastocysts. The single tumor cell contributed to the formation of all tissues of a chimeric mouse. This was final proof that single embryonal carcinomas were totipotent. After about 200 transplant generations as stem cells of a malignant tumor, these cells gave rise to normal benign differentiated tissues. Other experimental evidence demonstrated that the microenvironment of the normal embryonic cells apparently reversed the malignant state of the embryonal cells to normal. The nature of the influence of the normal cells is, as yet, unknown.

Illmensee [12] injected single cells from a transplantable ovarian teratocarcinoma into blastocysts of another genetically-marked strain and placed them in the uterus of pseudopregnant females. He obtained several chimeras derived from the normal embryonic cells and from the injected embryonal cells. One female produced functional eggs derived from an embryonal carcinoma cell. When these eggs were fertilized by sperm, they developed into normal offspring. The environment of the normal embryonic cells again caused the tumor cells to reverse to normality, and the tumor cell genes were transmitted to succeeding generations. This means that the genes in the embryonal carcinoma cell were functionally but not structurally different from normal.

Papaioannou et al [51] injected embryonal carcinoma from cell cultures, and they also obtained chimeric mice.

THE USE OF TERATOCARCINOMA CELLS FOR INTRODUCING MUTANT GENES INTO MICE

The development of the method of injecting embryonal carcinoma cells into blastocysts made it possible to introduce hybrid cells with rat and human chromosomal segments into mice. Illmensee et al [52, 53] transplanted single hybrid cells into genetically marked blastocysts, transferred them to the uteri of pseudopregnant females, and obtained healthy chimeric mice in which the rat and human gene products were identified.

Mintz [54] and others have suggested that cultured teratocarcinoma cells might be used as vehicles for introducing specific mutant genes such as those responsible for human diseases into mice. The possibility of converting hybrid teratocarcinoma stem cells into functional germ cells is a most exciting prospect.

ACKNOWLEDGMENTS

The investigations by the author were supported by grant number CA 02662, awarded by the National Cancer Institute, DHEW.

REFERENCES

1. Pierce GB, Stevens LC, Nakane PK: Ultrastructural analysis of the early development of teratocarcinoma. J Natl Cancer Inst 39:755–773, 1967.
2. Damjanov I, Katic V, Stevens LC: Ultrastructure of ovarian teratomas in LT mice. Z Krebsforsch 83:261–267, 1975.
3. Martin GR, Wiley LM, Damjanov I: The development of cystic embryoid bodies in vitro from clonal teratocarcinoma stem cells. Dev Biol 61:230–244, 1977.
4. Dewey MJ, Gearhart JD, Mintz B: Cell surface antigens of totipotent mouse teratocarcinoma cells grown in vivo: Their relation to embryo, adult, and tumor antigens. Dev Biol 55:359–374, 1977.
5. Jacob F: Mouse teratocarcinoma and embryonic antigens. Immunol Rev 33:3–32, 1977.
6. Stern P, Willison K, Kennox E, Galfre G, Milstein C, Secher D, Ziegler A, Springer T: Monoclonal antibodies as probes for differentiation and tumor-associated antigens: A Forssman specificity on teratocarcinoma stem cells. Cell 14:775–783, 1978.
7. Bernstine EG, Ephrussi B: Alkaline phosphatase activity in embryonal carcinoma and its hybrids with neuroblastoma. In Sherman MI, Solter D (eds): "Teratomas and Differentiation." New York: Academic Press, 1975, pp 271–287.
8. Dewey MJ, Filler R, Mintz B: Protein patterns of developmentally totipotent mouse teratocarcinoma cells and normal early embryo cells. Dev Biol 65:171–182, 1978.

9. Martin GR, Smith S, Epstein CJ: Protein synthetic patterns in teratocarcinoma stem cells and mouse embryos at early stages of development. Dev Biol 66:8–16, 1978.
10. Mintz B, Illmensee K: Normal genetically mosaic mice produced from malignant teratocarcinoma cells. Proc Natl Acad Sci USA 72:3585–3589, 1975.
11. Illmensee K, Mintz B: Totipotency and normal differentiation of single teratocarcinoma cells cloned by injection into blastocysts. Proc Natl Acad Sci USA 73:549–553, 1976.
12. Illmensee K: Reversion of malignancy and normalized differentiation of teratocarcinoma cells in chimeric mice. In Russell LB (eds): "Genetic Mosaics and Chimeras in Mammals." New York: Plenum Press, 1978, pp 3–25.
13. Stevens LC, Varnum DS: The development of teratomas from parthenogenetically activated ovarian mouse eggs. Dev Biol 37:369–380, 1974.
14. Stevens LC: The development of transplantable teratocarcinomas from intratesticular grafts of pre- and postimplantation mouse embryos. Dev Biol 21:364–382, 1970.
15. Stevens LC: Studies on transplantable testicular teratomas of strain 129 mice. J Natl Cancer Inst 20:1257–1275, 1958.
16. Damjanov I, Solter D: Experimental teratoma. Curr Top Pathol 59:69–130, 1974.
17. Stevens LC, Varnum DS: Data to be published.
18. Stevens LC: Comparative development of normal and parthenogenetic mouse embryos, early testicular and ovarian teratomas, and embryoid bodies. In Sherman MI, Solter D (eds): "Teratomas and Differentiation." New York: Academic, 1975, pp 17–32.
19. Stevens LC, Little CC: Spontaneous testicular teratomas in an inbred strain of mice. Proc Natl Acad Sci USA 40:1080–1087, 1954.
20. Stevens LC, Hummel KP: A description of spontaneous congenital testicular teratomas in strain 129 mice. J Natl Cancer Inst 18:719–747, 1957.
21. Stevens LC: The biology of teratomas. In Abercrombie M, Brachet J (eds): "Advances in Morphogenesis." New York: Academic Press, 1967a, pp 1–28.
22. Little CC: The relation of coat color to the spontaneous incidence of mammary tumors in mice. J Exp Med 59:229–250, 1934.
23. Heston WE, Deringer MK, Huges IR, Cornfield J: Interrelation of specific genes, body weight, and development of tumors in mice. J Natl Cancer Inst 12:1141–1157, 1952.
24. Law LW: The flexed-tail-anemia gene *(f)* and induced leukemia in mice. J Natl Cancer Inst 12:1119–1126, 1952.
25. Russell ES, Fekete E: Analysis of *W*-series pleiotropism in the mouse: Effect of $W^v W^v$ substitution on definitive germ cells and on ovarian tumorigenesis. J Natl Cancer Inst 21:365–381, 1958.
26. Heston WE, Vlahakis G: Elimination of the effect of the A^y gene on pulmonary tumors in mice by alteration of its effect on normal growth. J Natl Cancer Inst 27:1189–1196, 1961.
27. Stevens LC, Mackensen JA: Genetic and environmental influences on teratocarcinogenesis in mice. J Natl Cancer Inst 27:443–453, 1961.
28. Stevens LC: A new inbred subline of mice (129/terSv) with a high incidence of spontaneous congenital testicular teratomas. J Natl Cancer Inst 50:235–242, 1973.
29. Hummel KP, Chapman DB: Visceral inversion and associated anomalies in the mouse. J Hered 50:9–13, 1959.
30. Stevens LC: Experimental production of testicular teratomas in mice of strains 129, A/He, and their F_1 hybrids. J Natl Cancer Inst 44:923–929, 1970.
31. Stevens LC: Developmental biology of teratomas in mice. In McMahon D, Fox CF(eds): "Developmental Biology: Pattern Formation, Gene Regulation." Menlo Park, Cal., Reading, Mass., London, Don Mills, Ont., and Sidney: W.A. Benjamin, Inc., 1975, pp 186–204.
32. Stevens LC: Testicular teratomas in fetal mice. J Natl Cancer Inst 28:247–267, 1962.

33. Stevens LC: Origin of testicular teratomas from primordial germ cells in mice. J Natl Cancer Inst 38:549–552, 1967b.
34. Stevens LC, Bunker MC: Karyotype and sex of primary testicular teratomas in mice. J Natl Cancer Inst 33:66–78, 1964.
35. Stevens LC: Experimental production of testicular teratomas in mice. Proc Natl Acad Sci USA 52:654–661, 1964.
36. Diwan SB, Stevens LC: Development of teratomas from the ectoderm of mouse egg cylinder. J Natl Cancer Inst 57:937–942, 1976.
37. Damjanov I, Solter D: Host-related factors determine the outgrowth of teratocarcinomas from mouse egg cylinders. Z Krebsforsch 81:63–69, 1974.
38. Stevens LC, Varnum DS, Eicher EM: Viable chimaeras produced from normal and parthenogenetic mouse embryos. Nature 269:515–517, 1977.
39. Eicher EM, Hoppe PC: Use of chimeras to transmit lethal genes in the mouse and to demonstrate allelism of the two X-linked male lethal genes jp and msd. J Exp Zool 183:181–184, 1973.
40. Stevens LC: Totipotent cells of parthenogenetic origin in a chimeric mouse. Nature 276:266–267, 1978.
41. MacDowell EC: "Light" – A new mouse color. J Hered 41:35–36, 1950.
42. Linder D: Gene loss in teratomas. Proc Natl Acad Sci USA 63:699–704, 1969.
43. Eppig JJ, Kozak LP, Eicher EM, Stevens LC: Ovarian teratomas in mice are derived from oocytes that have completed the first meiotic division. Nature 269:517–518, 1977.
44. Eicher EM: Murine ovarian teratomas and parthenotes as cytogenetic tools. Cytogenet Cell Genet 20:232–239, 1978.
45. Eppig JJ: Granulosa cell deficient follicles. Occurrence, structure, and relationship to ovarian teratocarcinogenesis in strain LT/Sv mice. Differentiation 12:111–120, 1978.
46. Martin GR, Epstein CJ, Martin D: Use of teratocarcinoma stem cells as a model system for the study of X-chromosome inactivation in vitro. In Russell LB (ed): "Genetic Mosaics and Chimeras in Mammals." New York: Plenum Press, 1978, pp 269–296.
47. Martin GR, Epstein CJ, Travis B, Tucker G, Yatziv S, Martin DW, Clift S, Cohen S: X-chromosome inactivation during differentiation of female teratocarcinoma stem cells in vitro. Nature 271:329–333, 1978.
48. Kahan B: The stability of X-chromosome inactivation: Studies with mouse-human cell hybrids and mouse teratocarcinomas. In Russell LB: "Genetic Mosaics and Chimeras in Mammals." New York: Plenum Press, 1978, pp 297–328.
49. Kleinsmith LJ, Pierce GB: Multipotentiality of single embryonal carcinoma cells. Cancer 24:1544–1552, 1964.
50. Brinster RL: The effect of cells transferred into the mouse blastocyst on subsequent development. J Exp Med 140:1049–1056, 1974.
51. Papaioannou VE, Gardner RL, McBurney MW, Babinet C, Evans MJ: Participation of cultured teratocarcinoma cells in mouse embryogenesis. J Embryol Exp Morph 44:93–104, 1978.
52. Illmensee K, Hoppe PC, Croce CM: Chimeric mice from human – mouse hybrid cells. Proc Natl Acad Sci USA 75:1914–1918, 1978.
53. Illmensee K, Croce CM: Xenogeneic gene expression in chimeric mice derived from rat-mouse hybrid cells. Proc Natl Acad Sci USA 76:879–883, 1979.
54. Mintz B: Teratocarcinoma cells as vehicles for introducing mutant genes into mice. Differentiation 13:25–27, 1979.

Experimental Manipulation of the Mammalian Embryo: Biological and Genetic Consequences

Karl Illmensee

The experimental modification of the genome of mammalian cells and embryos might be envisaged by the following three microsurgical approaches: First, the removal of genetic information and its phenotypic consequences to differentiation; second, the addition of genetic material and its expression during development and neoplasia; and third, the entire replacement of the egg genome by a somatic nucleus and the production of genetically identical individuals. During the past few years, all three approaches have been realized in the mouse, the most suitable mammalian organism for genetic studies, and I should like briefly to summarize these new results, discuss their relevance and applicability to experimental embryology and cancer research, and finally remark upon some promising prospects in genetic manipulation of the mammalian embryo.

UNIPARENTAL MICE

Development of the fertilized egg in which only the maternal (gynogenesis) or paternal (androgenesis) genome is retained seems to occur infrequently in the animal kingdom. The same holds true for another type of reproduction that involves differentiation controlled exclusively by the maternal genes of the unfertilized but activated egg (parthenogenesis). Nevertheless, each event can happen spontaneously or be initiated experimentally in a number of vertebrates and insects [described in Beatty, 1967; White, 1978]. Although different in origin, these uniparental animals are of considerable interest for analysis of gene control during early development.

Various attempts to generate uniparental mice experimentally from fertilized and unfertilized eggs were limited to embryonic and early postimplantation stages. Similarly, most recent efforts using microsurgical techniques either to remove one pronucleus shortly after fertilization [Modlinski, 1975] or to bisect the fertilized egg at the pronuclear stage [Tarkowski, 1977] rarely resulted in the production of haploid blastocysts. Because mouse embryos with only one set of chromosomes are apparently nonviable, full restoration of the genome by diploidizing haploid embryos in the presence of cytochalasin B has recently been attempted. However, only blastocysts have been obtained, and no genetic markers were introduced to allow testing for homozygosity at the preimplantation stage [Markert and Petters, 1977].

Recently, we reported that viable homozygous-diploid mice can indeed be generated by microsurgically removing one pronucleus from the fertilized egg and diploidizing the residual one in the presence of cytochalasin B (Fig. 1). The resulting blastocysts are then implanted into the uterus of a pseudopregnant female in order to allow development to term. The isogeneic mice are expected to be females, because diploidized eggs left with only a Y-chromosome-bearing pronucleus should eventually develop as Y/Y embryos, which usually die during early cleavage [Tarkowski, 1977]. The X/X embryos, however, which grow normally, will give rise to females. The uniparental mice carry either the maternal or paternal genome, depending on whether the female or male pronucleus has been retained in the egg. Confirmation of such a process seems to be absolutely essential at the genetic level in order to reveal the gynogenetic or androgenetic origin of the experimental mice. Homozygosity for a number of gene loci positioned on different chromosomes has adequately been demonstrated in all the isogeneic females. Some of the uniparental females proved to be fertile and delivered healthy offspring [Hoppe and Illmensee, 1977].

In addition to using coat color as a genetic marker for identifying the isogeneic progeny, the blood of these mice was analyzed for glucosephosphate isomerase, hemoglobins, plasma proteins and esterases, and carbonic anhydrase; the urine was screened for the major urinary protein complex. All of the isogeneic mice exhibited only the enzymatic form or protein profile of one of the inbred parents. The occurrence of pure strain-specific variants of these different gene products clearly demonstrated that, after microsurgical removal of one pronucleus, the initial F_1 hybrid egg was still able to develop normally with the genome of the residual pronucleus. Furthermore, all seven uniparental mice showed normal diploid karyotypes with two X chromosomes, thus substantiating their diploid genetic constitution at the chromosomal level. Karyotypic analysis also revealed that X chromosome inactivation had occurred normally, as in regular females, irrespective of whether the X chromosomes had been inherited paternally or maternally [Eicher et al, in preparation].

The development to term of females that obtained the genome only from their father was comparable to that of females with genes inherited only from their mother indicating that one parental genome alone is fully capable of initiating and promoting normal development. The paternal genes must also have been active in extraembryonic tissues, which usually show a preferential inactivation of the paternal X chromosome [Takagi, 1978; West et al, 1978]. The apparently normal X-activation–inactivation process in our gynogenetic females as well as

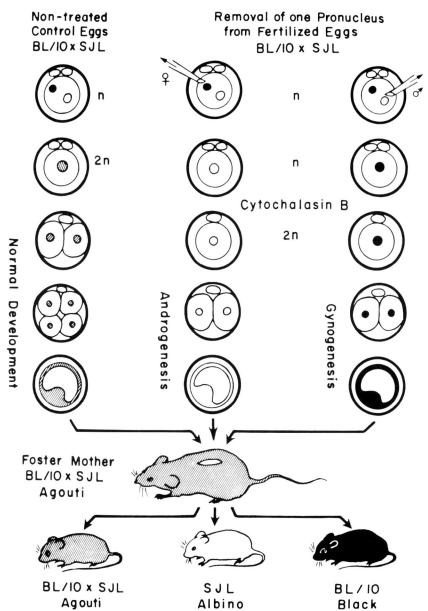

Fig. 1. Experimental scheme for the production of uniparental mice derived from either the paternal or maternal genome. Fertilized BL10SJLF$_1$ hybrid eggs at the pronuclear stage were placed in modified Whitten's medium containing cytochalasin B. Subsequently, either the female or male pronucleus was removed microsurgically. The resulting haploid eggs were diploidized with cytochalasin B and cultured in regular medium to the morula or blastocyst stage. These embryos, as well as nonmanipulated controls, were then transferred to the uteri of pseudopregnant foster mothers and allowed to develop to term. Live-born albino (SJL) or black (BL10) offspring indicated either androgenetic or gynogenetic development, depending on whether only the male or only the female pronucleus, respectively, had been left in the egg. The agouti (BL10SJLF$_1$) progeny originated from nonmanipulated eggs [Hoppe and Illmensee, 1977].

in parthenogenetic embryos [Kaufman et al, 1978] and human female patients [Latt et al, 1976] contradicts two current hypotheses [Cooper, 1971; Chandra and Brown, 1975] that would require two active X chromosomes in these individuals, which have not been observed. Furthermore, these hypotheses would demand inactive X chromosomes for our androgenetic females as well as for XO females, which both inherited the X chromosome from their father and therefore should not be viable.

The normal survival of uniparental mice raises again the question about the true origin of the prenatal developmental arrest during parthenogenesis frequently discussed as a result of deleterious genes or gene combinations being uncovered homozygously in these gynogenetic embryos (reviewed by Graham [1974]). The microsurgical production of homozygous-diploid mice provides a new means of analyzing the reasons and causes for the developmental failure of parthenotes, and may also be useful in studying the phenotypic consequences of homozygosity for recessive mutations, lethal or X-linked genes, and maternal versus paternal gene expression during mammalian differentiation.

XENOGENEIC GENE EXPRESSION IN CHIMERIC MICE

Genetic manipulation in vivo can be accomplished not only by removing genetic material from the organism but also by introducing cells, chromosomes, nuclei, or eventually genes into the early mouse embryo in order to study their phenotypic expression during in situ differentiation.

TERATOCARCINOMA CELL AS CARRIER

Recently it was found that, after microinjection into genetically different blastocysts, malignant cells of one particular tumor, a teratocarcinoma (described by Stevens [1975]; Graham [1977]), lost their neoplastic properties in the embryonic environment, participated in normal development, and clonally contributed to virtually all major adult tissues in chimeric mice, including those not seen before in the solid tumors [reviewed by Illmensee, 1978a]. It therefore appears as if the initially malignant teratocarcinoma cells remain developmentally totipotent and are able to express their genetic repertoire in an orderly sequence of differentiation into somatic and germ-line tissues. The reversion of malignancy was obviously a stable process, because when pieces of the teratocarcinoma-derived mosaic tissues were grafted under the skin of syngeneic mice, they never formed tumors. One could still argue that the teratocarcinoma cell population was heterogeneous with respect to malignant properties, so that one cell might be able to differentiate normally whereas another cell might be destined to form a tumor. An alternative explanation of the experimental results is that a single teratocarcinoma cell can either differentiate normally or give rise to a tumor, depending on the microenvironment into which it happens to be injected. In order to distinguish between these two possibilities, single teratocarcinoma cells were cultured in vitro in an appropriate medium, and some divided to form two daughter cells. One daughter cell was injected into a blastocyst, which was then transferred to the uterus of a foster mother; the other daughter cell was implanted under the skin of an adult mouse. In some cases the cell injected into a blastocyst participated in normal tissue differentiation, whereas its subcutaneously implanted sister cell gave rise to a teratocarcinoma [reviewed by Illmensee and Stevens, 1979]. It would appear that the embryonic environment plays an important role in bringing about the reversion of the malignant phenotype. At present, however, we do not know what signals cause a once-malignant cell to take part in normal development, nor do we have any insight as to the mode of action of these signals during cellular reprogramming.

During the past few years, several mouse teratocarcinoma cell lines have been established under culture conditions and selected for various cell phenotypes [Evans, 1975; Martin, 1977]. Since the in vitro assay systems are generally limited with respect to normal growth and differentiation, a more favorable environment is provided when teratocarcinoma cells are introduced into the living organism in order to reveal their full potential. In this way, cultured cells derived from various teratocarcinomas were shown to be capable of differentiating normally in the coat and several internal organs, although most of the chimeric mice additionally developed tumors in various anatomical sites [Papaioannou et al, 1978].

Considerable progress in clonally propagating teratocarcinoma cell lines has opened new possibilities of selecting for somatic mutations in vitro [described in Sherman and Solter, 1975]. It has therefore been proposed that teratocarcinomas might provide us with a unique kind of cell that can be selected in vitro as carrying a given somatic mutation and then cycled through mice via blastocyst injection for further in situ analysis [Mintz et al, 1975]. Following such an experimental scheme, teratocarcinoma cells deficient in hypoxanthine phosphoribosyltransferase (HPRT) apparently retained their developmental potential to a remarkable extent [Dewey et al, 1977]. Most recently, the thymidine kinase (TK) gene of herpes simplex virus together with the human β-globin gene have been transferred into TK-deficient mouse teratocarcinoma cells. After subcutaneous implantation of these transformed teratocarcinoma cells into syngeneic adult mice, viral TK enzyme activity but no human gene product has been detected in some of the developing tumors [Pellicer et al, 1980]. Another approach for introducing foreign genetic material into teratocarcinoma cells is to utilize somatic cell hybridization. Following the injection of such hybrid cells into genetically marked blastocysts, it should then be feasible to study xenogeneic gene expression during differentiation and to assay for the ontogenetic appearance, coexistence, and regulation of the foreign gene products in various tissues during mouse development.

Human × Mouse Hybrid Cells

In collaboration with Dr. C. Croce (Wistar Institute), the combination of cell hybridization techniques with our biological system has recently opened up a new line of research (Fig. 2). In his laboratory, cultured mouse teratocarcinoma cells first were selected for TK deficiency. After microinjection into mouse blastocysts, the TK-deficient cells became integrated during normal organogenesis and

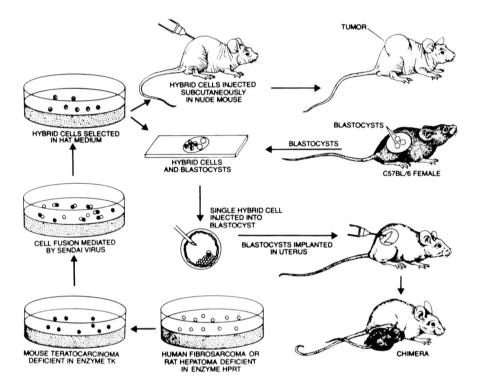

Fig. 2. Experimental scheme of cycling malignant hybrid cells through mice via microinjection into mouse blastocysts. Human fibrosarcoma or rat hepatoma cells, both deficient in hypoxanthine phosphoribosyltransferase (HPRT), were fused with mouse teratocarcinoma cells deficient in thymidine kinase (TK). Selection of hybrid cells was carried out in HAT medium using inactivated Sendai virus for cell fusion. To test for tumorogenicity, hybrid cells were injected subcutaneously into athymic nude mice. To test for developmental potential, single hybrid cells were injected into C57BL/6 blastocysts, bearing many genetic markers, in order to detect tissue differentiation derived from the injected cell. Shortly after micromanipulation, the blastocysts had to be surgically introduced into the uterus of a pseudopregnant foster mother to allow development to term. The live-born experimental offspring were then analyzed for hybrid-cell contributions in the coat and the various internal organs (modified after Illmensee and Stevens [1979]).

contributed substantially to several internal tissues, thus demonstrating their suitability as chromosome and gene carriers [Illmensee, 1978a]. The TK-deficient mouse teratocarcinoma cells were then fused in vitro with HPRT-deficient human fibrosarcoma cells using inactivated Sendai virus. Only the interspecific hybrid cells can grow in hypoxanthine/aminopterin/thymidine (HAT) selective medium. On the contrary, TK- and HPRT-deficient parental cells, which lack the enzymes required for the incorporation of thymidine and hypoxanthine, respectively, do not survive in this medium. Under these selective conditions, the viable hybrid cells, which quickly lose human chromosomes but not mouse chromosomes, retain at least human chromosome 17, which carries the locus for TK. This particular chromosome also carries a second known gene that is closely linked to TK and codes for galactokinase (GLK). The latter enzyme, with its characteristic electrophoretic mobility quite different from the equivalent mouse enzyme, serves as another useful biochemical marker for detecting the presence and normal expression of the human gene product in the hybrid cells.

After subcutaneous implantation into athymic nude mice, the human × mouse hybrid cells formed predominantly undifferentiated tumors. In contrast, when injected into genetically marked blastocysts, the malignant hybrid cells became integrated during embryogenesis and participated in orderly differentiation of the coat and seven internal organs (Fig. 3). Although the hybrid cells contributed up to 60% in some tissues of chimeric mice as judged from enzyme analysis, the human-specific gene product of GLK has only been detected in the heart of one chimera and the kidneys of another; both tissues, by the way, showed the highest participation of hybrid cells in the enzyme tests [Illmensee et al, 1978]. The failure to disclose human GLK in the other mosaic organs might have resulted from, among other possibilities, the loss of human chromosome 17 during development in the absence of any selective pressure to retain it.

Rat × Mouse Hybrid Cells

In order to facilitate a more extensive analysis of foreign gene expression in vivo, it would be desirable to use a hybrid cell line containing several xenogeneic chromosomes, thereby allowing the search for a number of different gene products. In this respect, interspecific hybrid cells between TK-deficient mouse teratocarcinoma and HPRT-deficient rat hepatoma, which had retained almost all of the mouse chromosomes and various numbers of rat chromosomes, seemed ideal for blastocyst injections. While the cell hybrids usually formed malignant tumors in athymic nude mice, at least some of these tumor cells were still capable of reverting to a normal phenotype in chimeric mice [Illmensee and Croce, 1979]. In contrast to previous results, tissue contributions of the rat × mouse hybrid cells remained limited to the liver and a few other organs of endomesodermal origin, perhaps because of the hepatoma origin of the rat cell used. Reversion of malignancy appeared more surprising because, on the one hand, the hybrid cells formed undifferentiated tumors in adult hosts and, on the other hand, differentiated normally into liver, lung, kidney, gut, and fat pad during in situ organogenesis. A

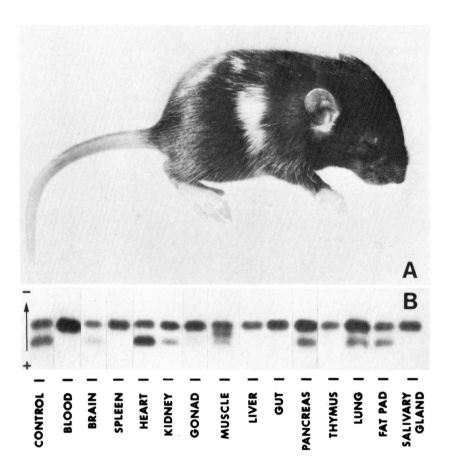

Fig. 3. Adult normal tissue differentiation from human fibrosarcoma × mouse teratocarcinoma hybrid cells after injection into C57BL/6 blastocysts. A. The chimeric male shows substantial coat mosaicism. The white patches originating from the hybrid cells contrast to the black coat of the C57BL/6 recipient. B. Cellulose-acetate electrophoresis of glucosephosphate isomerase (GPI) obtained from tissue homogenates of the chimeric male (shown in A). CONTROL represents a 50:50 GPI mixture of slow-migrating hybrid-cell type and fast-migrating C57BL/6 type. Clonal descendants of the injected human × mouse hybrid contributed significantly to seven of the 14 internal organs analyzed. They also fused normally with myoblasts of the C57BL/6 recipient, as judged from heterodimeric enzyme expression in skeletal muscle [Illmensee et al, 1978].

comparable situation occurred with the human X mouse hybrid cells, which produced undifferentiated tumors in nude mice but contributed normally to a number of different tissues in chimeric mice. It therefore seems as if the potential of the malignant hybrid cells remains limited in the adult host and will only be fully revealed in the environment of the embryo.

But what happened to the rat chromosomal genes in the chimeric mice? Because the rat X mouse hybrid cells retained several rat chromosomes, it was possible to detect nine different rat-specific enzymes in the various mosaic tissues [Illmensee and Croce, 1979]. The appearance in vivo of rat gene products not detectable in the hybrid cells in vitro and the formation of heteropolymers with the corresponding mouse enzymes indicate functional expression of the rat genes during mouse development. The synthesis of adult-specific enzyme variants further demonstrates the proper modulation of the foreign genes during organogenesis (Fig. 4).

Recently, we have obtained four chimeric fetuses, all of which revealed hybrid-cell participation in their livers, as well as in a few other organs such as intestinal tract and lungs. Xenogeneic gene expression was unequivocally documented by the presence of rat-specific enzyme variants of glucosephosphate isomerase (GPI), lactate dehydrogenase (LDH), glucerolphosphate dehydrogenase (GPD), and aldolase (ALD-B). After the chimeric organs were grown in vitro, rat chromosomes could be identified in these cultured cells. Although some of the enzymes reverted to a pattern similar to, but not identical with, the original hybrid cells, the overall protein synthesis remained organ-specific, as judged from 2-D gel electrophoresis. About one-third of the chimeric tissues formed tumors when subcutaneously injected into nude mice, indicating that some malignant hybrid stem cells might have been retained in these fetal organs, whereas others obviously lost their malignant properties during cellular differentiation [Duboule et al, in preparation]. At present, we do not know whether the rat-hepatoma origin of one of the parental cell types of the rat X mouse hybrid cells is causally related to the overt preference of these cells for liver development. It will be important to investigate this intriguing phenomenon further in order to elucidate the chromosomal and, eventually, genetic basis for such cell lineage preference.

NUCLEAR TRANSPLANTATION

In 1952 Briggs and King reported successful nuclear transplantations in the frog Rana pipiens, thereby opening up a new field in developmental biology of higher organisms. Subsequently, their technique of transplanting a nucleus from a somatic cell back into an unfertilized egg has been extensively applied to several other amphibian species, notably Xenopus laevis [reviewed by Gurdon, 1974]. It soon became apparent that such a bioassay system would provide new ways of attacking some of the long-standing questions in biology, among which belong the following: Are there changes in potentiality of nuclei during the course of development? Do nuclei become determined for a certain developmental fate and, if so, how and when does this happen in the organism? Is this "program," whatever its molecular nature may be, irreversibly restricting the nuclei to cell lineage-specific pathways?

Use of amphibia as the source of convenient experimental material for microinjection has tremendous advantages because they have been widely and effectively employed during the past two decades to study nucleocytoplasmic interactions in development [reviewed by DiBerardino, 1980]. However, the lack of a well-

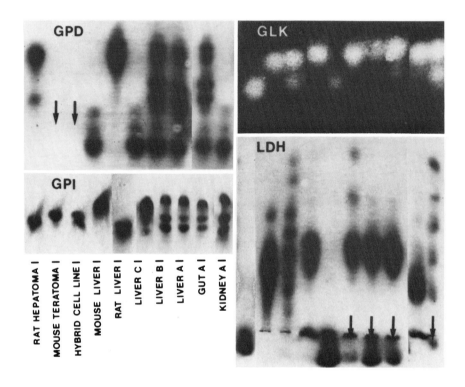

Fig. 4. Starch gel electrophoresis of rat- and mouse-specific enzyme variants in organs of chimeric mice. Cell extracts from liver, gut, and kidneys of chimeric mice A, B, and C (for their experimental production, see Fig. 2) reveal rat × mouse hybrid cell contributions, as judged from glycerolphosphate dehydrogenase (GPD), glycosephosphate isomerase (GPI), galactokinase (GLK), and lactate dehydrogenase (LDH). Note the absence of GPD activity (arrow) in the hybrid cells and mouse teratoma and its appearance in the liver B and A as well as in gut A. The formation of rat-mouse heteropolymers in these mosaic tissues documents functional cooperation of the interspecific gene products. The appearance of adult rat LDH-5 (arrow), not detectable in the hybrid cells, indicates differential modulation of the xenogeneic genes during in vivo development [Illmensee and Croce, 1979].

investigated genetic background amenable to gene manipulation and the long life cycle of these animals make it difficult to unravel the genetic mechanisms underlying differentiation and cellular diversification. In this respect, the most promising higher organisms from a genetic point of view appear to be the fruit fly and the house mouse. While nuclear transplantation in Drosophila melanogaster has already given some insight into the developmental fate and potential of embryonic nuclei and has demonstrated the epigenetic role of cytoplasm for cell commitment and cell lineage [reviewed by Illmensee, 1976, 1978b], similar experiments carried out in Mus musculus so far have met with rather limited success. Initial attempts at introducing somatic nuclei into mouse eggs, via inactivated Sendai virus-mediated cell fusion, resulted at best in a few abnormal cleavage divisions of the treated eggs [Graham, 1969; Baranska and Koprowski, 1970; Lin et al, 1973]. Recent studies have attempted to overcome the possible damage, attributable to the viral fusion procedure, by microsurgically injecting early embryonic nuclei into fertilized mouse eggs [Modlinski, 1978]. Some of the developing morulae and blastocysts revealed a tetraploid karyotype with the two T6 marker chromosomes of the donor nuclei, thus indicating functional participation of the transferred nuclei in preimplantation development. This experimental design, however, does not permit following the transplanted nucleus through normal embryogenesis, since this obviously is obscured by the continuing presence of the nucleus of the recipient egg.

Very recently, in collaboration with Dr. P. Hoppe (Jackson Laboratory), we successfully established a nuclear transplantation method in the mouse (Fig. 5) and first probed its biological applicability in the normal blastocyst. This preimplantation embryo is already differentiated into the external trophectoderm (TE) and the internal group of cells, called inner cell mass (ICM). Although the ICM cells will give rise to the embryo proper, each ICM cell is no longer able to develop independently into a complete embryo. The TE cells, however, do not contribute to the embryo, but rather to extraembryonic tissues [reviewed by Gardner, 1978]. Are these developmental restrictions at the cellular level also occurring simultaneously at the nuclear level?

In nuclear transplantation experiments we have demonstrated that nuclei from ICM cells of CBA/H-T6 blastocysts remain totipotent, since after their injection into fertilized but enucleated C57BL/6 eggs, they initiate normal development, giving rise to adult mice. Conversely, nuclei from TE cells appear to be restricted in their potential because of their failure to promote normal development [Illmensee and Hoppe, 1981].

In a second series of nuclear transplantations, we have continued and extended our ontogenetic analysis into the postimplantation period in an attempt to determine the developmental potential of nuclei derived from different cell types of the day 7 embryo. During this particular developmental stage, the mouse embryo is made up of an outer monocellular layer of distal endoderm, an inner layer of proximal endoderm, and finally a multicellular layer of embryonic and extraembryonic ectoderm [Snell and Stevens, 1968]. These three cell layers may conveniently be separated from one another by digestive enzyme treatment so that each can then be dissociated into single cells, which serve as donor material for nuclear transplantation. Are these different cell types also exhibiting different

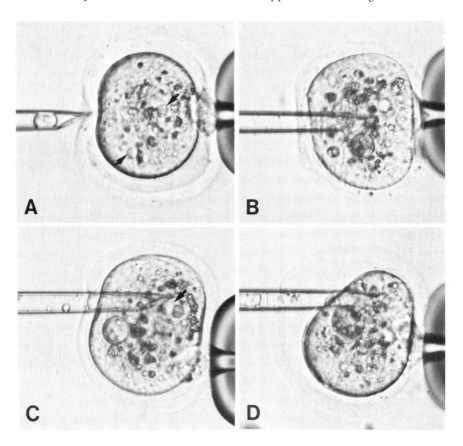

Fig. 5. Microsurgical procedure of nuclear transplantation into a fertilized mouse egg. A. The C57BL/6 egg (approx 70 μm diameter) is attached to a blunt holding pipet in the proper position so that the second polar body protects the egg surface from direct suction during the operation. The injection pipet containing the nucleus of a cell taken from the inner cell mass (ICM) of a LT/Sv blastocyst is introduced through the zona pellucida into the egg. At this developmental stage, the two pronuclei (arrows) are still separated, the female pronucleus being located near the second polar body. B. The nucleus is injected and the male pronucleus is slowly sucked into the transfer pipet. C. Subsequently, the female pronucleus (arrow) is removed from the egg by sucking it into the same pipet. D. Following this three-step operation, the micropipet is gently withdrawn from the egg. Note the changes in egg volume and shape during micromanipulation and the gradual swelling of the injected nucleus in the egg cytoplasm [Illmensee and Hoppe, 1981].

developmental potentials as far as their nuclei are concerned or, alternatively, do we observe similarities between the various nuclei, irrespective of the cell type?

In summary, nuclei from distal endoderm cells do not promote development of the enucleated eggs beyond a few cleavage divisions. Nuclei from either the proximal endoderm or the embryonic ectoderm, however, are capable of promoting development of the recipient eggs up to the adult stage [Illmensee and Hoppe, 1981]. Coat color phenotype, enzyme variants, and chromosomal T6 markers have confirmed the rather unusual origin of the nuclear-transplant mice from a somatic genome, which obviously has still retained the full developmental potential of the zygote genome despite the fact that cellular diversification has already taken place. In one instance, three nuclear-transplant mice have been derived clonally from embryonic ectoderm cells of a single day 7 donor embryo and are therefore considered to be genetically identical (Fig. 6).

The two experimental series have amply demonstrated that nuclear transplantation in the mouse may provide a new and useful approach to the study of nucleocytoplasmic interactions during mammalian differentiation.

Fig. 6. Three cloned nuclear-transplant mice with their albino ICR foster mother. From a single day 7 embryo of the CBA/H-T6 strain, cells of the embryonic ectoderm have been dissociated, and their nuclei injected into fertilized but enucleated C57BL/6 eggs. After culture in vitro and transfer into pseudopregnant females, the nuclear-transplant embryos are able to complete their development in utero. The resulting adult and fertile nuclear-transplant mice, which show the agouti coat and a diploid karyotype with the chromosomal T6 marker, characteristic of the nuclear donor CBA/H-T6 strain, demonstrate the developmental totipotency of somatic cell nuclei from day 7 embryonic ectoderm [Illmensee and Hoppe, in preparation 1981].

PROSPECTS

Recent progress in experimental embryology has provided new approaches to the genetic manipulation of the mammalian embryo. First, the production of uniparental mice, following microsurgical removal of one of the two pronuclei from a fertilized egg, will eventually allow us to study the expression of mutant genes, to compare maternal and paternal gene activity during development, and to elucidate further the problem of X chromosome inactivation.

Second, the injection of teratocarcinoma cells into mouse blastocysts has led to the striking observation that the stem cells of these malignant tumors contribute to normal tissue differentiation in chimeric mice, which has drawn attention to the fact that the neoplastic state of a cell can sometimes be reversed under appropriate conditions. In a manner of speaking, a teratocarcinoma resembles a mammalian embryo to some extent [reviewed by Jacob, 1975]. In our investigations we exploit this phenomenon to study a central problem in development: How does a single cell, the fertilized egg, differentiate to form all the cell types and tissues of the organism? Since the teratocarcinoma cell is potentially malignant and yet gives rise to normal tissues, we also expect to learn something from these tumors about the nature of neoplastic transformation. Now that it is possible to hybridize different kinds of cells with mouse teratocarcinoma cells, and thereby to make them carriers of foreign genetic material for integration into an excellent bioassay system – the tissues of a developing animal – the path should be open for advances in the study of mammalian differentiation and neoplasia.

Third, the transplantation of nuclei from somatic cells into fertilized but enucleated mouse eggs provides the most vigorous functional test for the developmental capacity of the nuclear genome. Another intriguing result refers to the totipotency of transplanted nuclei from embryonic ectoderm cells of the day 7 embryo. In future experiments, it will be important to reveal the nuclear potential of various cell types at more advanced embryonic and fetal stages and to determine the biological and, eventually, genetic consequences of nuclear changes during cell differentiation.

ACKNOWLEDGMENTS

I should like to thank Drs. C. M. Croce, P. C. Hoppe, L. C. Stevens, and my colleagues at the Laboratory of Cell Differentiation for continuous collaboration. Some of the experiments described in this article have been carried out during a very pleasant stay at The Jackson Laboratory. Our current research is supported by grants FN 3.442.0.79 from the Swiss National Science Foundation and CA27713-01 from the National Institutes of Health, as well as by appropriations from the Marc Birkigt and the Schmidheiny Foundations (Geneva).

REFERENCES

Baranska W, Koprowski H (1970): Fusion of unfertilized mouse eggs with somatic cells. J Exp Zool 174:1–14.
Beatty RA (1967): Parthenogenesis in vertebrates. In Metz CB, Monroy A (eds): "Fertilization," vol I. New York-London: Academic Press, pp 413–440.
Briggs R, King TJ (1952): Transplantation of living nuclei from blastula cells into enucleated frogs' eggs. Proc Natl Acad Sci USA 38:455–463.
Chandra HS, Brown SW (1975): Chromosome imprinting and the mammalian X-chromosome. Nature 253:165–168.
Cooper DW (1971): A direct genetic change model for X-chromosome inactivation in eutherian mammals. Nature 230:292–294.
Dewey MJ, Martin DW, Martin GR, Mintz B (1977): Mosaic mice with teratocarcinoma-derived mutant cells deficient in hypoxanthine phosphoribosyltransferase. Proc Natl Acad Sci USA 74:5564–5568.
DiBerardino MA (1980): Genetic stability and lability of metazoan nuclei transplanted into eggs and oocytes. Differentiation 17:17–30.
Duboule D, Illmensee R, Petzoldt U, Croce CM, Illmensee K (in preparation): Tissue preference and differentiation of malignant hybrid cells during mouse development.
Eicher EM, Washburn LL, Hoppe PC, Illmensee K (in preparation): X-Chromosome inactivation in gynogenetically and androgenetically derived female mice.
Evans MJ (1975): Studies with teratoma cells in vitro. In Balls M, Wild AE (eds): "The Early Development of Mammals." London: Cambridge University Press, pp 265–284.
Gardner RL (1978): The relationship between cell lineage and differentiation in the early mouse embryo. In Gehring W (ed): "Results and Problems in Cell Differentiation, vol IX. Heidelberg: Springer-Verlag, pp 205–241.
Graham CF (1969): The fusion of cells with one- and two-cell mouse embryos. In Defendi V (ed): "Heterospecific Genome Interaction." Symposium Monograph vol IX, Philadelphia Wistar Institute Press, pp 13–35.
Graham CF (1974): The production of parthenogenetic mammalian embryos and the use in biological research. Biol Rev 49:399–422.
Graham CF (1977): Teratocarcinoma cells and normal mouse embryogenesis. In Sherman MI, Graham CF (eds): "Concepts in Mammalian Embryogenesis." Cambridge: MIT Press, pp 315–394.
Gurdon JB (1974): "The Control of Gene Expression in Animal Development." Cambridge: Harvard University Press.
Hoppe PC, Illmensee K (1977): Microsurgically produced homozygous-diploid uniparental mice. Proc Natl Acad Sci USA 74:5657–5661.
Illmensee K (1976): Nuclear and cytoplasmic transplantation in Drosophila. In Lawrence PA (ed): "Insect Development." London: Blackwell, pp 76–96.
Illmensee K (1978a): Reversion of malignancy and normalized differentiation of teratocarcinoma cells in chimeric mice. In Russell LB (ed): "Genetic Mosaics and Chimeras in Mammals." New York: Plenum Press, pp 3–25.
Illmensee K (1978b): Drosophila chimeras and the problem of determination. In Gehring W (ed): "Results and Problems in Cell Differentiation," vol IX, Heidelberg: Springer-Verlag, pp 51–69.
Illmensee K, Croce CM (1979): Xenogeneic gene expression in chimeric mice derived from rat-mouse hybrid cells. Proc Natl Acad Sci USA 76:879–883.
Illmensee K, Hoppe PC, Croce CM (1978): Chimeric mice derived from human-mouse hybrid cells. Proc Natl Acad Sci USA 75:1914–1918.

Illmensee K, Hoppe PC (1981): Nuclear transplantation in Mus musculus: I. Developmental potential of nuclei from preimplantation embryos. Cell (in press).

Illmensee K, Hoppe PC (in preparation): Nuclear transplantation in Mus musculus: II. Developmental potential of nuclei from postimplantation embryos.

Illmensee K, Stevens LC (1979): Teratomas and chimeras. Sci Am 240:87–98.

Jacob F (1975): Mouse teratocarcinoma as a tool for the study of the mouse embryo. In Balls M, Wild AE (eds): "The Early Development of Mammals." London: Cambridge University Press, pp 233–241.

Kaufman MH, Guc-Cubrilo M, Lyon MF (1978): X-chromosome inactivation in diploid parthenogenetic mouse embryos. Nature 271:547–549.

Latt S, Willard HF, Gerald PS (1976): BrdU-33258 Hoechst analysis of DNA replication in human lymphocytes with supernumerary or structurally abnormal X chromosomes. Chromosoma 57:135–153.

Lin TP, Florence J, Oh JO (1973): Cell fusion induced by a virus within the zona pellucida of mouse eggs. Nature 242:47–49.

Markert CL, Petters RM (1977): Homozygous mouse embryos produced by microsurgery. J Exp Zool 201:295–302.

Martin GR (1977): The differentiation of teratocarcinoma stem cells in vitro: Parallels to normal embryogenesis. In Karkinen-Jaaskelainen M, Saxen L, Weiss L (eds): "Cell Interactions in Differentiation." London: Academic Press, pp 59–75.

Mintz B, Illmensee K, Gearhart JD (1975): Developmental and experimental potentialities of mouse teratocarcinoma cells from embryoid body cores. In Sherman MI, Solter D (eds): "Teratomas and Differentiation." New York-London: Academic Press, pp 59–82.

Modlinski JA (1975): Haploid mouse embryos obtained by microsurgical removal of one pronucleus. J Embryol Exp Morphol 33:897–905.

Modlinski JA (1978): Transfer of embryonic nuclei to fertilized mouse eggs and development of tetraploid blastocysts. Nature 273:466–467.

Papaioannou VE, Gardner RL, McBurney MW, Babinet C, Evans MJ (1978): Participation of cultured teratocarcinoma cells in mouse embryogenesis. J Embryol Exp Morphol 44: 93–104.

Pellicer A, Wagner EF, El Kareh A, Dewey MJ, Reuser AJ, Silverstein S, Axel R, Mintz B (1980): Introduction of a viral thymidine kinase gene and the human β-globin gene into developmentally multipotential mouse teratocarcinoma cells. Proc Natl Acad Sci USA 77:2098–2102.

Sherman MI, Solter D (eds) (1975): "Teratomas and Differentiation." New York-London: Academic Press.

Snell GD, Stevens LC (1968): Early embryology. In Green EL (ed): "Biology of the Laboratory Mouse." New York: Dover Publications, pp 205–245.

Stevens LC (1975): Teratocarcinogenesis and spontaneous parthenogenesis in mice. In Markert LC, Papaconstantinou J (eds): "The Developmental Biology of Reproduction." New York-London: Academic Press, pp 93–106.

Takagi N (1978): Preferential inactivation of the paternally derived X chromosome in mice. In Russell LB (ed): "Genetic Mosaics and Chimeras in Mammals." New York: Plenum, pp 341–360.

Tarkowski NK (1977): In vitro development of haploid mouse embryos produced by bisection of one-cell fertilized eggs. J Embryol Exp Morphol 38:187–202.

West JD, Papaioannou VE, Frels WI, Chapman VM (1978): Preferential expression of the maternally derived X chromosome in extraembryonic tissues of the mouse. In Russell LB (ed): "Genetic Mosaics and Chimeras in Mammals." New York: Plenum, pp 361–377.

White M (1978): "Modes of Speciation." San Francisco: Freeman.

SESSION III. INHERITED DISEASES OF MOUSE AND MAN

VICTOR A. McKUSICK, Chairman

Introduction to Session III

Many single-gene-induced diseases are now recognized in man and mouse, as are a smaller array of defects in other experimental mammals. Although the ways of identifying mutants and methods used to determine the manner of their inheritance differ between man and mouse, many apparently homologous hereditary diseases are known in the two species. Advances have resulted from analysis of mouse conditions which, although not precise interspecies homologs, nevertheless contribute significantly to understanding the etiology and pathology of many inherited human syndromes.

Chairman Victor McKusick's account of 20 years of progress in recognizing and understanding human genetic diseases clearly demonstrates that rapid advances have been made in this field as is shown by the increasing size of each successive edition of his "Mendelian Inheritance in Man." Another fruitful contribution has been his promotion of a successful series of Johns Hopkins–Jackson Laboratory Short Courses in Medical Genetics. McKusick's account of landmarks in clinical genetics, of growth of the human linkage map, and his organized listing of known molecular defects in hereditary diseases provides orientation especially useful to many of our readers who are not clinicians.

Douglas Coleman provides several excellent examples of contributions from studies of mouse mutants to understanding of the etiology of inherited diabetes and obesity. Mutants at any one of five different genetic loci cause, in young mice, a syndrome similar to human maturity-onset diabetes. If the condition, including hyperglycemia, progresses relatively slowly (as in all A^y/a, A^{vy}/a, *fat/fat*, or *tub/tub* mice), this relatively benign clinical course persists through a near-normal life-span. If, however, the condition progresses rapidly (as in *db/db* or *ob/ob* mice), prognosis depends on the rest of the genome. When congenic with C57BL/6J, both *db/db* and *ob/ob* mice continue to show maturity-onset diabetes. However, C57BL/Ks-*db/db* and C57BL/Ks-*ob/ob* mice, with different genetic background, shift as young adults to a much more severe condition (similar to human juvenile diabetes), because their islet cells cannot proliferate rapidly enough to keep up with increasing hyperglycemia.

Coleman's parabiosis experiments demonstrate clearly that actions of both the *ob* and *db* genes involve a satiety center in the hypothalamus. The diabetic mouse produces, but cannot respond to, a factor that prevents overeating. The obese mouse responds to this factor, but cannot produce it. Coleman's beautifully designed experiments also show that obese and diabetic mice metabolize even restricted amounts of food differently from their normal counterparts, with more energy diverted to lipid synthesis and fat storage and can, under fasting conditions, survive longer than +/+ mice.

Samuel Lux and co-workers have made excellent use of erythrocytes of mutant mice to study normal vs abnormal structure of the red cell membrane. Durable yet very flexible erythrocytes are essential for preservation and continuing circulation of hemoglobin. Defective fragile erythrocytes are responsible for a variety of hemolytic anemias in both mouse and man. Each of four distinct hemolytic-anemia-producing mutants in the mouse has been shown to have very unstable red cell membranes (containing markedly less than the normal amount of spectrin), the extent of deficiency correlating with severity of anemia. The fact that mice of four distinct genotypes show spectrin deficiency implies the existence of four different mechanisms for reducing spectrin content. Erythrocyte membrane skeleton makeup also has been investigated for four different types of human hereditary hemolytic anemia. In these, spectrin defects appear to be qualitative rather than quantitative. It may well be that both quantitative and qualitative spectrin defects will eventually be found in both species. Meanwhile, up to the present, mouse models of hemolytic anemias have proved useful as analogs, rather than homologs, of human conditions.

Charles Scriver's account of "Garrod's Legacy to the Nations of Mice and Men." —discussing errors of metabolism — provides a delightful climax to this session on inherited diseases. He speaks of the special usefulness of homologs, citing mutants (histidinemia; ornithine transcarbamylase deficiency; and hypophosphatemia) as mouse homologs of human genetic defects. Each of these may have potential for revealing the nature of a basic defect common to man and mouse. He cites the deficiency of proline dehydrogenase activity in the PRO/Re mouse as an analog of human hyperprolinemia, useful as a probe into biologic processes relating to proline transport.

Scriver points out that achievement of control of chemical traffic across lipid bilayer membranes was a giant step in the evolution of living cells. In the epithelia of multicellular organisms, solute transport is transcellular, not paracellular, and transport of high molecular weight molecules involves activity of many different genes. Membrane transport and intracellular metabolism are independent functions. The mendelian patterns for inheritance of distinct inborn errors of membrane transport show that a diversity of specific gene products is responsible for the observed phenotypes. Transport mutations expressed in only one tissue — often in only one membrane of a cell — demonstrate the high degree of specificity of transport control, as does the involvement of alleles at two or more loci in control of transport of a single substance. Scriver exemplifies these principles in his elegant presentation of transport of proline in normal and prolinemic (PRO/Re) mice and transport of phosphate in normal mice and mice with sex-linked hypophosphatemia.

Scriver ends by predicting great future interest in membrane structure and membrane events — the stuff of cellular organization and compartmentation — and recommends greater attention to the important problem of evolutionary origin of membranes.

The Last Twenty Years: An Overview of Advances in Medical Genetics

Victor A. McKusick

The year 1979 is not only the 50th anniversary of the founding of the Jackson Laboratory, but also the 20th anniversary of the conception of the short course in medical genetics conducted jointly by the Jackson Laboratory and The Johns Hopkins University, with the financial support of the National Foundation—March of Dimes. Exactly 20 years ago this month, I first visited the Jackson Laboratory; and it was at lunch at a well-known Bar Harbor restaurant that the idea of a course first came up in conversation with Dr. John L. Fuller, then Assistant Director for Training. Those of us in medical genetics and the scientists at the Jackson Laboratory have all along had similar goals; both are interested in determining the genetics and basic gene-determined defect of the mutations they encounter as well as in determining the role of genetic factors in cancer, longevity, and behavior. The methodology then used by each was quite different, but each had much to learn from the other.

The short course in medical genetics, in essentially the same format that it has used throughout its 20-year history, was first given in the summer of 1960. From the beginning, the course was supported by the March of Dimes as part of the foundation's then new venture into the area of birth defects (a term that, as far as I know, was first devised by the National Foundation to denominate congenital malformations; certainly, it was the National Foundation that gave the word the wide currency it now enjoys). From the beginning, the objective of the course was to "teach the teachers." A great vacuum then existed in genetics instruction in medical schools; and the largest part of the "students" in the short course consisted of members of medical school faculties, although the faculties of dental and other professional schools and persons teaching human genetics at the university level were also represented. We insisted that this was a course and not a seminar or symposium. An attempt was made to cover the field in a systematic manner. In addition, new material was introduced each year, much of which had not yet been published, thereby conveying a sense of immediacy and excitement. This was easy to do because of the rapid advances being made throughout the last 20 years and the active research of the faculty. Thus, although the primary objective of the course was educational, research inevitably was fostered, as was, of course, clinical application.

Each year, in addition to eight to twelve members of The Johns Hopkins University faculty and about an equal number from the staff of the Jackson Laboratory, the faculty has consisted of six or more guest lecturers. Dr. Park S. Gerald of the Boston Children's Hospital enjoys the unique distinction of having been a member of the guest faculty each year. Rotation of the guest lecturers has introduced useful enrichment of the course and has made the experience additionally worthwhile for students and faculty alike. The course was described in some detail in a note in the journal Science in 1970 [1].

Beginning in 1967, the short course in medical genetics alternated with a short course in experimental mammalian genetics (primarily mouse genetics) in odd-numbered years. By good rights, 1979 would be a year of the mammalian genetics course, but a combined course seemed indicated for this anniversary year. A convergence of mouse and man as objects of genetic study has occurred during the last 20 years. The body of genetic information in the two species is now approximately on a par, and some of the experimental methodology is now the same.

Advances in medical genetics in the last 20 years include all of human chromosome mapping, all of our understanding of the major histocompatibility complex of man, all of the developments in several areas of biochemical genetics (eg, lysosomal diseases, vitamin-responsive inborn errors of metabolism, receptor defects), and all of our understanding of the genetic code. Cliometrics would be difficult. I suspect that more than 95% of the material presented in the 1979 course was unknown at the time of the 1960 course. Perhaps, not unexpectedly, the growth has been nonlinear. An acceleration in the last decade is evident. Methodologic advances, as always, have made this scientific progress possible. These technical advances have included chromosome banding, somatic cell hybridization, and other methods that provide surrogate approaches to the genetics of man.

During the past two decades, human genetics has become much more an experimental science, via the study of cultured cells as a surrogate for the whole organism as well as via the study of cell-free systems. The last 20 years have seen less exclusive reliance on statistical methods in human genetics and medical genetics. Human genetics has, furthermore, become highly "medicalized" during this period. This has been, in part, because of pressure to understand man's ailments and, in part, because insight is best provided into the normal by a study of the abnormal. I think it is fair to say that mouse genetics has, likewise, become in part "medicalized," perhaps through the influence of the short course in medical genetics and the contacts that it facilitated.

CYTOGENETICS

Chromosomal aberrations have been identified in seven categories of clinical abnormalities (Table I). Two of them — mental retardation syndromes (of which the archetype is Down syndrome) and sexual abnormalities (of which the archetypes are Turner (XO) syndrome and Klinefelter (XXY) syndrome) — were the only ones known by the summer of 1959, the point of initiation of this review. In both areas, knowledge has expanded greatly.

The body of cytogenetic information that has accumulated since 1959 includes the poly-X states, Robertsonian and reciprocal translocations, isochromosomes, deletions, inversions, and mosaicism. Description of the cri-du-chat syndrome (resulting from deletion of the short arm of chromosome 5) was notable from a clinical point of view, as was description of familial balanced karyotypes as the basis for segregation of unbalanced karyotypes with deficiencies and/or duplications. Also noteworthy was description of familial inversions leading to the generation of anomalous gametes from inversion loops in meiosis.

"Banding" techniques uncovered a previously unknown, and almost unsuspected, order of polymorphism in the human chromosomes. These methods also permitted a degree of precision in the definition of chromosomal aberrations, which was previously quite impossible. Karyotypically "pure culture" groups of cases could be compared phenotypically with the clinical delineation of "new" chromosomal syndromes [2]. An example of this is the trisomy-8 syndrome, which previously was undefined, largely because chromosome 8 could not be confidently distinguished from other members of the C group.

In addition to the classic example of a neoplastic disease [3] associated with a specific chromosomal aberration — chronic myeloid leukemia, the discovery of deletion of a specific small segment of the long arm of chromosome 13 (13q14) with retinoblastoma [4, 5] and of the short arm of chromosome 11 (11p13) with Wilms tumor (often with aniridia) [6] is of great theoretic and practical importance.

TABLE I. Clinical Abnormalities Causally Related to Chromosomal Aberrations*

1. Mental retardation syndromes, eg, T21 (Down syndrome). Jan. 1959.
2. Abnormalities of sexual development, eg, XO Turner syndrome, XXY Klinefelter syndrome. Early 1959.
3. Complex malformation syndromes, eg, trisomy 13 (Patau syndrome). Early 1960.
4. Malignant transformation, eg, Philadelphia chromosome in chronic myeloid leukemia 21q–, 1960; revised to t(22q–;9q+), 1974.
5. Behavioral abnormalities, eg, XYY syndrome. About 1964.
6. Spontaneous abortion in first trimester. In 1960s.
7. Chromosomal breakage in autosomal recessive disorders, eg, Fanconi syndrome, 1964.

*Discovered mainly since 1959.

Progress in Understanding the Molecular Basis of Mendelian Disease (Table II)

In 1959, the enzyme deficiency in only nine disorders had been demonstrated, validating the hypothesis of Archibald Garrod in his "Inborn Errors of Metabolism." (Methemoglobinemia is not, *strictu sensu*, an inborn error of metabolism.) Today, 20 years later, deficiencies of over 185 different enzymes have been identified as the basis of mendelian abnormalities [7].

Lysosomal diseases, which now number over 30, have been known only since 1964. Vitamin-responsive inborn errors of metabolism first came to attention with pyridoxine-dependent convulsion in the 1950s. Interestingly, this earliest-to-be-described vitamin-dependent disorder is probably the one least understood at the enzymatic level. Homocystinuria and the several forms of methylmalonic-aciduria are examples that are better understood.

It has come to be realized that the deficiency may involve an enzyme involved in the posttranslational processing of a protein. The clearest example is the defect in collagen in Type VI Ehlers-Danlos syndrome, occasioned by deficiency of lysyl hydroxylase. Some enzyme deficiencies may, in fact, be deficiencies of processing or activating enzymes, although no clear examples are known in man. The discovery of enzyme deficiencies underlying neoplastic disease (deficiency in DNA repair enzymes) and immunodeficiencies (ADA and PNP deficiencies) was noteworthy.

TABLE II. Molecular Defects in Hereditary Diseases* (Discovered Mainly Since 1959)

I. Enzymopathies (inborn errors of metabolism)
   A. Nine enzyme deficiencies known by 1959
      1. Diaphorase in methemoglobinemia (1948)
      2. Alkaline phosphatase in hypophosphatasia (1948)
      3. Catalase in acatalasemia (1949)
      4. Glucose-6-phosphatase in GSD I (1952)
      5. Phenylalanine hydroxylase in PKU (1953)
      6. G-6-PD in hemolytic anemia (1956)
      7. Gal-1-P-uridyl transferase in galactosemia (1956)
      8. Pseudocholinesterase in succinylcholine sensitivity (1956)
      9. Homogentisic acid oxidase in alkaptonuria (1958)
   B. Over 185 known by 1979, including
      1. Vitamin-responsive inborn errors of metabolism, eg, B_6-responsive homocystinuria
      2. Defects in 15 red-cell enzymes leading to hemolytic anemia [40]
      3. Four autosomal-dominant porphyrias, with defects in heme synthesis pathway [57]
      4. Lysosomal diseases: defects in each of over 30 specific acid hydrolases (Tay-Sachs disease)
      5. Deficiency of enzymes involved in posttranslational processing of protein (lysyl hydroxylase in Ehlers-Danlos syndrome VI)
      6. Enzymic basis of hereditary neoplastic disease (deficiency of DNA repair enzyme(s) in xeroderma pigmentosa)

(continued next page)

TABLE II. Continued

  7. Enzymatic basis of immunodeficiency (adenosine deaminase or nucleoside phosphorylase deficiency)
  8. Fetal abnormality due to inborn error of metabolism in mother (phenylketonuria)
  9. Abnormal pregnancy due to placental enzyme deficiency determined by fetal genotype (placental steroid-sulfatase deficiency)
  10. Defects in coagulation cascade (factors VII, IX, X, XI, XII, XIII)
  11. Defects in early stages of complement pathway
II. Defects in nonenzymatic proteins: hemoglobinopathies
  A. Amino acid substitutions
    1. Two known by 1959: β-6val (HbS); β-6lys (HbC)
    2. By 1979: about 100 alpha, 200 beta, 15 gamma, 10 delta
  B. Other defects (all discovered since 1959)
    1. Hybrid nonalpha-globin chain (Hb Lepore) from nonhomologous pairing and unequal crossing-over
    2. Interstitial deletion, probably through same mechanism (Hb Gun Hill)
    3. Deletion leading to frame shift (Hb Wayne)
    4. Extra long polypeptide due to mutation in "stop" codon (Hb Constant Spring)
    5. Nonsense mutation (at beta-codon-17 in β^0-thalassemia)
    6. Deletion of one or more genes (several forms of thalassemia and hereditary persistence of fetal hemoglobin)
III. Defects in other nonenzymatic proteins (or peptides)
  A. Phenylalanine to leucine as AA 23 or 24 in B chain of insulin in diabetic [14]
  B. Hyperproinsulinemia, due to change at cleavage site (?)
  C. Coagulopathies (factor VIII deficiency in hemophilia A)
  D. Deficiency of specific nonenzymatic complement components
  E. Deficiency of immunoglobulins
  F. Deficiency of peptide hormones (growth hormone)
IV. Receptor diseases
  A. Defects in cell-surface receptor (eg, LDL in hypercholesterolemia)
  B. Defects in cytosol-nuclear receptor (that of dihydrotestosterone in testicular feminization)
V. Molecular basis of congenital malformations
  A. Defect in dynein in immobile cilia syndrome (Kartagener syndrome)
  B. Defect in X-linked H-Y regulator in XY gonadal dysgenesis
  C. Deficiency of testosterone reductase in pseudovaginal perineoscrotal hypospadias
  D. Defect in cortisol synthesis in congenital adrenal hyperplasia, with masculinized external genitalia in female
VI. Transmembrane-transport defects
  A. For AAs (in cystinuria)
  B. For phosphate (in hereditary hypophosphatemia)
  C. For copper (in Menkes disease)
  D. For hydrogen ion (in renal tubular acidosis)
  E. For glucose-6-phosphatase (into lumen of endoplasmic reticulum) in GSD Ib
  F. For cobalamin (into mitochondrion) in a form of methylmalonicaciduria
VII. Defect in a Specific Differentiated Cell Type (of osteoclasts in osteopetrosis; of specific immunocytes in certain immunodeficiencies)

*For additional bibliographic references, see Reference 11.

Fetal abnormality of genetic causation arising, not from defect in the genotype of the fetus, but rather from a genetic enzyme deficiency in the mother, found illustration in defects occurring in most offspring of women with phenylketonuria [8]. Conversely, recurrent abnormalities in pregnancy – earlier misinterpreted as resulting from a defect in the maternal genotype – proved to be the result of placental steroid-sulfatase deficiency, determined by a mutant fetal genotype. An X-linked recessive, the deficiency is manifested by delay in the onset of labor, relative refractoriness to oxytocic agents, increased frequency of stillbirth, and low estriol levels in urine and plasma of the mother [9]. The steroid-sulfatase deficiency is expressed in postnatal life as ichthyosis [10].

Defects in nonenzymatic proteins are illustrated, in particular, by the hemoglobinopathies. By 1959, the amino acid change had been determined in only two variant hemoglobins, S and C. By 1979, about 100 alpha-, 200 beta-, 15 gamma-, and 10 delta-globin amino acid substitutions had been identified [11]. Other types of mutational change in hemoglobin were deduced from the study of amino acid sequences of the globins or, in more recent times, by studies of the genes themselves by the methodology of recombinant DNA and restriction enzyme digestion [12, 13]. Deletions lead to: 1) production of hybrid nonalpha-globin chains, as in hemoglobin Lepore; 2) interstitial deletions, as in hemoglobin Gun Hill; or 3) deletion of the entirety of one or more closely linked nonalpha-globin genes, as in several forms of thalassemia and in hereditary persistence of fetal hemoglobin. Because of mutation in the terminator codon, extra long nonalpha-globin polypeptides are formed with reading of a terminal flanking sequence, which is ordinarily not translated, eg, hemoglobin Constant Spring. Conversely, nonsense mutation in a codon to a "stop" was found to be the basis of failure of globin synthesis; for example, mutation in beta-codon-17 to a terminator codon produces β^0-thalassemia.

Understanding of the enzymatic basis of inborn errors of metabolism was greatly advanced by the use of cell cultures. It came as a pleasant surprise in the 1960s that the lowly fibroblast has an unexpectedly wide enzymatic repertoire, making the study of cells from affected individuals a useful approach. In this way, methods similar to those used in the study of bacterial metabolism could be brought to bear, and many human diseases were opened up for study by a wide range of laboratory scientists.

Defects have been delineated in other nonenzymatic proteins, although, in many of these, not at the level of precise amino acid structure. Examples include deficiency of specific clotting factors (in fact, several of the clotting factors are enzymatic in nature), deficiency of specific complement components, deficiency in immunoglobulins, and deficiency of peptide hormones such as growth hormone. The recently described amino acid substitution in insulin [14], which renders it metabolically ineffective, is an example of the precise change in nonenzymatic proteins, of which one can expect many more examples in the next few years.

Receptor diseases, like all mendelian disorders, illustrate a special case of Murphy's law: If anything can go wrong, it will go wrong. They also are powerful examples of the Harvey-Garrod principle [15]: The abnormal teaches much about the normal. Examples are defects in the cell-surface receptor for LDL in hypercholesterolemia [16] and those for the specific cytosol-nuclear receptor [17] for dihydrotestosterone in the testicular feminization syndrome (androgen insensitivity).

The molecular basis of congenital malformations is not easily determined because the human embryo is not readily accessible to study at pertinent stages. But, here too, some progress has been made. The axonemal defect in the immobile cilia syndrome (Kartagener syndrome) is a particularly noteworthy example [18] because, although the sinusitis, bronchitis with bronchiectasis, and male infertility due to immobile sperm are readily observable and explicable postnatal features, dextrocardia is not so obvious in its pathogenesis. A defect in an X-linked H-Y regulator is inferred in XY gonadal dysgenesis. Because of the importance of dihydrotestosterone to the normal fetal development of the external genitalia, deficiency of testosterone reductase leads to the form of male pseudohermaphroditism, which is referred to as pseudovaginal perineoscrotal hypospadias [19]. Defects in cortisol synthesis in congenital adrenal hyperplasia lead to masculine external genitalia in the female because of the excessive androgens that develop as a side effect.

Defects in the transmembrane transport of a variety of metals, ions, enzymes, provitamins, and amino acids are now known. In some cases, these are defects in transport across the plasma membrane. In other cases, these are defects of transport into the mitochondria or endoplasmic reticulum. (Testicular feminization might be classified as a transmembrane-transport defect — a defect in transport across the nuclear membrane. It is not, however, a defect in an active transmembrane-transport system.)

Although perhaps not properly discussed as examples of molecular defects, the delineation of the basic abnormality in some disorders has focused on failure of development of specific differentiated cell types. A cardinal example is the failure of development of osteoclasts in osteopetrosis of the "malignant" autosomal recessive type [20]. Failure of development of a specific immunocyte in certain immunodeficiencies represents another example.

Immunogenetics

Although white-cell typing was just beginning in 1959, it is accurate to say that almost all understanding of the major histocompatibility complex in man came within the last 20 years. In few areas of human genetics is as much owed to the previous and concurrent work in the mouse as in this field. Studies of the H2 "locus" go back to its initial description by Peter Gorer in the 1930s. The description of multiple loci in the H2 region and the discovery of immune-response genes in the mouse led the way to parallel discoveries in man. The importance of the human counterpart, HLA, in transplantation forced rapid progress in man [21]. Widespread work in service laboratories in many parts of the world led to the accumulation of extensive information on man in the last 15 years, comparable to the vast amount of work on the blood groups of man coming out of blood banks in the period 1940–1965.

The part of immunogenetics concerned with humoral immunity likewise advanced greatly, with some parallels to the advances in the study of the hemoglobinopathies, eg, the description of molecular change associated with mutation. Consistent with the complexity of the immune system, definition of the normal and the abnormal is not yet in as satisfactory a state of understanding as that of the hemoglobins. Defects in polymorphonuclear leukocytes in their host-defense function have also been delineated.

In the decades to come, we may see, in man, a delineation of a system homologous to the "T locus" of the mouse. Information in the mouse, much of it coming from the study of congenital malformations resulting from point mutations in the T region, will predictably be a valuable guide in studies of the human fetus.

The Formal Genetics of Hereditary Disease in Man (Table III)

The Lyon hypothesis (the concept of the one inactive X chromosome), Ohno's law of evolutionary conservatism of the mammalian X, and the human chromosome map, insofar as we now know it, are almost exclusively advances of the last 20 years. The description by Lyon and Hawkes [22] of X-linked testicular feminization in the mouse, taken together with Ohno's law of the evolutionary conservatism of the mammalian X, effectively settled the long-standing dilemma as to whether this disorder in man is an X-linked recessive or an autosomal-dominant male-limited trait. More direct proof of X-linkage was provided by Meyer et al [23], when they succeeded in cloning two classes of cells from cultures made from women heterozygous for testicular feminization: One class with absent dihydrotestosterone receptor and one with normal receptor.

Paternal age effect in mutation was referred to by Weinberg in 1912 (in relation to dwarfism) and was demonstrated by Penrose in the 1950s. Elaboration on these earlier studies has been achieved in the last few years. In the last 20 years it has been demonstrated that the recombination frequency usually is greater in females than in males. This had been noted in mice by R. A. Fisher in the early 1920s. Contradictory evidence has been presented on the question of a higher rate of X-linked mutation in males than in females. Linkage disequilibrium demonstrated in man by Boyer et al [24], in relation to the delta- and beta-globin loci, has become a frequently observed phenomenon, particularly in relation to HLA.

Deletion mapping of loci first was achieved in man by Ferguson-Smith et al [25], who used this approach to place the locus for red-cell acid phosphatase (ACP-1) on the distal part of the short arm of chromosome 2. Centromere mapping by the study of ovarian teratomas was introduced in the 1970s by Linder, Hecht, Ott and others [26, 27].

TABLE III. Formal Genetics of Hereditary Disease in Man (Discovered Mainly Since 1959 – A Partial Listing)

1. Lyon's hypothesis of one inactive X chromosome
2. Ohno's law of evolutionary conservatism of X chromosome
3. Allelic series and genetic compounds as the bases of phenotypic variability resulting from deficiency of a given enzyme [29]
4. Mutation rate in X chromosome: ?M > F
5. Recombination frequency: F > M
6. Linkage disequilibrium
7. Ethnic distribution of genetic disease
8. Deletion mapping
9. Centromere mapping from study of ovarian teratomas
10. Surrogate methods (Haldane: "alternatives to sex"; Pontecorvo: "parasexual")
    a. Sequencing
        i. AAs in protein gene product
        ii. Nucleic acids in gene or mRNA
    b. Interspecies somatic cell hybridization for gene mapping
    c. Man-man somatic cell hybridization for recognitions of genetic heterogeneity
    d. Cloning of two populations of cells from heterozygous females as proof of X linkage
11. Confusion of slow virus and mendelian gene (Creutzfeld-Jakob disease)
12. Mitochondrial genetics
13. The genetic map of human chromosomes [30, 58]

The confusion of slow viruses and mendelian genes was illustrated by the work of Gajdusek and colleagues [28] in Creutzfeldt-Jakob disease. This disorder often shows typical autosomal dominant pedigree patterns, but a filtrable agent can be transmitted to primates from familial cases, and transmissibility in man has been demonstrated from the development of the disease in a patient given a corneal graft from an affected person. Similar transmission is suspected as an occupational hazard in neurosurgeons. A heuristic hypothesis is that the virus becomes incorporated into one chromosome of an autosomal pair and, thereafter, is transmitted as an autosomal dominant mendelian trait. Linkage studies (difficult to perform because of the late development of the clinical picture and the death of affected individuals in earlier generations) might settle the question of a specific chromosomal integration site. Cellular genetic studies could assign the agent to a specific chromosome but, again, would have difficulties because of the cumbersome test system for demonstration of the virus. Probably, molecular DNA methods hold the greatest hope for definition of the basis of this viral disease.

The recognition of multiple allelic series and of genetic compounds as the bases of phenotypic variability resulting from one and the same enzyme deficiency [29] was a contribution of the last 10–15 years.

In the last two decades, progress in the formal genetics of man has been possible because of the devising of surrogate methods, ie, what Haldane referred to as "alternatives to sex" and what Pontecorvo referred to as "parasexual" methods. In the 1960s, these were mainly methods of sequencing the amino acids in protein gene products and arriving at genetic inferences therefrom. In the 1970s, surrogate methods have involved, particularly, somatic cell hybridization, either interspecies hybridization for purposes of gene mapping or intraspecies hybridization for identification of genetic heterogeneity (nonallelism). Late in the 1970s, sequencing of nucleic acids in the gene itself or in the messenger RNA (mRNA) has been the main surrogate method.

The human gene map (Fig. 1), perhaps the single most important development in the formal genetics of man, is almost exclusively an advance of the 1970s [30]. In 1969, a map comparable to that in Figure 1 would convey very little information: The Duffy blood-group locus had been assigned to chromosome 1 by family studies; the alpha-haptoglobin locus had been assigned to chromosome 16, also by family study; thymidine kinase locus had been assigned tentatively to an E-group chromosome, possibly 17, by interspecies somatic cell hybridization; and about 85 loci had been assigned to the X chromosome. Several autosomal linkage groups were known as well as two linkage groups on the X chromosome. The specific autosome carrying these groups was not known, and the regional localization of the genes on the X chromosome was also not known. Today, over 110 loci have been assigned to the X chromosome with confidence and over 230 loci to specific autosomes, and the rather precise regional localization of many of these genes is known. Through all of this development, continual correlation with the map in the mouse has been valuable as a guide in the search of homologies.

The ethnic distribution of genetic disease, although long a matter of interest, has come in for both expanded description in the last two decades and some increase in basic understanding. In a parallel way, the basis of genetic polymorphism has become better understood. In both areas, controversy concerning the relative importance of selection and random events has occupied workers in the field. Following Allison's classic hypothesis of the interrelationship between malaria and the sickle-cell gene in the 1950s, other examples of the role of selection have recently appeared (with mechanistic explanation). Especially convincing is the role of vivax malaria in determining the high frequency of the Duffy null gene in West Africans [31]. The description of the functional significance of particular polymorphic proteins has provided a plausible basis for the observed distribution of that polymorphism, eg, the vitamin-D binding role of the Gc protein [32]. In the case of many mendelian diseases, founder effect and random genetic drift provide a plausible explanation for their distribution in particular ethnic groups, eg, the Ellis-van Creveld syndrome in the Amish [33] and, I think, Tay-Sachs disease [34] in the Ashkenazim. In other instances — ie, cystic fibrosis — selection remains a more plausible explanation, but one which is unproved both in terms of a demonstrated reproductive advantage and in terms of a physiologic superiority of heterozygotes.

Embryonic beginnings have been made on the mitochondrial genetics of man, with the description of mutations in the mitochondrial DNA of cultured human cells [35–37].

Clinical Genetics [38]

Five landmarks, perhaps, characterize the development of clinical genetics in the last 20 years (Table IV). Expansion in genetic nosology (classification of genetic diseases, or better, delineation of separate genetic diseases) has occurred, especially through the identification of genetic heterogeneity [39]. Progress has been made, not so much in the identification of new phenotypes, as in identification of heterogeneity within phenotypes such as congenital deafness, mental retardation, chondrodystrophies, chorioretinopathies, and categories of skin disease such as epidermolysis bullosa and ichthyosis. Biochemical approaches, of course, have helped greatly in the delineation of heterogeneity; for example, hereditary chronic nonspherocytic hemolytic anemia was, for all practical purposes, viewed as a single entity before 1959. Now, about 15 different enzyme defects, each a separate entity, are known to lead to this syndrome [40].

A second landmark of the last two decades has been the development of new diagnostic techniques, particularly the chromosomal methods and fibroblast culture, both of which have already been discussed.

I would list prenatal diagnosis as a third and separate landmark because, although an advance in diagnosis, it is a new dimension thereof. The methods of prenatal diagnosis are of two main types: imaging and amniocentesis. The imaging methods include sonography and x-ray and, perhaps, can be considered to include fetoscopy also. The fluid obtained by amniocentesis has been mainly useful for study of the contained cells, although determination of alpha-fetoprotein in the fluid itself has had some usefulness in relation to neural-tube defects. Enzyme deficiencies and chromosomal aberrations can be detected by study of amniotic cells.

TABLE IV. Five Landmarks of Clinical Genetics in the Last Twenty Years

1. Genetic nosology, mainly identification of genetic heterogeneity
2. Expanded diagnostic methods
    a. Cytogenetics
    b. Fibroblast culture
    c. Biochemical methods
3. Prenatal diagnosis
    a. Enzyme deficiency
    b. Chromosomal aberration
    c. Diagnosis by the linkage principle
        1. Hemophilia-G-6-PD
        2. Myotonic dystrophy-secretor
        3. Sickle-cell anemia-Hpa I
        4. Congenital adrenal hyperplasia-HLA
4. Definition of basis of genetic susceptibility to disease
    a. HLA-B27 (ankylosing spondylitis)
    b. Alpha-1-antitrypsin deficiency (emphysema, cirrhosis)
    c. Transketolase defect (Wernicke-Korsakoff syndrome)
    d. Duffy null (tertian malaria)
5. New methods of treatment: eg, vitamin supplementation in vitamin-responsive inborn errors of metabolism: replacement of defective tissue or organ, eg, kidney in congenital cystic disease, bone marrow or thymus in immunodeficiencies, bone marrow in osteopetrosis.

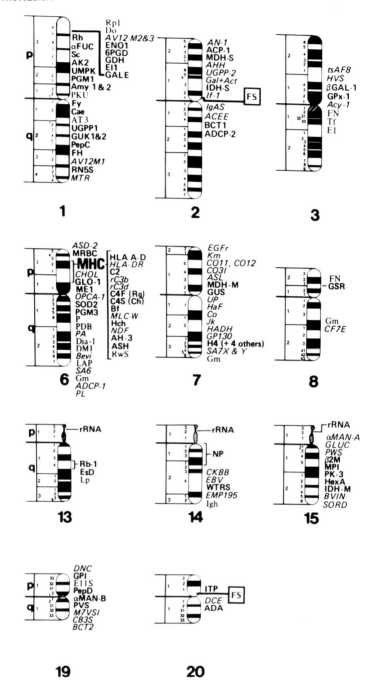

Fig. 1. The human gene map. See Reference 58 for key.

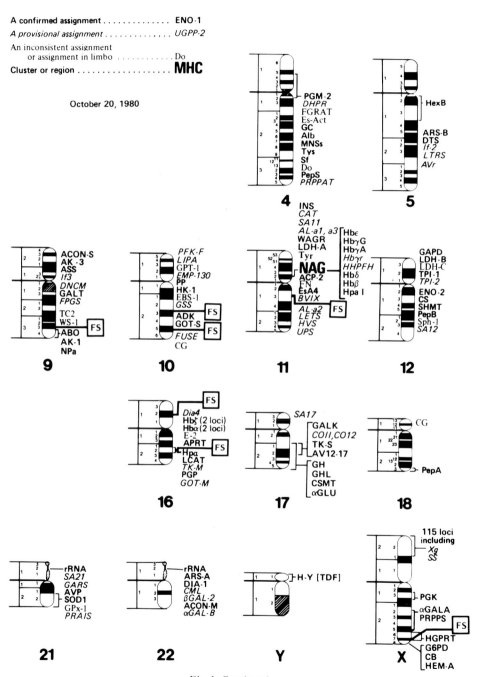

Fig. 1. Continued.

In 1956 Klaus Fuchs suggested that fetal sex determination is possible by studying the material obtained by amniocentesis for Barr bodies in the amniotic cells. John H. Edwards immediately suggested that the principle of linkage could be applied [41]. The principle has, indeed, been put to use for the prenatal diagnosis of four disorders: 1) hemophilia A (linkage to G-6-PD [42]; 2) myotonic dystrophy (linkage to ABH secretion) [43]; 3) sickle-cell anemia (linkage to the HpaI restriction-enzyme polymorphism) [44, 45]; and 4) congenital adrenal hypoplasia (linkage to HLA) [46, 47]. The most effective of these is the last, because of very tight linkage and because of the high order of polymorphism of HLA; but this approach may not be needed because of the availability of diagnosis by the content of steroid hormones in the amniotic fluid [48].

A fourth landmark in clinical genetics relates to the understanding of genetic susceptibility to disease. Two cardinal examples are the association between HLA B27 and ankylosing spondylitis [21], and the relation between Duffy null bloodgroup and tertian malaria [31], as previously mentioned. The first of these is such a strong association that testing for B27 is used diagnostically (although considerations of sensitivity and specificity impair its usefulness). (The second provides an explanation for an old clinical observation; namely, that in the days when fever therapy with induced malaria was used in the treatment of central nervous system syphilis, tertian malaria could not be used in black patients because it was known that, in most cases, the infection would not "take." Quartan malaria was used instead.) Two other examples are alpha-1-antitrypsin deficiency in emphysema [49] and in cirrhosis [50], and transketolase variant in Wernicke-Korsakoff syndrome [51].

A fifth landmark in clinical genetics is the development of treatment of some disorders. This area, in general, has not advanced as much as other areas; but one can point to the use of vitamin supplementation in the treatment of certain inborn errors of metabolism such as vitamin B_6 in one form of homocystinuria, and to the transplantation of tissues or organs to replace defective or missing tissues. Examples of the latter are kidney transplantation in polycystic renal disease; thymus or bone-marrow transplantation in immunodeficiencies; and bone-marrow transplantation for replacement of stem cells and, therefore, osteoblasts in osteopetrosis [20]. (The advance in osteopetrosis was based on the findings in the homologous disorders of the mouse [52].) The time is now right for the development of other new forms of treatment. Enzyme replacement has had modest success in the case of lysosomal diseases. The infusion of red cells in ADA deficiency is a form of enzyme replacement [53].

SUMMARY

By way of summary, as well as comparison between man and mouse, I present in Figure 2 a count of reasonably firmly established loci (actually phenotypes, but no locus has wittingly been represented more than once) in man and mouse in 1968 and in 1979. At both times, in man, dominants exceed recessives, whereas in the mouse, the converse is true. The difference between man and mouse in the relative frequency of dominants and recessives presumably is attributable to patterns of mating; recessives come more readily to attention under conditions of close mating in the laboratory mouse.

Inborn errors of metabolism are underrepresented in the catalog of mouse mutations, as compared with man. Ascertainment of many human inborn errors of metabolism have been through the phenotype of mental retardation, which is not recognizable in the mouse, and/or the phenotype of "failure to thrive," which has usually been disregarded (and discarded) in the mouse. (Ornithine transcarbamoyl-transferase deficiency [54, 55] and proline-oxidase deficiency [56] have been identified in both man and mouse.) As pointed out by Knudsen (personal communication), who would find it useful to have such mutations for study, autosomal-dominant neoplasia syndromes comparable to neurofibromatosis, retinoblastoma, Wilms tumor, intestinal polyposis, multiple endocrine adenomatosis, basal-cell nevus syndrome, and tuberous sclerosis of man are unknown in mice, as are spontaneous mutations to abnormal hemoglobin structure.

The state of knowledge of the X chromosome is noteworthy in comparing man and mouse. The first X linkage to be specifically so interpreted in man was color blindness, in 1911 by E. B. Wilson. By the early 1950s, when the first X-linked loci — tabby (Ta) and jimpy (jp) — were identified in the mouse by Falconer in Edinburgh, over 35 X-linked loci had already been demonstrated in man. Man has continued to maintain a substantial lead over the mouse, illustrating one advantage of our species for genetic study, namely, the massive body of information — morphologic, physiologic, biochemical, pathologic, immunologic — that is accumulated on large numbers of people. The situation as far as autosomal

	—1968—		
	Dominant	Recessive	X
Man	344	280	68
Mouse	99	207	12

	—1979—		
	Dominant	Recessive	X
Man	780	540	110
Mouse	230	248	23

Fig. 2. Numbers of identified phenotypes, relating to separate loci, in mouse and man. The dominant counts include codominant. Counts in the mouse are from Dr. M.C. Green (1968) and Dr. T.H. Roderick (1979).

linkage is concerned has been precisely the converse. The first autosomal linkage in a mammal was demonstrated in the mouse, the linkage between albino and pink-eye, by J. B. S. Haldane and colleagues in 1915. By the early 1950s, when the first human autosomal linkage was discovered by Mohr — that between secretor and Lutheran blood group — 12 autosomal linkage groups were known in the mouse.

In the future, genetics of the mind and genetics of development should be the leading areas of advance in human genetics. Defects in these areas should, again, play an important role, and comparisons between man and mouse will be useful. There will, however, be need for development of ingenious surrogate methods for the study of these areas in man.

This review of 20 years in the study of genetic disease in man should also indicate, to some extent, the convergence of mouse and man as objects of genetic study. Hybridization of human and mouse cells is symbolic of what has happened in the fields of human genetics and mouse genetics in the last 20 years. In the early years of the Bar Harbor short courses, the content of the Medical Genetics course was very different from that of the Course on Experimental Mammalian Genetics but, in recent times, many portions of the two courses have become interchangeable.

REFERENCES

(Bibliographic references are provided mainly for those topics that are not discussed and referenced in McKusick [11] and in Stanbury et al [7], including references to publications since those books went to press.)

1. McKusick VA: The Bar Harbor course in medical genetics. Science 176:820–821, 1972.
2. Yunis JJ (ed): "New Chromosomal Syndromes." New York: Academic Press, 1977.
3. Sandberg AA: "The Chromosomes in Human Cancer and Leukemia." New York: Elsevier North-Holland, 1979.
4. Yunis J, Ramsay N: Retinoblastoma and subband deletion of chromosome 13. Am J Dis Child 132:161–163, 1978.
5. Hoegerman SF: Chromosome 13 long arm interstitial deletion may result from maternal inverted insertion. Science 205:1035–1036, 1979.
6. Riccardi VM, Sujansky E, Smith AC, Francke U: Chromosomal imbalance in the aniridia-Wilms tumor association: 11p interstitial deletion. Pediatrics 61:604–610, 1978.
7. Stanbury JB, Wyngaarden JB, Fredrickson DS (eds): "The Metabolic Basis of Inherited Disease," ed 4. New York: McGraw-Hill, 1978.
8. Frankenburg WK, Duncan BR, Coffelt RW, Koch R, Coldwell JG, Son CD: Maternal phenylketonuria: Implications for growth and development. J Pediatr 73:560–570, 1968.
9. France JT, Seddons RJ, Liggins GC: A study of a pregnancy with low estrogen production due to placental sulfatase deficiency. Biochem Med 10:167–174, 1974.
10. Shapiro LJ, Weiss R, Webster D, France JT: X-linked ichthyosis due to steroid-sulphatase deficiency. Lancet 1:70–72, 1978.
11. McKusick VA: "Mendelian Inheritance in Man: Catalogues of Autosomal Dominant, Autosomal Recessive and X-linked Phenotypes," ed 5. Baltimore: Johns Hopkins University Press, 1978.

12. Weatherall DJ, Clegg JB: Recent developments in the molecular genetics of human hemoglobin. Cell 16:467–479, 1979.
13. Forget BG: Molecular genetics of human hemoglobin synthesis. Ann Intern Med 91:605–616, 1979.
14. Tager H, Given B, Baldwin D, Mako M, Markese J, Rubenstein A, Olefsky J, Kobayashi M, Kolterman O, Poucher R: A structurally abnormal insulin causing human diabetes. Nature 281:122–125, 1979.
15. McKusick VA: Preface. "Heritable Disorders of Connective Tissue," ed 4. St. Louis: C.V. Mosby, 1972, p x.
16. Goldstein JL, Brown MS: The LDL receptor locus and the genetics of familial hypercholersterolemia. Annu Rev Genet 13:259–289, 1979.
17. Keenan BS, Meyer WJ III, Hadjian AJ, Jones HW, Migeon CJ: Syndrome of androgen insensitivity in man: Absence of 5-alpha-dihydrotestosterone binding protein in skin fibroblasts. J Clin Endocrinol 38:1143–1146, 1974.
18. Afzelius BA: A human syndrome caused by immobile cilia. Science 193:317–319, 1976.
19. Imperato-McGinley J, Guerrero L, Gautier T, Peterson RE: Steroid 5-alpha-reductase deficiency in man: An inherited form of male pseudohermaphroditism. Science 186:1213–1215, 1974.
20. Ballet W, Griscelli C, Coutris C, Milhand G, Maroteaux P: Bone marrow transplantation in (human) osteopetrosis, letter. Lancet 2:1137, 1977.
21. Svejgaard A, et al: The HLA system: An introductory survey. In Hauge et al (eds): "Monographs in Human Genetics," vol 7. Basel: Karger, 1979.
22. Lyon MF, Hawks SG: X-linked gene for testicular feminization in the mouse. Nature 225:1217–1219, 1970.
23. Meyer WJ III, Migeon BR, Migeon CJ: Locus on human X chromosome for dihydrotestosterone receptor and androgen insensitivity. Proc Natl Acad Sci USA 72:1469–1472, 1975.
24. Boyer SH, Rucknagel DL, Weatherall DJ, et al: Further evidence of linkage between the beta and delta loci governing human hemoglobin and the population dynamics of linked genes. Am J Hum Genet 15:438–448, 1963.
25. Ferguson-Smith MA, Newman BF, Ellis PM, Thomson DMG, Riley ID: Assignment by deletion of human red cell acid phosphatase gene locus to the short arm of chromosome 2. Nature 243:271–273, 1973.
26. Linder D, McCaw BK, Hecht F: Parthenogenic origin of benign ovarian teratomas. N Engl J Med 292:63–66, 1975.
27. Ott J, Linder D, McCaw BK, Lovrien EW, Hecht F: Estimating distances from the centromere by means of benign ovarian teratomas in man. Ann Hum Genet 40:191–196, 1976.
28. Roos R, Gajdusek DC, Gibbs CJ Jr: The clinical characteristics of transmissible Creutzfeldt-Jakob disease. Brain 96:1–20, 1973.
29. McKusick VA: Phenotypic diversity of human diseases resulting from allelic series. Am J Hum Genet 25:446–456, 1973.
30. McKusick VA, Ruddle FH: The status of the gene map of the human chromosomes. Science 196:390–405, 1977.
31. Miller LH, Mason SJ, Clyde DF, McGinnis MH: The resistance factor to Plasmodium vivax in blacks: The Duffy blood group genotype, FyFy. N Engl J Med 295:302–304, 1976.
32. Daiger SP, Schanfield MS, Cavalli-Sforza LL: Group-specific component (Gc) proteins bind vitamin D and 25-hydroxyvitamin D. Proc Natl Acad Sci USA 72:2076–2080, 1975.
33. McKusick VA, Egeland JA, Eldridge R, Krusen DE: Dwarfism in the Amish. I. The Ellis-van Creveld syndrome. Bull Johns Hopkins Hosp 115:306–336, 1964.

34. Chase GA, McKusick VA: Controversy in human genetics: Founder effect in Tay-Sachs disease. Am J Hum Genet 24:339–340, 1972.
35. Borst P, Grivell LA: The mitochondrial genome of yeast. Cell 15:705–723, 1978.
36. Fine PEM: Mitochondrial inheritance and disease. Lancet 2:659–662, 1978.
37. Yatscoff RW, Goldstein S, Freeman KB: Conservation of genes coding for proteins synthesized in human mitochondria. Somatic Cell Genet 4:633–645, 1978.
38. McKusick VA: The growth and development of human genetics as a clinical discipline. Am J Hum Genet 27:261–273, 1975.
39. McKusick VA: Genetic nosology: Three approaches. Am J Hum Genet 30:105–122, 1978.
40. Beutler E: Red cell enzyme defects as nondiseases and as diseases. Blood 54:1–7, 1979.
41. Edwards JH: Antenatal detection of hereditary disorders, letter to the editor. Lancet 1:579, 1956.
42. McCurdy PR: Use of genetic linkage for the detection of female carriers of hemophilia. N Engl J Med 285:218–219, 1971.
43. Schrott HG, Omenn GS: Myotonic dystrophy: Opportunities for prenatal prediction. Neurology 25:789–791, 1975.
44. Kan YW, Dozy AM: Antenatal diagnosis of sickle-cell anemia by DNA analysis of amniotic-fluid cells. Lancet 2:910–911, 1978.
45. Kan YW, Dozy AM: Polymorphism of DNA sequence adjacent to human beta-globin structural gene: Relationship to sickle mutation. Proc Natl Acad Sci USA 75:5631–5635, 1978.
46. Levine LS, Zachmann M, New MI, Prader A, Pollack MS, O'Neill GJ, Yang SY, Oberfiel DSE, Dupont B: Genetic mapping of the 21-hydroxylase-deficiency gene within the HLA linkage group. N Engl J Med 299:911–915, 1978.
47. Pollack MS, Levine LS, Pang S, Owens RP, Nitowski HM, Maurer D, New MI, Duchon M, Merkatz IR, Sachs G, Dupont B: Prenatal diagnosis of congenital adrenal hyperplasia (21-hydroxylase deficiency) by HLA typing. Lancet 1:1107–1108, 1979.
48. Warsof SL, Larsen JW, Kent SG, Rosenbaum KN, August GP, Migeon CJ, Schulman JD: Prenatal diagnosis of congenital adrenal hyperplasia. Obstet Gynecol 55:751–754, 1980.
49. Falk GA, Briscoe WA: Alpha-1-antitrypsin deficiency in chronic obstructive pulmonary disease, editorial. Ann Intern Med 72:427–429, 1970.
50. Sharp HL, Bridges RA, Krivit W, Freier EF: Cirrhosis associated with alpha-1-antitrypsin deficiency: A previously unrecognized inherited disorder. J Lab Clin Med 73:934–939, 1969.
51. Blass JP, Gibson GE: Abnormality of a thiamine-requiring enzyme in patients with Wernicke-Korsakoff syndrome. N Engl J Med 297:1367–1370, 1977.
52. Walker DG: Bone resorption restored in osteopetrotic mice by transplants of normal bone marrow and spleen cells. Science 190:785–786, 1975.
53. Polmar SH, Stern RC, Schwartz AL, Wetzler EM, Chase PA, Hirschhorn R: Enzyme replacement therapy for adenosine deaminase deficiency and severe combined immunodeficiency. N Engl J Med 295:1337–1343, 1976.
54. DeMars R, LeVan SL, Trend BL, Russell LB: Abnormal ornithine carbamyltransferase in mice having the sparse-fur mutation. Proc Natl Acad Sci USA 73:1693–1697, 1976.
55. Qureshi IA, Letarte J, Oellet R: Ornithine transcarbamylase deficiency in mutant mice. I. Studies on the characterization of enzyme defect and suitability as an animal model of human disease. Pediatr Res 13:807–811, 1979.
56. Blake RL: Animal model for hyperprolinemia: Deficiency of mouse proline oxydase activity. Biochem J 129:987, 1972.
57. Romeo G: Enzymatic defects of hereditary porphyrias: An explanation of dominance at the molecular level. Hum Genet 39:261–276, 1977.
58. McKusick VA: The anatomy of the human genome. J Hered 71:370–391, 1980.

Inherited Obesity-Diabetes Syndromes in the Mouse

D. L. Coleman

INTRODUCTION

Although numerous studies have been conducted by many investigators in attempts to understand the various syndromes, no one has as yet defined the primary lesion involved in any one of these conditions. For comprehensive bibliographies and excellent reviews of the various studies carried out with each of these syndromes, see references [1–4]. Although the abnormalities associated with each diabetes obesity type have many similarities with regard to their overall development, the documentation of several different single genes being involved in each obesity condition makes it unlikely that the various syndromes will all be reduced to disturbances in a single metabolic pathway. Those mutations maintained on inbred strains are particularly useful in that appropriate matings can be made that will produce predictable numbers of unaffected and diabetic mice, all of which are of the same strain, differing only by a single gene. The availability of congenic histocompatible controls permits organ and tissue transplantation without concern for tissue rejection, enabling one to transplant specific organs to establish their role in the development of the syndrome.

That several different mutations are known that cause similar obesity-diabetes syndromes in mice suggests that similar, if not identical, genetic insults may be responsible for diabetes in human beings. A knowledge of the defects caused by these mutations in mice should lead to much more insight into the possible defects involved in human disease. The finding that the mutant gene interacts differently with different inbred backgrounds suggests reasons why the disease expression in the human condition often varies from individual to individual. A thorough knowledge of the nature of these interactions with the background genome should increase our knowledge of similar interactions in human beings.

TABLE I. Single Gene Mutations Causing Diabetes-Obesity Syndromes in Mice

Gene name	Inheritance	Chromosome
Yellow (A^y, A^{vy}, A^{iy})	Autosomal dominant	2
Obese (ob)	Autosomal recessive	6
Diabetes (db, db^{2J}, db^{3J}, db^{ad})	Autosomal recessive	4
Fat (fat)	Autosomal recessive	?
Tubby (tub)	Autosomal recessive	7

MOUSE MODELS

The diabetes syndromes in mice vary in the degree rather than in the developmental profile of the disease. In most cases the syndrome includes hyperphagia, some degree of hyperinsulinemia, hyperglycemia (either transient or sustained), and marked obesity. Thermoregulatory defects, hypogonadism, and functional sterility are typical features. The activities of many enzymes in many pathways (lipogenesis, gluconeogenesis) have been studied, and any abnormalities observed usually reflect the degree of hyperinsulinemia. The most notable exceptions occur with those enzymes involved in gluconeogenesis that remain elevated in spite of high plasma insulin concentrations. Hyperplasia and hypertrophy of the islets of Langerhans are associated with all of the syndromes in mice. Insulin resistance associated with a loss of receptor sites causes excessive insulin demand. Whether the increase in number and size of the islets can keep up with the abnormal metabolic demand for insulin to control the diabetes depends, in some instances, on the inbred background on which the mutation is maintained [5, 6].

The Obese Mouse

Obese (ob) — an autosomal recessive mutation (Chromosome 6) — occurred in a non-inbred stock [7], but was established later, and has been maintained, in the C57BL/6J (BL/6) strain. BL/6 obese mice are characterized by marked obesity, hyperphagia, transient hyperglycemia, glucose intolerance, and markedly elevated plasma-insulin concentrations associated with an increase in number and size of the beta cells of the islets of Langerhans. Homozygous mutants of both sexes are infertile, and stock must be maintained by mating known heterozygotes. The rate of liver and adipose tissue lipogenesis is more than doubled, although this increase is not strictly proportional to the circulating insulin concentration (10–20 times normal). Marked insulin resistance is associated with a loss of insulin receptors in several tissues. This loss of receptor sites in diabetes-obesity states seems general in all models studied and may be an attempt by the animal to protect itself against the hypoglycemic effect of massive levels of circulating insulin [2]. Those enzymes involved in gluconeogenesis that normally are decreased in the hyperinsulinemic state remain elevated in obese mice. This abnormality occurs early in development in most diabetes mutants and may be associated with the loss of insulin receptor sites in liver cells, or may represent a primary defect.

The Diabetes Mouse

Diabetes (db) — an autosomal recessive mutation (Chromosome 4) — occurred in the C57BL/KsJ (BL/Ks) inbred strain and on this background is characterized by obesity, hyperphagia, and a severe diabetes with marked hyperglycemia [4, 8]. Increased plasma-insulin concentration is observed as early as 10 days of age [9]. The concentration of insulin peaks at 6–10 times normal by 2–3 months of age, then drops precipitously to near-normal levels. During the same time, blood-glucose concentration rapidly rises to over 400 mg per 100 ml and remains elevated until death at 5–8 months. Prior to the drop in plasma-insulin concentration, the most consistent morphological feature of the islets of Langerhans is hyperplasia and hypertrophy of the beta cells in an attempt to produce sufficient insulin to control blood-sugar concentration at physiological levels. The drop in plasma-insulin concentration is concomitant with typical degenerative changes and atrophy of the islets. One striking finding in islets from diabetes mice is a topological relocation of alpha and delta cells away from their normal position at the periphery of the islet and surrounding the beta cell mass. Delta cells, in particular, appear to migrate toward the center of the islet, where much more delta-to-beta cell contact is established. This increased number of contacts is apparently undertaken in order to provide more somatostatin (insulin-secretion inhibitor) to the overly reactive beta cells. Compared with the phenotypically similar obese mouse, the diabetic condition is more severe and the life-span is markedly decreased. Other alleles (db^{2J}, db^{3J}, db^{ad}) at this locus are known, and when maintained on the BL/Ks background, db^{2J} and db^{ad} produce the identical syndrome as does the original db mutation; and when these alleles are introduced to the BL/6 background, phenotypes are produced that are identical to that seen in BL/6-ob/ob mice (marked obesity and mild diabetes).

The Agouti Locus

Several alleles (A^y, A^{vy}, and A^{iy}) at the agouti (A) locus (Chromosome 2) change the normal agouti (black-and-yellow banded) hair pattern to a complete yellow. These alleles also cause mild hyperglycemia and excess accumulation of adipose tissue [10, 11]. The condition is inherited as an autosomal dominant. Homozygous (A^y/A^y) mice are lethal, such embryos developing only to the blastocyst stage. In contrast, viable yellow (A^{vy}) and intermediate yellow (A^{iy})

are viable in the homozygous state. These latter two alleles produce progeny with coat-color patterns ranging from a clear yellow, to a yellow splotched with typical agouti patches, or to a complete agouti pattern [12]. The degree of adiposity and diabetes that develops in genetically yellow mice is proportional to the percentage of yellow in the coat. Yellow obese mice have only moderate hyperphagia and hyperglycemia, coupled with mild hyperinsulinemia and some hypertrophy and hyperplasia of the islets of Langerhans.

The Fat Mouse

A recessive mutation (*fat*, linkage unknown) occurred in the HRS/J stock at the Jackson Laboratory that produces a slowly developing but severe obesity [13]. Pronounced hyperglycemia is not typical of any stage of the disease. Hyperinsulinemia is marked, and the islets of Langerhans show enlargement coupled with marked degranulation of the beta cells. Metabolic studies with this mutant have not been undertaken.

The Tubby Mouse

Tubby (*tub*) — an autosomal recessive mutation (Chromosome 7) — occurred in the BL/6 inbred strain [14]. These mice are characterized by slowly developing obesity, hypertrophy of the islets of Langerhans, and hyperinsulinemia. Apparently, increased synthesis and secretion of insulin are sufficient to maintain blood-sugar concentrations at normal or even hypoglycemic levels. These mice, if mated prior to becoming markedly obese, are fertile, permitting a limited production of afflicted animals by matings of homozygous mutants. Metabolic studies of this mutation have not been undertaken.

Effects of Inbred Background

The mutations diabetes and obese are particularly interesting to diabetologists because the two conditions are identical when the mutant genes are maintained on the same inbred background and, when maintained on a different inbred background, the syndrome changes from a severe obesity to a life-shortening, fulminating diabetes [5, 6]. These two divergent syndromes are not produced by the mutant gene action itself but, rather, result from an interaction between the mutant gene and the background genome. The effects of background on the different mutants are shown in Table II. The rapidly developing syndromes (obese and diabetes) on the BL/Ks background are characterized by severe diabetes and early attempts to increase insulin supply, followed by islet atrophy and beta cell degeneration. In contrast, both mutations on the BL/6 background have mild glucose intolerance, transient hyperglycemia, and severe hyperinsulinemia produced by marked overactivity of the pancreatic islets. Islet hypertrophy and hyperplasia, rather than degenerative changes, are characteristic of either mutation in the BL/6 background. Histological evaluation of the changes that take place in the islets of

TABLE II. Effect of Genetic Background on the Diabetic Syndromes

Locus	Strain	
	C57BL/6J	C57BL/KsJ
Obese (*ob*)	Marked obesity, mild diabetes, islet hyperplasia, hyperinsulinemia	Obesity, severe diabetes, transient hyperinsulinemia, islet atrophy
Diabetes (*db*, db^{2J}, db^{ad})	Marked obesity, mild diabetes, islet hyperplasia, hyperinsulinemia	Obesity, severe diabetes, transient hyperinsulinemia, islet atrophy
Agouti (A^y, A^{vy})	Moderate obesity, minimal diabetes, some islet hyperplasia	Same as on BL/6 background
Fat (*fat*)	HRS/J background: obesity, normal or hypoglycemia, hyperinsulinemia, islet hypertrophy and beta-cell degranulation	Same as on HRS/J background (BL/6 type)
Tubby (*tub*)	Slowly developing obesity, little or no diabetes, islet hyperplasia	?

Langerhans suggests that the initial response of both mutants on either background is to increase insulin supply by rapid hypertrophy and hyperplasia of the islets of Langerhans. However, at about 12 weeks of age, the islet-proliferating capacity of the islets in BL/Ks mice ceases, and islet atrophy and insulinopenia ensues. In contrast, islet proliferation and increased insulin secretion continue indefinitely in the BL/6 strain. It seems that the rapid rate of islet proliferation and insulin secretion required in these rapidly developing syndromes cannot be sustained indefinitely in the BL/Ks mutants. However, the more slowly developing syndromes do not put so much pressure on the proliferative capacity of the islet cells, and proliferation, at a slower rate, can continue indefinitely on both backgrounds. Thus, both yellow (A^y) and fat (*fat*) on the BL/Ks background remain only mildly diabetic, similar to the condition seen on the BL/6 background. The newer mutation, tubby (*tub*), presumably would respond in a similar fashion on the BL/Ks background. It appears that islet atrophy and severe diabetes occur only when the progression of the disease is sufficiently rapid to exceed the islet-proliferative capacity of the particular strain. Factors that slow the rate of disease development (illness, food restriction, high protein diets), if they interact at a critical developmental stage, have been found to convert an incipient severe diabetic (BL/Ks) condition to the milder (BL/6) obesity. The markedly different obesity-diabetes states exhibited when obese and diabetes mice are on different backgrounds point out the importance of strict genetic control in studies with all types of obese-hyperglycemic mutants. Genetic studies [15] have shown that the modifiers leading to islet hypertrophy and well-compensated diabetes compatible with a nearly normal life-span are dominant to those factors causing severe diabetes. The mode of action of these modifying genes in mice has tremendous significance in relation to diabetes in human beings, where susceptibility to diabetes is probably caused by a few major genes interacting with a susceptible genotype.

POTENTIAL PRIMARY DEFECTS

Although studies with these diabetes-obesity models have been extensive and exhaustive in scope, no primary lesion has been defined for any of the syndromes described. Several possibilities have been suggested, and many remain acceptable in light of present data. One primary defect applicable to several mutants involves a hypothalamic abnormality. Most of the models are infertile, have defects in thermoregulation, and are hyperphagic, all of which facts support the concept of a hypothalamic defect. The best evidence for a hypothalamic defect is found in the diabetes (db) mouse, which may have a defective satiety center [16]. Parabiotic unions of diabetes (db/db) with normal mice consistently resulted in the death of the normal mouse, apparently of starvation within 3 weeks after surgery. Our interpretation of the finding is that the diabetes partner produces, but cannot respond to, a factor that prevents overeating. In parabiosis, this factor crosses into the circulation of the normal partner, where it acts on the normal partner's satiety center to inhibit eating, producing starvation and death. Although these studies appear very clear-cut, they must be interpreted with caution since the syndrome of hypothalamic obesity in the diabetes mouse differs in many respects from the classical syndrome, in which the ventromedial nucleus is destroyed by electrolytic lesioning. The genetic lesion probably involves a defect in a much more discrete area of the hypothalamus than is destroyed by chemical or electrolytic methods, and many of the functions of the region other than satiety may be preserved. In this respect, the diabetes mouse presents with many of the features typical of the Prader-Willi syndrome, which is associated with a satiety-center defect that produces hyperphagia, massive obesity, hypogonadism, and diabetes.

The obese (ob) mouse is phenotypically identical to the diabetes mouse when maintained on the same genetic background, and this syndrome might also be expected to be caused by a defective hypothalamus. When obese (ob/ob) mice were parabiosed to diabetes (db/db) mice, the obese partner lost weight and died of starvation, while no abnormal changes were observed in the diabetes partner. This implies that obese mice are like normals in that they have normal satiety centers responsive to satiety factor. In unions of obese with normal mice, both partners survive. However, obese mice, in such pairs, eat less and fail to gain weight as fast as expected, when compared with the weight gain typical of obese-with-obese pairs [17]. This observation suggests that the normal partner provides some humoral (satiety) factor that regulates weight gain in the obese partner. These parabiosis experiments, taken together, provide strong evidence that the obese mouse is unable to produce sufficient satiety factor to regulate its food consumption, whereas the diabetes mouse produces satiety factor but cannot respond appropriately to it because of a defective satiety center. These studies suggest that the primary lesion in both mutants is associated with similar defects in satiety. Such closely related defects would explain the identical phenotypes observed when either mutant is maintained on the same inbred backgrounds. Strautz [18] has reported that transplantation of normal islets to obese mice stabilized rate of weight gain and reduced both hyperglycemia and hyperinsulinemia. These studies imply that the missing satiety factor may be pancreatic in origin. Unpublished studies by others have been unable to confirm this result. Somatostatin, pancreatic polypeptide, and serotonin — all compounds found in various cells of the pancreas — have been found to be ineffective in both obese and diabetes mice.

The yellow mouse represents the only single-gene mutation that has multiple pleiotropic effects affecting viability and pigmentation, and produces mild diabetes and obesity. This model is less severe than either diabetes or obese and has only moderate hyperphagia and hyperglycemia (and then, only for discrete periods in its development). Body weight may be regulated more by the control of energy utilization than by the control of energy (food) intake. The yellow mouse is particularly intriguing because the yellow (A^y) gene is lethal, and homozygotes do not develop past the blastocyst stage. Thus, we can conclude that the gene controls a metabolic pathway essential to normal development as well as being involved in producing glucose intolerance. Very subtle environmental changes in utero influence the phenotype expressed. Genetically yellow ($A^{vy}/?$) mice may be phenotypically wild type (agouti) on maturing. If the phenotype is agouti, regardless of the genotype, insulin secretion is normal, hyperinsulinemia is not a factor, and the mouse does not become obese or diabetic. In those mice that are phenotypically yellow or mottled, the degree of hyperinsulinemia and of obesity correlates with the extent of yellow pigmentation of the pelage. Defining the nature of the environmental insults that can occur in utero to change subsequent hair development from yellow to agouti, and also prevent obesity and glucose intolerance, should elucidate factors involved in the control of both parameters.

The two new obesity syndromes, tubby and fat, have not been studied sufficiently to allow any conclusions to be drawn regarding the potential primary lesions in these conditions.

All of the syndromes described have varying degrees of hyperinsulinemia coupled with hyperphagia. Either could be primary. However, preweaning studies

(10–12 days) in my laboratory suggest that hyperinsulinemia occurs before hyperphagia. This hyperinsulinemia would cause the hyperphagia, and the increased eating would provoke further hyperinsulinemia and, eventually, the marked insulin resistance. Ultimately, the insulin-synthesizing capacity would be either expanded or exhausted, depending on the inbred background. Our studies of the expression of the diabetes gene on the BL/Ks background indicate that the beta cell is not intrinsically defective but, instead, it is the alpha cell (glucagon-synthesizing cell) that shows hypersecretion prior to beta-cell deterioration. This excessive alpha-cell activity very early in life could predispose the islets to the subsequent beta-cell degeneration later in life. Thus many possible primary defects – hyperphagia, beta-, alpha-, or even delta-cell defects could lead to the sequence of events observed, which culminates in marked obesity and severe diabetes.

METABOLIC EFFICIENCY OF OBESITY MUTANTS

In attempts to see whether hyperphagia was an important prerequisite to the development of either obesity or diabetes, several investigators [4, 19] have pair-fed mutants the same amount of food that is eaten by normal mice. Under these conditions, obese mutants gained an additional 2–3 grams of body weight, most of which was fat. Since feeding a limited amount of food to hyperphagic mutants once a day leads to an obesity-producing stuff-starve regimen, we pair-fed mutants (either obese or diabetes) under more natural feeding conditions. A normal mouse was trained to press a bar to obtain a single pellet of food. This system was coupled to food dispensers in other cages where mutants were housed. Under these conditions the mutants would receive exactly the same amount of food on the identical time schedule, and the stuff-starve regimen would be avoided. Even with yoked controls, the mutant mice still gained an additional 1–2 g of weight in a 1-month period over that seen in normal mice [4, 19]. Carcass analysis demonstrated that, even when fed normal amounts of food, most of the extra weight gain consisted of abnormal depositions of fat (Table III). When restricted to two-thirds of the food intake of normal mice, weight gain remained about the same, and the mutants still retained the obese body composition (line 4, Table III). Even when fed only 50% of the amount of food eaten by the normal mice (line 5), the obese mutants were still able to maintain normal body weights and retain the obese body composition. The synthesis of fat is nearly twice as energy-expensive as the synthesis of any other body constituent. This extraordinary accretion of body fat in spite of severe food restriction represents an increase in metabolic efficiency of considerable magnitude. Histological examination of the pancreas under all conditions of underfeeding revealed incipient islet atrophy similar to that seen in young mutants fed ad libitum. These studies establish that hyperphagia, although a major contributory factor in the rapid rate and amount of fat accre-

TABLE III. Body Weight and Body Fat on Various Feeding Schedules

Genotype	Starting weight	Weight after 4 weeks		% Fat	
		Fed ad lib	Pair-fed	Fed ad lib	Pair-fed
BL/Ks-+/+	14.9 g	26.1 g (11.5)[a]	21.4 g (6.8)	19.1	14.9
BL/Ks-ob/ob	17.0 g	38.6 g (21.6)	25.6 g (8.6)	42.3	43.7
BL/Ks-db/db	16.8 g	38.2 g (21.4)	24.3 g (7.5)	36.8	43.7
BL/Ks-db/db[b]	15.3 g	—	26.2 g (10.9)	—	47.4
BL/Ks-db/db[c]	25.5 g	—	24.3 g (−1.2)	—	37.0

[a]Figures represent average values obtained from four mice in each group. Weight gain per 4-week period is in parentheses.
[b]Mice fed two-thirds the amount of food eaten by normal mice on the same schedule.
[c]Mice fed one-half the amount of food on the same schedule.

tion, is not essential to the development of either diabetes or obesity. Even severe under-feeding will not prevent the abnormal diversion of energy (food) to the synthesis of lipid and its storage in abnormally large fat stores. This marked increase in efficiency would be of great advantage to mice living in the wild that have to subsist on a limited food supply, but obviously becomes a liability in affluent conditions.

We have shown that homozygous mutants (either ob/ob or db/db), in addition to being more efficient in storing food in lipid stores for subsequent use, can utilize their large fat stores to survive prolonged periods of fast up to 40 days, as compared with 10 days for normal (+/+) mice (Table IV). Before death, the mutants utilized up to 80% of their body weight (mostly fat), whereas the normal mice died after only 40% of their reserves had been utilized and had some fat stores remaining at autopsy. Apparently mutants can totally metabolize their entire lipid and tissue reserves in order to prolong survival. Theoretically, lipid metabolism should cease when the animal's glucose and gluconeogenic reserves are exhausted. The mutant mice, although grossly obese, have no more gluconeogenic precursors than normal mice and should die when these glucose reserves are gone, prior to exhausting their entire fat reserves. The total utilization of lipid in mutants suggests that they may have unique pathways for the total utilization of fat.

The increased metabolic efficiency typical of mutant homozygotes extends to heterozygous carriers, which carry only a single dose of either the obese or diabetes gene. Heterozygotes are able to withstand a total fast for up to 3 days longer than normal (+/+) homozygotes [20]. The increment in survival time varies somewhat with the mutant gene and with the inbred background on which it is maintained. This increased survival time represents one of the rare cases of a single dose of a normally deleterious gene having a beneficial effect.

Acetone is normally considered to be an unmetabolizable end product of fatty-acid metabolism, which accumulates when the gluconeogenic reserves required for complete oxidation of fatty acids are depleted. If acetone could be metabolized and, better, if it were gluconeogenic, its metabolism could provide the glucose both

TABLE IV. Survival Time and Body-Weight Changes on Fasting*

Genotype	Weight (g)		Survival time (days)
	Starting	At death	
C57BL/6J-+/+	36.8 ± 0.4	20.2 ± 0.4	10.8 ± 0.4
C57BL/6J-db/db	43.9 ± 0.7	12.0 ± 0.4	35.8 ± 0.9
C57BL/6J-ob/ob	44.2 ± 0.6	12.9 ± 0.4	37.2 ± 0.5

*Mice were subjected to total fast, with only water provided ad libitum. Figures represent average values obtained from 7 to 8 mice per group.

to permit the total fat metabolism and to prolong survival. We found that acetone-2-^{14}C injected into fasted mice of any genotype was metabolized to $^{14}CO_2$ (30%) in 3 hours. Furthermore, the radioactivity from the injected acetone was incorporated into both blood glucose and liver glycogen. When comparing normal homozygotes with heterozygotes, it was seen that the heterozygotes converted 50% more acetone into glycogen than the normals (Table V). Regardless of the pathway used in the metabolism of acetone, the heterozygous mutants utilized it more efficiently. Our further studies (in vitro) with mitochondria-free supernatant fluids from liver incubated in the presence of oxygen and fortified with an NADPH-generating system have shown that acetone is converted to a nonvolatile substance that has been identified as lactic acid. Lactic acid is gluconeogenic and can be rapidly converted to glucose. This ability to convert metabolically produced acetone to glucose could increase the survival time on fasting by providing sufficient new glucose to permit the total utilization of lipid reserves.

Other mechanisms might contribute to the increased metabolic efficiency of heterozygotes and mutants as well as to a more effective utilization of acetone. Obese mutants do not thermoregulate [21, 22]. Thermoregulation requires energy. The energy saved by not thermoregulating could be diverted to extra fat synthesis during food abundance and would decrease the energy requirement during a prolonged fast. Normal mice consume much energy in the continuous synthesis and breakdown of tissues (proteins and triglycerides). Insulin, being an anabolic hormone, would suppress degradation and favor synthesis. Thus tissue synthesis would predominate in the hyperinsulinized mutants, and less turnover would occur. The energy conserved by the nonfunctioning of these energy-consuming, futile cycles could produce what would appear to be greater metabolic efficiency in obese mutants. Regardless of the mechanisms used by these mutants to conserve energy, they are excellent models to define metabolic efficiency mechanisms and to explore new pathways of metabolite utilization.

DISCUSSION

All of these models have aspects of both maturity-onset and juvenile diabetes, and each has some application to studies of both types of diabetes. Varying the background on which the mutation is maintained has profound effects on the type

TABLE V. Incorporation of Acetone-^{14}C Into Liver Glycogen (cpm/g liver)*

C57BL/6J		C57BL/KsJ	
+/+	ob/+	+/+	db/+
168,600 ± 55	247,700 ± 52	104,800 ± 60	183,400 ± 78

*Mice were fasted for 24 hours, injected with glucose to initiate glycogen synthesis, injected with 2-^{14}C acetone; and liver glycogen was isolated, purified, and counted.

of diabetes produced. Investigators wishing to study a model of juvenile-type diabetes could choose either the diabetes or obese mutations on the BL/Ks background, whereas those more interested in obesity and a maturity-onset diabetes could choose these same mutants on the BL/6 background. An understanding of the mode of action of these modifying genes, which change the course of the disease from a severe juvenile type to a more benign obesity, or maturity-onset-type diabetes, would be an important contribution to the understanding of human diabetes variants.

Some degrees of hyperphagia and hyperinsulinemia are associated with the early stages of all of the diabetes syndromes. Either abnormality would lead to obesity, beta-cell hyperactivity, and compensatory hypertrophy and hyperplasia of the beta cells. The events leading to most of the obesity-diabetes states follow a similar sequence. A vicious cycle is established, with hyperinsulinemia encouraging hyperphagia, and hyperphagia favoring further secretion of insulin. The target organs for insulin become maximally stimulated. Insulin resistance intervenes, probably to protect the animal against the hypoglycemic effect of the hyperinsulinemia. The pancreatic output of insulin either is sustained, thus maintaining grossly elevated concentrations of insulin and producing a massive obesity; or islet atrophy occurs, causing eventual insulinopenia, severe diabetes, and premature death.

In most of these models, the mutation or mutations that have occurred fit the definition of thrifty genes [23]. All mutations have produced metabolic changes that favor more efficient utilization and conversion of food into fat. Under conditions of deprivation, these metabolic changes have obvious advantages. However, under laboratory conditions, the increased ability to stockpile energy into fat reserves becomes a liability. Similar evolutionary changes have been suggested to explain the persistence of diabetes in human populations in spite of negative selection pressure. Primitive man had to forage for a limited supply of food and was subject to periods of plenty followed by indefinite periods of food limitation, and even famine. Those individuals (hyperphagic) who could eat the most when food was available, and who could store this excess food more efficiently as fat for subsequent use when food was unavailable, had a distinct advantage. However, when food became plentiful, or when these individuals moved to more affluent societies, they overate, became obese and finally diabetic. Thus what was an advantage in primitive societies became a liability in affluent circumstances, by conferring a susceptibility to diabetes on the thrifty individuals. Studies on those factors causing thriftiness are simpler in inbred mutants maintained on standard inbred backgrounds than in human populations that are probably segregating for many different genes that could be contributing to metabolic efficiency in both negative and positive ways. More detailed studies with these mutants could provide fundamental information regarding the association of thriftiness and diabetes susceptibility in human populations.

SUMMARY

Several different single-gene mutations are known to cause varying degrees of diabetes and obesity in mice. The severity of the diabetes produced depends on both the mutation itself and the interaction of the mutant gene with the inbred background. Establishing the nature of these gene-background interactions should aid us in our understanding of similar interactions that occur in human diabetes. The documentation of several different genes that produce similar, if not identical, diabetes-obesity syndromes suggests that lesions in many pathways can cause diabetes. An understanding of these defects in mice should help us to understand similar defects involved in the human disease. The developmental stages in each mutant are similar. The early symptoms include hyperphagia, hyperinsulinemia, and hypertrophy and hyperplasia of the beta cells of the islets of Langerhans. Hyperglycemia, obesity, and severe diabetes are secondary features that result from insulin resistance and the failure to sustain the secretion of massive amounts of insulin. All models appear to be able to utilize their food in a more efficient manner than normal. Even when restricted to 50% of that amount of food eaten by a normal mouse, mutants are able to maintain their weight and still remain obese. On fasting, the stored fat is utilized more efficiently. One cause of this efficiency in obese and diabetes mice is the ability to convert acetone (the end product of fatty-acid metabolish) to lactate which, in turn, can be converted to glucose, which can sustain continued lipolysis. The occurrence of increased efficiency in obesity and diabetes mutants lends credence to the thrifty-genotype hypothesis regarding the maintenance of the deleterious diabetes genes in human populations.

ACKNOWLEDGMENTS

This manuscript was prepared from a talk given at the 50th Anniversary Symposium of The Jackson Laboratory, "A Century of Mammalian Genetics and Cancer, 1929–2029. A View in Mid-Passage." This research was supported in part by research grants AM 14461 and AM 20725 from the National Institutes of Arthritis, Metabolism and Digestive Diseases, and by a grant from the Juvenile Diabetes Foundation. The Jackson Laboratory is fully accredited by the American Association for the Accreditation of Laboratory Animal Care.

The author wishes to dedicate this paper to the staff of the Jackson Laboratory, both past and present, who spent much time away from their studies to teach him, a chemist, the fundamentals of genetics.

REFERENCES

1. Bray CA, York DA: Genetically transmitted obesity in rodents. Physiol Rev 51:598, 1971.
2. Assimocopoulos-Jeannet F, Jeanrenaud B: The hormonal and metabolic basis of experimental obesity. Clin Endocrinol Metabol 5:337, 1976.

3. Herberg L, Coleman DL: Laboratory animals exhibiting obesity and diabetes syndromes. Metabolism 26:59, 1977.
4. Coleman DL: Obese and diabetes: Two mutant genes causing diabetes-obesity syndromes in mice. Diabetologia 14:141, 1978.
5. Hummel KP, Coleman DL, Lane PW: The influence of genetic background on expression of mutations at the diabetes locus in the mouse. I. C57BL/KsJ and C57BL/6J strains. Biochem Genet 7:1, 1972.
6. Coleman DL, Hummel KP: The influence of genetic background on the expression of the obese (*ob*) gene in the mouse. Diabetologia 9:287, 1973.
7. Ingalls AM, Dickie MM, Snell GD: Obese, a new mutation in the mouse. J Hered 41:317, 1950.
8. Coleman DL, Hummel KP: Studies with the mutation, diabetes, in the mouse. Diabetologia 3:238, 1967.
9. Coleman DL, Hummel KP: Hyperinsulinemia in pre-weaning diabetes (*db*) mice. Diabetologia 10:607, 1974.
10. Cuénot L: Les races pures and leur combinaisons chez les souris. Arch Zool Exp Gen 3:123, 1905.
11. Danforth CH: Hereditary adiposity in mice. J Hered 18:153, 1927.
12. Wolff GL: Composition and coat color correlation in different phenotypes of "viable yellow" mice. Science 147:1146, 1965.
13. Coleman DL: Unpublished.
14. Eicher EM, Coleman DL: Unpublished.
15. Coleman DL, Hummel KP: Influence of genetic background on the expression of mutations at the diabetes locus in the mouse. I. Studies on background modifiers. Isr J Med Sci 11:708, 1975.
16. Coleman DL, Hummel KP: Effects of parabiosis of normal with genetically diabetic mice. Am J Physiol 217:1298, 1969.
17. Coleman DL: Effects of parabiosis of obese with diabetes and normal mice. Diabetologia 9:294, 1973.
18. Strautz RL: Studies of hereditary obese mice (*ob/ob*) after implantation of pancreatic islets in Millipore filter capsules. Diabetologia 6:306, 1970.
19. Cox JE, Powley TL: Development of obesity in mice pair-fed with lean siblings. J Comp Physiol Psychol 91:347, 1977.
20. Coleman DL: Obesity genes: Beneficial effects in heterozygous mice. Science 203:663, 1979.
21. Trayhurn P, James WPT: Thermoregulation and non-shivering thermogenesis in the genetically obese (*ob/ob*) mouse. Pfluegers Arch 373:189, 1978.
22. James WPT, Trayhurn P: An integrated view of the metabolic and genetic basis for obesity. Lancet 2:770, 1976.
23. Neel JV: Diabetes mellitus, a thrifty genotype rendered detrimental by "progress." Am J Hum Genet 14:353, 1962.

Hemolytic Anemias Due to Abnormalities in Red Cell Spectrin: A Brief Review

Samuel E. Lux, Lawrence C. Wolfe, Barbara Pease,
Mary Beth Tomaselli, Kathryn M. John, and Seldon E. Bernstein

INTRODUCTION

Comparative hematology suggests that the development of the red-cell membrane was critical in the evolution of red-cell oxygen-carrying capacity [1]. The red-cell membrane allowed hemoglobin concentrations to rise to meet the needs of mammalian metabolism and permitted the red cell to control the environment of hemoglobin, keeping it reduced and functional. As a consequence, hemoglobin molecules are able to survive for months, instead of minutes, in the circulation. However, the membrane must meet certain structural requirements if the red cell is to complete its prescribed four-month journey in humans. In particular, it must be sufficiently durable to survive nearly half a million circulatory circuits and yet be sufficiently flexible to squeeze through the $3-4\,\mu$ fenestrations that separate the splenic cords and sinuses. These structural requirements are controlled by a mesh of fibrillar proteins that laminate the inner-membrane surface, which are referred to as the "membrane skeleton." This structure was discovered by Yu et al [2], who observed that an insoluble, ghost-shaped residue remained after extraction of membrane lipids and integral membrane proteins with nonionic detergents such as Triton X-100. In recent years the organization and function of the skeleton have been intensively investigated in a number laboratories. These studies recently have been reviewed [3, 4] and will not be discussed in detail here. Spectrin, the principal skeletal protein, is a large molecule composed of two enormous polypeptide chains, bands 1 (240,000 daltons) and 2 (220,000 daltons), which are similar, but not identical [5]. Recent studies indicate that these two subunits associate as heterodimers and heterotetramers [6, 7]. In the dimer, the two parallel chains are variably entwined about each other and form a long, thin, highly flexible structure [8]. Dimers associate head to head to form tetramers, the probable physiological species. Current concepts (Fig. 1) [9] suggest that spectrin tetramers are joined by short strands of polymerized actin [10, 11], which complexes with spectrin in the presence of a third protein, band 4.1 [11], to form the core of the skeleton. The membrane skeleton is attached to the overlying lipid bilayer by ankyrin (band 2.1) [12–14], a protein that connects spectrin to band 3 [15], the transmembrane anion-exchange channel.

Studies of membrane skeletons and skeletal proteins of fragile, rigid, or mishapen red cells suggest that the skeleton is a major determinant of red-cell membrane shape, strength, and flexibility. We previously have shown that red cell membranes (ghosts) and membrane skeletons of irreversibly sickled cells are fixed in a sickle shape, which indicates that an acquired skeletal defect is responsible for the distortion of this rigid cell [16]. In this brief review, we focus on four types of inherited defects: spherocytic hemolytic anemias of mice and human hereditary elliptocytosis, pyropoikilocytosis, and spherocytosis.

Spherocytic Anemias of Mice

Four mutants of the common house mouse, Mus musculus, with recessively inherited hemolytic anemias have been identified: Normoblastosis *(nb/nb)*, Hemolytic anemia *(ha/ha)*, Spherocytosis *(sph/sph)*, and Jaundiced *(ja/ja)* [3, 17–19]. All of the mutants have severe hemolysis. Reticulocyte counts range from 90 100% *(ja/ja, sph/sph,* and *ha/ha* mutants) to 60% *(nb/nb)*. Judging from the degree of anemia and the effect of the mutation on survival, the order of severity is: *ja/ja* > *sph/sph* > *ha/ha* [18]. Peripheral blood smears and scanning

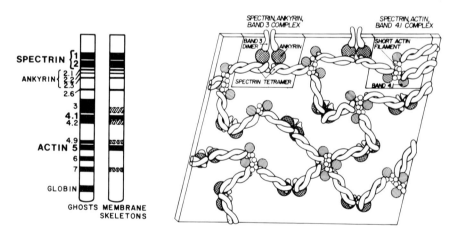

Fig. 1. Left: SDS polyacrylamide gel electrophoretic patterns of red-cell membranes (ghosts) and membrane skeletons. Right: Schematic illustration (not to scale) of the structural organization of the red-cell membrane skeleton (reprinted from Reference 9, with permission).

electron microscopy show intense spherocytosis and in vivo red cell budding (Fig. 2), which suggests that the mutant red cells have very unstable membranes [17, 19, 20]. This impression is confirmed by the fact that, unlike normal, mutant ghosts are extremely fragile and spontaneously vesiculate during preparation.

All of the mutants are spectrin deficient (Fig. 2). Interestingly, the degree of deficiency correlates closely with clinical severity: *ja/ja* ghosts have no detectable spectrin; *sph/sph* ghosts lack band 1, but have small amounts of band 2; *ha/ha* ghosts have about 20–30% of the normal amount of both bands 1 and 2; and *nb/nb* ghosts have approximately 50% of the normal amount of spectrin [19]. Extraction studies show that the bands identified as "spectrin" in each of the mutants are completely extractable at low ionic strength, like normal spectrin, and cross-react with an antihuman spectrin antibody. These results confirm and extend the recent report by Greenquist et al [17] of severe spectrin deficiency in the *sph/sph* mutant and clearly demonstrate the structural importance of spectrin in the membrane skeleton to the red cell. Without this structure, mouse red cells cannot withstand circulatory stresses, rapidly fragment, and are destroyed. In contrast, when normal mouse spectrin is reinserted into spectrin-deficient *sph/sph* red cells by exchange hemolysis, membrane stability is considerably improved and membrane fusion, which occurs readily in untreated *sph/sph* red cells, is virtually eliminated [20].

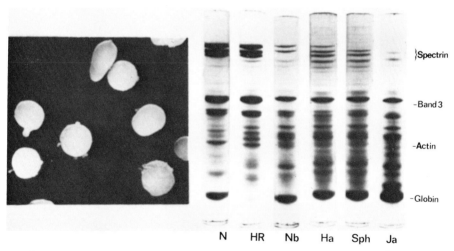

Fig. 2. Spectrin-deficient mutant mice. Left: Scanning electron micrograph of *sph/sph* erythrocytes. Note the intense spherocytosis in membrane budding. Right: Polyacrylamide gel electrophoresis in sodium sulfate of normal (N), high reticulocyte control (HR), *nb/nb* (Nb), *ha/ha* (Ha), *sph/sph* (Sph), and *ja/ja* (Ja) mouse red-cell membranes (reprinted from Reference 3, with permission).

Hereditary Elliptocytosis (HE)

Hereditary elliptocytosis in man is a relatively common, dominantly inherited disorder in which more than 20% of the circulating red cells are elliptical or rod shaped. The disease is clinically heterogeneous. In most patients, elliptocytes have nearly-normal life spans, and the disease is clinically innocuous. A minority (10–15%) have overt hemolysis. We find that elliptocytes from both types of patients are uniformly converted to elliptical ghosts and elliptical membrane skeletons, which implies that the shape defect in HE red cells is due to a defective membrane skeletal component [3, 21]. The protein composition of HE ghosts and skeletons is quantitatively normal on SDS-polyacrylamide gel electrophoresis (the standard assay technique for red-cell membrane proteins), which suggests that the basic skeletal defect is a qualitative rather than a quantitative one. Initial studies [21] have shown that, in some families, this defect appears to reside in spectrin and is detectable as a change in its thermal stability. When highly purified normal spectrin dimers are heat denatured, two subcomponents are detected (Fig. 3): 65% is heat labile and denatures at $49.0° \pm 0.3°C$; the remainder is heat stable to at least $54°C$. This heterogeneity also is seen when spectrin denaturation is monitored by circular dichroism [22], but it remains unexplained. The denaturation temperature of the heat-labile subcomponent corresponds to the temperature at which intact red cells fragment when heated, a phenomenon first noted by Schultze [23] more than a century ago, and subsequently confirmed by others [24–27]. Unlike normal, spectrin from some HE kindreds is almost entirely heat labile and abnormally heat sensitive [21]. It denatures at $48.0° \pm 0.1°C$ instead of at $49°C$, a highly significant difference. In these kindreds, red-cell fragmentation in spherocytosis also occurs at $48°C$ instead of at $49°C$. In other HE kindreds, spectrin and red cells denature and fragment normally.

These observations, which reemphasize the correlation between spectrin and red-cell shape evident in other studies, suggest that defects in the membrane skeleton are a common feature of HE in all kindreds and indicate that at least two genetically distinct forms of HE exist. HE patients with heat-sensitive spectrin must have a molecular abnormality that decreases the conformational stability of the spectrin molecule. At present, it is not clear whether this is a primary-sequence defect or a defect in a posttranslational process such as spectrin phosphorylation. Spectrin phosphorylation of elliptical ghosts is reportedly normal [28], but it is not known whether the patients tested had heat-sensitive spectrin.

In the limited number of kindreds tested, the presence of heat-sensitive spectrin does not correlate with the degree of elliptocytosis or hemolysis. This is not surprising since instability of spectrin at the unphysiologic temperature of $48°C$ probably does not contribute to the pathogenesis of HE. It simply reflects a decline in the conformational stability of the molecule. Presumably in the rare HE kindreds with hemolysis and heat-sensitive spectrin, the conformational change affects sensi-

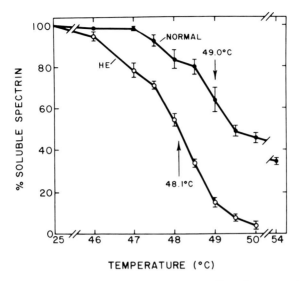

Fig. 3. Effect of heat on the solubility of purified spectrin dimer from normal (●) or hereditary elliptocytosis (○) erythrocytes. The vertical bars indicate ± 1 SE above the mean. The arrows designate the median of each solubility curve (reprinted from Reference 3, with permission).

tive functional regions of the molecule such as the binding sites for actin, band 4.1, or ankyrin. Similarly, the lack of heat instability in some HE kindreds does not exclude a spectrin defect, since functionally important molecular modifications may presumably occur without jeopardizing conformational stability. Future comparisons of the interactions between HE and normal skeletal proteins should permit differentiation of these possibilities.

Hereditary Pyropoikilocytosis (HPP)

HPP is a rare, recessively inherited disorder, characterized by severe hemolytic anemia and bizarre red cell morphology. Peripheral blood smears show large numbers of budding red cells, fragments, spherocytes, elliptocytes, triangulocytes, and other bizarre poikilocytes. This morphology is replicated in ghosts and membrane skeletons (Lux SE, John KM, Zarkowsky HS: Unpublished observations) and strongly suggests a defect in skeletal durability. The pathognomonic feature is the unusual thermal sensitivity of the red cells. HPP red cells fragment at 45 to 46 degrees instead of at 49 degrees after short incubations (10–15 minutes), and fragment at 37 degrees on prolonged incubations [29]. This observation suggests a defect in spectrin. The content of spectrin and other membrane proteins is normal on SDS gels [29]; however, recent studies [30, 31] indicate that spectrin is structurally abnormal. Both isolated [30] and membrane-associated [30, 31] spectrin are unusually heat sensitive and denature at temperatures at which hereditary pyropoikilocytes fragment. In addition, spectrin phosphorylation is markedly depressed in HPP ghosts in spite of increased spectrin-kinase activity [31], suggesting an abnormality in the structure or accessibility of the phosphorylated end of the molecule. The exact location of the molecular abnormality and its effects on the interaction of spectrin with neighboring molecules remains to be determined.

Hereditary Spherocytosis (HS)

Hereditary spherocytosis is a common, dominantly inherited trait and is the paradigm of red-cell membrane disorders. Peripheral blood smears characteristically show numerous partially spherical red cells that appear smaller in diameter and more dense than their normal counterparts. The basic defect in HS red cells appears to be an inherent instability of the membrane. Hereditary spherocytes fragment more easily than normal when directly [32] or metabolically stressed [33–36] and presumably lose membrane fragments more rapidly than normal during circulatory travel. This suggests a defect of the membrane skeleton. With the exception of a few unique cases (discussed in Reference 4), the protein composition of HS red-cell membranes and membrane skeletons is normal [37, 38], except for a subtle increase in membrane-associated catalase and hemoglobin [39]. There is, however, increasing evidence favoring a qualitative defect of the membrane skeleton.

Several years ago, Jacob and his co-workers reported that solubilized but unfractionated HS membrane proteins aggregate less than normal in the presence of cations or vinblastine [40]. Subsequently, they discovered that many of the characteristics of hereditary spherocytes can be induced in normal red cells by treatment with vinblastine or colchicine [41] and that these deleterious effects are blocked by cyclic nucleotides such as cAMP or cGMP, but not by their noncyclic analogs [42]. Studies on isolated ghosts and spectrin extracts suggest that vinblastine selectively precipitates spectrin and inhibits spectrin phosphorylation and that these actions are prevented by cyclic nucleotides [42]. These observations imply that spectrin may function abnormally in HS red cells. Studies from our lab support this premise [43]. We prepared membrane skeletons by extracting ghosts with the nonionic detergent Triton X-100 and monitored their disintegration during titration with urea. Membrane skeletons from 15 HS patients in 13 kindreds uniformly fragmented and dissociated at concentrations of urea much lower than those required to dissociate normal skeletons. However, recent observations indicate that this effect is evident only in skeletons prepared from Triton that has been stored for long periods of time at room temperature (Wolfe LC, Lux SE: Unpublished observations). Apparently, normal skeletons are modified by contaminants (eg, oxidants) in aged Triton in a way that makes them more resistant than HS skeletons to dissociation by urea. The nature of this effect and the component or components in this skeleton that are altered are currently under investigation.

SUMMARY

In summary, it is now clear that spectrin, actin, and band 4.1 form a submembranous red-cell skeleton and that this structure is a major determinant of red-cell membrane shape, strength, and flexibility. In mice, hereditary deficiencies of spectrin produce very fragile red cells that spontaneously fragment in the circulation. The degree of spectrin deficiency correlates with hemolytic severity. In hereditary elliptocytosis and pyropoikilocytosis, the shape defect is localized to the membrane skeleton. Some families with these disorders have an abnormal, thermally sensitive spectrin. In hereditary spherocytosis, membrane skeletons are probably also abnormal, since skeletons prepared under certain conditions (ie, using aged Triton X-100) disintegrate at urea concentrations that do not perturb similarly treated normal skeletons.

These findings represent some of the recent forays by physicians and molecular biologists, ourselves included, into a new area of hematology: the study of red cell membrane protein disorders. For reasons that are beyond the scope of this review (but are discussed elsewhere [4]), it is likely that most of these maladies will prove to be due to defects of the red-cell membrane skeleton. The search for such disorders is by no means limited to the diseases discussed above. Evidence is accumulating that membrane skeletal defects also may be present in irreversibly sickled cells [16], in oxidant-damaged erythrocytes [44–48], in disorders of cation permeability [36, 49, 50], and in a number of other abnormally shaped red cells [3]. Detailed structural analysis of these abnormal skeletons should soon be possible, as the structure of the normal skeleton is rapidly being dissected [9]. It will be surprising if this does not prove to be one of the most active areas of hematology in the coming decade.

REFERENCES

1. Lehmann H, Huntsman RG: Why are red cells the shape they are? The evolution of the human red cell. In Macfarlane RB, Robb-Smith AHT (eds): "Functions of the Blood." New York: Academic Press, 1961, pp 73–148.
2. Yu J, Fischman DA, Steck TL: Selective solubilization of proteins and phospholipids of red blood cell membranes by nonionic detergents. J Supramol Struct 1:233–248, 1973.
3. Lux SE: Spectrin-actin membrane skeleton of normal and abnormal red blood cells. Semin Hematol 16:21–51, 1979.

4. Lux SE, Glader BE: Disorders of the red cell membrane. In Nathan DG, Oski FA (eds): "Hematology of Infancy and Childhood," ed2. Philadelphia: WB Saunders, in press.
5. Anderson JM: Structural studies on human spectrin. Comparison of subunits and fragmentation of native spectrin. J Biol Chem 254:939–944, 1979.
6. Ralston GB: The isolation of aggregates of spectrin from bovine erythrocyte membranes. Aust J Biol Med Sci 28:259–266, 1975.
7. Ungewickell E, Gratzer W: Self-association of human spectrin. A thermodynamic and kinetic study. Eur J Biochem 88:379–385, 1978.
8. Shotton DM, Burke BE, Branton D: The molecular structure of human erythrocyte spectrin. Biophysical and electron microscopic studies. J Mol Biol 131:303–329, 1979.
9. Lux SE: Dissection of the red cell membrane skeleton. Nature 281:426–429, 1979.
10. Cohen CM, Branton D: The role of spectrin in erythrocyte membrane-stimulated actin polymerization. Nature 279:163–165, 1979.
11. Ungewickell E, Bennett PM, Calvert R, Ohanian V, Gratzer WB: In vitro formation of a complex between cytoskeletal proteins of the human erythrocyte. Nature 280:811–814, 1979.
12. Bennett V, Stenbuck PJ: Identification and partial purification of ankyrin, the high affinity membrane attachment site for human erythrocyte spectrin. J Biol Chem 254:2533–2541, 1979.
13. Luna EJ, Kidd GH, Branton D: Identification by peptide analysis of the spectrin-binding protein in human erythrocytes. J Biol Chem 254:2526–2532, 1979.
14. Yu J, Goodman S: Syndeins: The spectrin-binding protein(s) of the human erythrocyte membrane. Proc Natl Acad Sci USA 76:2340–2344, 1979.
15. Bennett V, Stenbuck PJ: The membrane attachment protein for spectrin as associated with band 3 and human erythrocyte membranes. Nature 280:468–473, 1979.
16. Lux SE, John KM, Karnovsky MJ: Irreversible deformation of the spectrin-actin lattice in irreversibly sickled cells. J Clin Invest 58:955–963, 1976.
17. Greenquist AC, Shohet SB, Bernstein SE: Marked reduction of spectrin in hereditary spherocytosis in the common house mouse. Blood 51:1149–1155, 1978.
18. Bernstein SE: Inherited hemolytic disease in mice. Lab Anim Sci 30:197–205, 1980.
19. Lux SE, Pease B, Bernstein SE: Spectrin-deficient red blood cells in four genetically distinct mouse mutants with hereditary spherocytic hemolytic anemias. Submitted for publication.
20. Shohet SB: Reconstitution of spectrin-deficient, spherocytic mouse erythrocyte membranes. J Clin Invest 64:483–494, 1979.
21. Tomaselli MB, Lux SE: Elliptical spectrin-actin skeletons and heatsensitive spectrin in hereditary elliptocytosis, abstracted. 25:519A, 1977.
22. Brandts JF, Erickson L, Lysko K, Schwartz AT, Taverna RD: Calorimetric studies of the structural transition of the human erythrocyte membrane. The involvement of spectrin in the A transition. Biochemistry 16:3450–3454, 1977.
23. Schultze M: Ein heizbarer objectisch und seine verwendung bei unter suchungen des blutes. Arch Mikrosk Anat 1:1–42, 1865.
24. Ham TH, Shen SC, Fleming EM, Castle WB: Studies on the destruction of red blood cells. IV. Thermal injury: Action of heat in causing increased spheroidicity, osmotic and mechanical fragilities and hemolysis of erythrocytes; observations on the mechanisms of destruction in such erythrocytes in dogs and in a patient with a fatal thermal burn. Blood 3:373–403, 1948.
25. Kimber RJ, Lander H: The effect of heat on human red cell morphology, fragility and subsequent survival in vivo. J Lab Clin Med 64:922, 1964.

26. Mohandas N, Greenquist AC, Shohet SB: Effects of heat and metabolic depletion on erythrocyte deformability, spectrin extractability and phosphorylation. In Brewer GJ (ed): "The Red Cell." New York: Alan R. Liss, 1978, pp 435–472.
27. Lux SE, John KM, Ukena TE: Diminished spectrin extraction from ATP-depleted human erythrocytes. Evidence relating spectrin to changes in erythrocyte shape and deformability. J Clin Invest 61:815–827, 1978.
28. Greenquist AC, Shohet SB, Bernstein SE: Marked reduction of spectrin in hereditary spherocytosis in the common house mouse. Blood 51:1149–1155, 1978.
29. Zarkowsky HS, Mohandas N, Speaker CB, Shohet SB: A congenital haemolytic anaemia with thermal sensitivity of the erythrocyte membrane. Br J Haematol 29:537–543, 1975.
30. Chang K, Williamson J, Zarkowsky H: Altered circular dichroism of spectrin in hereditary pyropoikilocytosis, abstracted. Blood 52(suppl 1):95, 1978.
31. Walter T, Mentzer W, Greenquist A, Schrier S, Mohandas N: RBC membrane abnormalities in hereditary pyropoikilocytosis, abstracted. Blood 50(suppl 1):98, 1977.
32. LaCelle PL, Evans EA, Hochmuth RM: Erythrocyte membrane elasticity, fragmentation and lysis. Blood Cells 3:335–350, 1977.
33. Reed CF, Swisher SN: Metabolic dependence of the critical hemolytic volume of human erythrocytes: Relationship to osmotic fragility and autohemolysis in hereditary spherocytosis and normal red cells. J Clin Invest 45:1137–1149, 1966.
34. Weed RI, Bowdler AJ: Metabolic dependence of the critical hemolytic volume of human erythrocytes: Relationship to osmotic fragility and autohemolysis in hereditary spherocytosis and normal red cells. J Clin Invest 45:1137–1149, 1966.
35. Cooper RA, Jandl JH: The selective and conjoint loss of red cell lipids. J Clin Invest 48:906–914, 1969.
36. Snyder LM, Lutz HU, Suaberman N, Jacobs J, Fortier NL: Fragmentation and myelin formation in hereditary xerocytosis and other hemolytic anemias. Blood 52:750–761, 1978.
37. Wolfe LC, Lux SE: Membrane protein phosphorylation of intact normal and hereditary spherocytic erythrocytes. J Biol Chem 253:3336–3342, 1978.
38. Johnsson R: Red cell membrane proteins and lipids in spherocytosis. Scand J Haematol 20:341–350, 1978.
39. Allen DW, Cadman S, McCann SR, Finkel B: Increased membrane binding of erythrocyte catalase in hereditary spherocytosis and in metabolically stressed normal cells. Blood 49:113–123, 1977.
40. Jacob HS, Ruby A, Overland ES, Mazia D: Abnormal membrane protein of red blood cells in hereditary spherocytosis. J Clin Invest 50:1800–1805, 1971.
41. Jacob H, Amsden T, White J: Membrane microfilaments of erythrocytes: Alteration in intact cells reproduces the hereditary spherocytosis syndrome. Proc Natl Acad Sci USA 69:471–474, 1972.
42. Jacob HS, Yawata Y, Matsumoto N, Ahman S, White J: Cyclic nucleotide-membrane proteins interaction in the regulation of erythrocyte shape and survival: Defect in hereditary spherocytosis. In Brewer GJ (ed): "Erythrocyte Structure and Function." New York: Alan R Liss, 1975, pp 235–247.
43. Wolfe LC, Lux SE: Diminished stability of red cell membrane skeletons in hereditary spherocytosis, abstracted. Blood 52(suppl 1):106, 1978.
44. Allen DW, Johnson GJ, Cadman S, et al: Membrane polypeptide aggregates in glucose-6-phosphate dehydrogenase-deficient and in vitro aged red blood cells. J Lab Clin Med 91:321–327, 1978.
45. Johnson GJ, Allen DW, Cadman S, Fairbanks VF, White JG, Lampkin BC, Kaplan ME:

Red cell membrane polypeptide aggregates in glucose-6-phosphate dehydrogenase mutants with chronic hemolytic disease. N Engl J Med 301:522–527, 1979.
46. Kahane I, Polliack A, Rachmilewitz EA, et al: Distribution of sialic acids on the red blood cell membrane in β-thalassaemia. Nature 271:674–675, 1978.
47. Kahane I, Shifter A, Rachmilewitz EA: Cross-linking of red blood cell membrane proteins induced by oxidative stress in β-thalassemia. FEBS Lett 85:267–270, 1978.
48. Salhany JM, Swanson JC, Cordes KA, Gaines SB, Gaines KC: Evidence suggesting direct oxidation of human erythrocyte membrane sulfhydryls by copper. Biochem Biophys Res Commun 82:1294–1299, 1978.
49. Mentzer WC Jr, Smith WB, Goldstone J, et al: Hereditary spherocytosis: Membrane and metabolism studies. Blood 46:659–669, 1975.
50. Mentzer WC Jr, Lubin BH, Emmons S: Correction of the permeability defect in hereditary stomatocytosis by dimethyl adipimidate. N Engl J Med 294:1200–1204, 1976.

Garrod's Legacy to the Nations of Mice and Men

Charles R. Scriver

Inborn errors of metabolism have fascinated students of mammalian genetics throughout the 20th century. The human component of the topic is quite enough, without the challenge to discuss murine and human counterparts together. However, Dr. Elizabeth Russell's invitation was irresistable and, I find myself, a novice student of the mouse (and then only in a small corner of his domain) forced to think about the lessons I have been learning at mouse school and how they inform me about human biochemical genetics. I admit to a familiarity with a certain chronicle of Amos,* and I know how well I can be instructed by the principal residents of the Jackson Laboratory to improve my understanding of the human counterparts of murine inborn errors of metabolism. However, under the circumstances, it would be advisable to raise a note of caution on how mouse "models," or murine mutants of medical interest – as Dr. Russell calls them – are used to elucidate human disease, and yet communicate with genuine enthusiasm, the lessons we are learning from them.

GARROD'S LEGACY

Garrod [1] introduced the "inborn error of metabolism" to human biology and medicine as a new concept early in this century. His understanding of the biochemical problems he described was profound and far ahead of his time; in fact, we have spent most of the intervening years catching up with his insights [2].

---

*Amos – a mouse – "reports" on the real story of Benjamin Franklin in "Ben and Me: A New and Astonishing Life of Benjamin Franklin, As Written by His Good Mouse Amos, Lately Discovered, Edited, and Illustrated by Robert Lawson" (Boston: Little, Brown and Co, 1939). The narrative, which I read as an impressionable child, is a revisionist view of Franklin's achievements (including the founding of "The Montreal Gazette," 201 years ago, in my place of birth and dwelling). Amos claims – in his role as friend and advisor – that he was, in great part, responsible for the remarkable achievements commonly attributed to Franklin. And, indeed, so might have been the case, because Franklin was an excellent student. I have tried to emulate Franklin in my relationship with Amos' descendants: *PRO/Re* and *Hyp*.

Garrod insisted, as early as 1902 [3], that a disease such as alcaptonuria was merely an extreme example of the biochemical individuality that characterize all humans with the exception of monozygotic twins. Harris and Hopkinson [4] eventually proved Garrod right with their delineation of human biochemical polymorphisms and the quantitation of heterozygosity in man. Meanwhile, others had revealed that extensive heterogeneity often characterized the rare mutant alleles at the loci involved in the individual inborn errors [5]. These two important biochemical themes — heterogeneity of rare alleles and polymorphism of common alleles — are important for our interpretation of disease and well-being [6, 7]. An understanding of polymorphisms may reveal risks to health for particular individuals in specific environment. Awareness of heterogeneity in the inborn errors allows physicians to interpret screening and diagnostic tests with greater precision and to titrate euphenic treatment to the exact needs of the particular patient.

The rate of enumeration and growth of knowledge about mendelian birth defects [8] and of inborn errors of metabolism, in particular [9], have risen nearly exponentially in recent years. Nonetheless, a basic challenge remains, since the sequence of events linking mutation to all aspects of phenotype is rarely fully understood. Various strategies have been used to meet the challenge, particularly in the case of inborn errors of metabolism. Intensive investigation of patients themselves, within ethical limits, has often been a path to enlightenment, but it is impossible to learn everything by this method. Tissue biopsy and cell cultivation in defined media have been welcomed as supplemental approaches, but this combination reductive method lacks an essential element: The integration of components such as enzymatic and metabolic phenotypes into a comprehensive view of the total clinical phenotype. For this reason a third approach has appeared: The use of an animal model or analog of the human problem that can, ideally, provide both reductive and integrative solutions.

The nonhuman models of inborn errors of metabolism play twin roles. Primarily, they can serve as probes of biological processes, so that the integrated events in the human counterparts can be understood by analogy. Secondly, if they are true homologues of the human disease, the counterparts can reveal the basic defect and, consequently, guide us in how to relax selection against mutant alleles in the human patient and apply our knowledge to prediction and prevention of inborn disease. Therefore, we must be concerned with the validity of the animal probes; and if they are mutants, ask these questions: How truly does the mutant of medical interest reflect its human counterpart? Is it an analogue or a homologue? Such questions are not academic: The answers determine how the information derived from the counterpart can be used for the interpretation of human disease and how society will perceive its investment in the nonhuman line of research. Garrod was too long misunderstood, because those who would not perceive his larger message did not understand his analogy. We dare not repeat the scenario by misunderstanding the significance of animal mutants of medical interest.

ON THE USE OF GENETICALLY DEFINED MURINE INBORN ERRORS OF METABOLISM

More than 500 mutations are known in the mouse [10] and almost six times that number are known in man [8]. Seventy-four biochemical variants have been identified among inbred strains of mice; these are catalogued in Part III of the FASEB Biological Handbooks [10] and more have been described since the catalogue was compiled in 1978. However, of the murine inborn errors of metabolism, relatively few (only about twenty) have clearly recognized human counterparts.* From these, I have selected four (Table I) to illustrate how mutations of medical interest in the mouse, on the one hand, can be useful analogues that point the way to interpretation of biological processes in man, and on the other hand, may be true homologues of human disease processes that elucidate the counterpart in ways impossible to achieve in man.

The *His* Mouse

The *His*-bearing mouse [11–13] has severe hepatic deficiency of histidine-ammonia lyase activity (EC 4.3.1.3; histidase), with histidine accumulation in body fluids and tissues and overproduction of minor histidine metabolites including imidazole pyruvic acid, which yields a green ferric chloride reaction in urine. The *His* allele is autosomal. Homozygotes (*His/His*) can have a balance defect (phenotype, BLA), manifested in circling behaviour, head tilting, spatial disorientation while swimming and after spinning, and poor maze learning. The behavioural disorder is related to defective development of the otoliths. Whereas there is no direct relationship between genotype and BLA phenotype, it is apparent

---

*Some inborn errors of metabolism with counterparts (analogues or homologues) in man and mouse are *autosomal* [10]; maturity-onset diabetes mellitus, pituitary dwarfism, Chédiak-Higashi syndrome, albinism, acatalasemia, purine nucleoside phosphorylase deficiency, galactokinase deficiency, hyperhistidinemia, hyperprolinemia. Others are *X-linked* homologues [10 and Table I; Table XXXVII in Reference 8]: glycogen storage disease (phosphorylase kinase deficiency), Menkes' syndrome, ornithine transcarbamylase deficiency, X-linked hypophosphatemia, testicular feminizing syndrome (receptor deficient or negative), phosphoglycerate kinase deficiency (erythrocyte, kidney, liver), α-galactosidase deficiency, G-6-PD deficiency, and HGPR transferase deficiency. X-linked ichthyosis (steroid sulfatase deficiency) and placental steroid sulfatase deficiency in man, and the scurfy mutant in mouse may be homologues, as may be Pelizaeus-Merzbacher disease in man, and the jimpy mutant in mouse.

TABLE I. Some Murine Counterparts of Problematic Human Inborn Errors of Metabolism

	Mouse counterpart			Enzyme or protein activity	Human counterpart		McKusick cat. no.[a]
Gene symbol	Gene name	Linkage	Phenotype		Phenotype	Linkage assignment	
His	Histidine	(A)[b]	Hyperhistidinemia ± balance defect (BLA)	Histidine-ammonia lyase (EC4.3.1.3)	Hyperhistidinemia ± mental retardation or learning defect	A[d]	23,580
pro 1	Proline	(A)	Hyperprolinemia & prolinuria (benign)	Proline dehydrogenase (Component 1)	Hyperprolinemia, iminoglycinuria (benign)	A	23,950
spf	Sparse fur	X[c]	Hyperammonemia, high postnatal mortality in males	Carbamyl phosphate L-ornithine carbamyl-transferase (EC 2.1.3.3)	Hyperammonemia, usually fatal in male infants	X	31,125
Hyp	Hypophos-phatemia	X	Hypophosphatemia, rickets, dwarfism	Na$^+$-dependent phosphate transport system (renal brush-border membrane)	Hypophosphatemia, rickets, dwarfism	X	30,780

[a]McKusick catalogue numbers taken from Reference 8.
[b](A) = autosomal.
[c]X = X chromosome.
[d]A = autosomal.

that interaction between the maternal genotype, the maternal metabolic phenotype, the genotype of offspring, and timing of exposure to maternal environment determines penetrance of the BLA phenotype in the offspring. The *His* mouse is an important model because it shows that the metabolically compromised pregnancy can be teratogenic to offspring, the risk for a developmental lesion in the latter being influenced by its genotype.

The murine model is of particular interest, because the human histidinemia counterpart [14] is homologous in enzyme deficiency, metabolic phenotype, and mode of inheritance. However, there is controversy whether human histidinemia is a "disease": Some patients are mentally retarded or have behavioural abnormalities, while most are normal. Resolution of the uncertainty about the medical significance of human histidinemia is important, because newborn screening programs discover infants with the condition. Whether these infants should be treated with a low-histidine diet remains uncertain, in which case prognosis and counseling are indecisive.

Murine and human histidinemia share the common problem of maternal histidinemia. Human maternal histidinemia is sometimes [15], but not always [16], associated with a behavioural disorder in offspring. The variability in effect of maternal histidinemia on human offspring is reminiscent of the murine BLA phenotype, and it suggests that maternal genotype, maternal metabolic phenotype, and timing of exposure in human histidinemia come to influence behavioural development in man, as they do in the mouse. Because it is not possible to study this problem fully in the human, the *His* mouse, whether a homologue or not, is a powerful probe of the general problem of the metabolically compromised pregnancy. Accordingly, it may shed light on the particular controversy concerning the significance of human histidinemia and, parenthetically, on the larger and more pressing problem of maternal phenylketonuria [17].

The *spf* Mouse

The sparse-fur (*spf*-bearing) mouse has X-linked deficiency of hepatic ornithine transcarbamylase (OTC) activity [18a, b]. In the absence of any dissociation between the sparse-fur and abnormal OTC phenotypes, it is believed that the *spf* locus determines OTC activity in the mouse. Mutant male hemizygotes (*spf*/Y) have hyperammonemia; female heterozygotes (*spf*/+) have less severe and more variable expression of OTC deficiency. In biochemical terms, murine and human OTC deficiencies essentially are identical counterparts, although no analogue of sparse fur is seen in human patients with OTC deficiency [19]. Nevertheless, we expect the two species of mutants to have homologous OTC gene product deficiency, because the X chromosome has been conserved in mammalian evolution, and homology is evident for the products of corresponding X-linked genes in placental mammals [20]. Therefore, the *spf* mutant, seen as a homologue of human OTC deficiency, could serve to answer numerous questions about the latter; in particular, those pertaining to efficacy of the newer forms of therapy utilizing keto acids, prenatal expression of the mutant allele in affected males, and identification of female heterozygotes. The *spf* mouse illustrates how the other X-linked mutants listed in the footnote below could be used to elucidate the human counterparts.

The *PRO/Re* Mouse

The murine *pro-1* allele [21–23] is associated with deficiency of proline dehydrogenase activity and hyperprolinemia in mutant homozygotes of the PRO/Re inbred strain. The condition is benign in the mouse. Autosomal recessive hyperprolinemia (Type I – proline dehydrogenase deficiency), the human counterpart [24], is also a benign condition – or so we believe, until proved otherwise. The *pro-1* phenotype in mouse differs from its human counterpart in the extent of prolinuria observed in the homozygote: massive in the mouse, modest in man. From this single observation have come certain insights into the process of transepithelial transport, which are discussed in detail later. The PRO/Re mouse has served as a novel probe of a normal cellular process in biology.

The *Hyp* Mouse

The *Hyp* allele is X linked [25]. Accordingly, the mutant gene product is believed to be homologous with that involved in human X-linked hypophosphatemia (XLH) [26]. The latter, a mendelian form of vitamin-D-resistant rickets, has long been the focus of vivid controversy as to the cause of hypophosphatemia, rickets, and dwarfism. The *Hyp* mouse yields insight into the primary defect in XLH (see below) and has, thus, directed our attention not only to the primary mode of disease expression in the human disease, but also to its treatment. At the same time, it has been an elegant probe of a basic biological process, namely, transepithelial and membrane transport of phosphate anion.

ON INBORN ERRORS OF MEMBRANE TRANSPORT

Garrod had few apparent lapses in his acute perception of the inborn error of metabolism. However, one of them was in his understanding of cystinuria: He considered the problem as an undisclosed abnormality of cystine metabolism, that is, in the catalytic conversion of cystine by the pathway serving transsulfuration. It was not until 1951 that Dent and Rose [27] offered an alternate interpretation, which accommodated an important finding in cystinuria, namely, the renal loss of three dibasic amino acids (lysine, ornithine, and arginine) in addition to cystine. Dent and Rose [27] and all subsequent studies of the disease [28] have considered cystinuria to be an inborn error of membrane transport belonging to that class of disorders first brought to attention by hereditary renal glucosuria [29]. The remainder of this essay is largely devoted to a brief appreciation of the plasma membrane and its inborn errors of transport functions and an appreciation of two murine mutants that illustrate, in some detail, how mouse models can amplify our understanding of transport processes and diseases thereof in man.

Cells, Membranes, and Transport

Living cells have membranes composed largely of lipids and proteins. Thus, when cells collect water-soluble nutrients, a hydrophobic domain is encountered, which must be passed during the vectorial process, when solutes pass from one aqueous environment to another. Biological membranes play many functional roles; for example, receptor or binding activity, maintenance of chemical gradients, uptake and partition of fuels and building blocks. Such biological functions have extraordinary specificity, that is, they can select unique sets of reactants from larger populations of molecules. Only proteins can confer this property to biological functions.

The control exerted on chemical traffic across lipid bilayer membranes is one of the important achievements of evolution. When the prokaryotic cell gained the ability to accumulate solutes and nutrient across its plasma membrane, organization of the transport process began, refinement of which has been an important characteristic of eukaryotic evolution. Membranes establish compartments whose chemical composition can be different, both one from another and from the external environment, depending on the control exerted by the membrane on the flow of physical and chemical information. Accordingly, the appearance of functional membranes was a giant step in the evolution of living cells [30].

When lipids come into contact with an aqueous environment, an organizing force is exerted on them. That force is the hydrophobic effect, and it determines the assembly and organization of lipid bilayers in all biological systems [31]. Lipids are repulsed from the aqueous phase and, under thermodynamic control, they aggregate in micelles, bilayers, or vesicles — according to their chemical nature. In the absence of strong attractive forces between its constituent molecules, the lipid membrane is fluid and deformable and serves as a permeability barrier between aqueous phases.

Low-Molecular-Weight Substances: Transport

Penetration of most low-molecular-weight nutrients into intact living cells deviates from that expected of their oil-water partition coefficients. Passive diffusion across the lipid bilayer membrane does not account for the uptake of L-amino acid, D-glucose, or phosphate, for example, by cells; mediation of such processes must be considered. Facilitated diffusion is one possibility, energized transport is another. Both involved specificity of the process, in which the transport protein apparently can differentiate the functional groups on the nutrient, strip the nutrient of its water of hydration, and position it properly for a vectorial translocation across ~ 75Å of lipid bilayer [32]. Integration of functioning transport proteins in membranes implies that they span the lipid bilayer [33].

Nutrient and solute transport take place across a single plasma membrane in prokaryotes. Transport in eukaryotes often involves at least two membranes — plasma and mitochondrial. The situation is even more complicated in metazoa, where transport across epithelial is a fact of life.

Epithelial cells are characterized by structural asymmetry: "Apical" and "basilar" poles have different morphology. In the epithelium of nephron and intestine, for example, the apical pole faces the lumen, where solute is outside the body; the basilar pole is bathed by interstitial fluid, which is inside the body. Nutrition of the epithelial cell, to sustain its own work, is served by uptake from either pole; but vectorial uptake across the cell is essential to serve nutrition of the whole organism.

Certain features of epithelial cell structure obviously serve the vectorial process. The apical (brush border) membrane is enlarged with microvilli [34], while the basolateral (basilar) membrane possesses infoldings, which may intrude around cellular organelles. This structural asymmetry, in some way, accomodates the functional vector in transepithelial transports, perhaps through imbalanced distribution of transport systems among the apical and basolateral membranes.

Epithelia function as effective permeability barriers because they possess tight junctions between the cells [35]. Solute transport is, accordingly, transcellular rather than paracellular, and two sets of plasma membranes are traversed in the process [36]. However, we should not assume that the tight junctions are obligatory for vectorial transport or that, in themselves, they confer the property of imbalanced fluxes. Transcellular concentration of solute can occur, under certain conditions, across thin layers of simple cells lacking tight junctions between them [37], indicating that intracellular events probably contribute to transport asymmetry.

Net transcellular flux of solute usually takes place against an electrochemical gradient. In the case of net reabsorption of solute from ultrafiltrate in the nephron, for example, the inward flux must exceed the outward flux. However, only influx at the brush-border membrane must exceed backflux from the cell, presumably on the same carrier, to accomplish a net flux is maintained at the brush-border membrane is of special interest. The thermodynamics of the system merely require that there be a carrier mediating the solute fluxes across the luminal membrane and that it be energized in some way [38].

Two possibilities for the maintenance of a net flux of low-molecular-weight substances deserve consideration [36]. First, the effective cytoplasmic concentration of solute close to the inner face of the membrane must be kept lower than the steady-state concentration that would allow backflux to equal influx; second, the carrier must constrain efflux in the direction of the gradient at the luminal membrane in order to establish a gradient.

For the first case, involving the effective intracellular concentration of solute, two mechanisms can be visualized. In one, metabolic "runout" removes solute from the apical cytoplasmic zone, thus lowering the effective concentration generating basic flux at the apical membrane; in the other, asymmetry of fluxes at the basolateral membrane permits transport "runout," so that there is a differential permeability of basolateral and apical membranes to the solute. It follows that the agencies that permit vectorial imbalance at luminal and basolateral membranes must be of dissimilar nature in the two locations.

For the second case, involving carrier kinetics controlling backflux of solute, too little is known yet to postulate meaningfully how relative affinities of substrate for the carrier-serving net flux may influence the gradients that permit influx to be greater than backflux. Independent agencies in plasma membranes, with differing functional characteristics, can achieve dissimilar transports of the same solute [39–41]. If one type of carrier achieves a steep uphill gradient in solute distribution across a membrane, and another working in parallel in the same membrane generates a less steep gradient, they can complement each other under differing conditions of the environment; or they could work in opposing fashion to control flux and counterflux across the same membrane; or, if placed in series in different membranes that are metabolically intercommunicated, they could achieve "transport runout" along the vector during transcellular flux. The different specificities and characteristics of such carriers would of course, be genetically controlled.

High-Molecular-Weight Substances: Uptake

Membranes combine their receptor and binding activities with a transport role in the process of receptor-mediated endocytosis [42]. This recently defined activity of the plasma membrane is an important general mechanism by which animal cells take up high-molecular-weight substances such as nutritional and regulatory proteins from the extracellular fluid. Receptor-mediated endocytosis is a coupled process by which proteins or peptides are first bound to specific cell-surface receptors, which are aggregated in subcellular organelles (known as coated pit regions) and are then internalized by the cell through invagination of the coated pits.

Four stages of receptor-mediated endocytosis have now been defined: 1) binding of an endogenous ligand at a specific receptor to achieve a particular physiological effect; 2) clustering of bound ligands in coated pits in the plasma membrane; 3) rapid internalization of the bound ligand as coated vesicles; 4) processing of the internalized vesicles – usually in lysosomes, occasionally, in other cellular structures. The importance of receptor-mediated absorptive endocytosis – analogous to pinocytosis in the amoeba – gained momentum from studies of familial hypercholesterolemia and the uptake of the low-density lipoprotein-cholesterol complex in normal and mutant fibroblasts and lymphocytes. Twelve additional systems involving receptor-mediated endocytosis in nonphagocytic cells recently have been identified.*

GENERAL LESSONS FROM THE INBORN ERRORS OF MEMBRANE TRANSPORT

Mutation is a useful and powerful probe of plasma-membrane transport functions serving low-molecular-weight substances [43–46]. While specific enumeration of the inborn errors of transport (IET) is no longer as interesting as it was when these disorders were first recognized for what they were, a catalogue (Table II) is still useful for the insight it offers about the nature of human membrane transport systems. We learn that:

1. The IETs are all mendelian disorders. Therefore, specific gene loci are associated with each IET, from which it follows that specific gene products in, or associated with, membranes are responsible for the various phenotypes observed.

2. Mutation may be expressed in one cell type (eg, the nephron) and not in another (eg, the erythrocyte) and, yet, under normal conditions, transport of a specific solute takes place in both tissues, from which it follows that the transporters in the two tissues must be different.

3. A particular mutation may be expressed in only one membrane, and not in another, of a given cell. This is not surprising when the structures differ in their function, such as plasma and mitochondrial membranes; but it is most interesting when different regions of the same membrane are involved, as, for example, the apical vs basilar regions of the plasma membrane of epithelial cells; or in the coated pit vs the nondifferentiated region of the plasma membrane.

4. Mutations, when fully expressed as autosomal recessives (in homozygotes), "codominants" (in genetic compounds), and X-linked recessives or dominants in hemizygotes may still permit residual function to be retained for transport of a particular substance. One conventional explanation for this phenomenon postulates a "Km mutation" at the structural gene locus causing catalytic properties of

---

*The following systems of receptor-mediated absorptive endocytosis in animal cells have been described (see Goldstein et al [43] for details): a) Transport proteins: low-density lipoproteins in fibroblasts, lymphocytes, smooth muscle cells, endothelial cells, adrenal cells; yolk proteins in oocytes; transcobalamin II in kidney, liver, and fibroblast; transferrin in erythroblasts and reticulocytes; b) Protein hormones: epidermal growth factor in fibroblasts; nerve growth factor in sympathetic ganglion cells; insulin in hepatocytes, adipocytes, and lymphocytes; chorionic gonadotropin in ovarian luteal cells; β-melanotropin in melanoma cells; c) other proteins: asialoglycoproteins in hepatocytes; lysosomal enzymes in fibroblasts; α-2-macroglobulin in fibroblasts and macrophages; and maternal immunoglobulin (IgG) in fetal yolk sac and neonatal intestinal cells.

TABLE II. Mendelian Disorders of Membrane Transport in Man*

Condition	McKusick cat. no.	Substance(s) affected	Tissues affected	Apparent inheritance
Hyperdibasic-aminoaciduria	12,600 & 22,270[a]	Lysine, ornithine arginine (probably different transport systems affected in 12,600 & 22,270)	Kidney, intestine (liver? in 22,270), brain (?)	AD/AR
Cystinuria	22,010	Lys, orn, arg, cystine	Kidney, intestine brain	AR
Hypercystinuria	23,820	Cystine	Kidney	AR(?)
Iminoglycinuria	24,260	Proline, hydroxy-pro, glycine	Kidney, intestine	AR
Dicarboxylic aminoaciduria	23,165	Glutamic acid aspartic acid	Kidney, intestine	AR
Hartnup disease	23,450	Neutral amino acids (excl. pro, gly)	Kidney, intestine	AR
Methionine malabsorption syndrome	25,090	Methionine	Intestine	AR(?)
"Blue-diaper" syndrome	21,110	Tryptophan	Intestine	AR(?)
B_{12}-malabsorption syndrome	26,110	Vitamin B_{12}	Intestine, kidney	AR(?)
Renal glucosuria	23,310	D-glucose	Kidney	AR
Glucose-galactose malabsorption syndrome	23,160	D-glucose D-galactose	Kidney, intestine	AR
Bartter's syndrome	24,120	Na^+, (K^+)	Kidney, erythrocyte, muscle	AR
Hereditary spherocytosis	18,290	Na^+	Erythrocyte	AD
Pseudohypoparathyroidism(s)	30,080	Calcium	Kidney, bone	XL dominant or sex-influenced AD
Familial hypophosphatemia	30,780	Phosphorus (as phosphate)	Kidney, intestine, bone(?)	XLD
Hypophosphatemic bone disease	14,635	Phosphorus (as phosphate)	Kidney	AD
Renal tubular acidosis (Type II)	31,240	HCO_3^-	Kidney, intestine	XL recessive (or AR)

(Continued next page)

TABLE II. Mendelian Disorders of Membrane Transport in Man* (Continued)

Condition	McKusick cat. no.	Substance(s) affected	Tissues affected	Apparent inheritance
Renal tubular acidosis (Type I)	17,980	H^+	Kidney	AD
RTA variant with deafness	26,730	H^+	Kidney, erythrocyte	AR
Familial chloride diarrhea	21,470	Cl^-	Intestine	AR
Diabetes insipidus (ADH resistant)	30,480	H_2O	Kidney	XL recessive
Fanconi's syndromes				
a) idiopathic	22,770 22,780	Many solutes and water	Kidney, other tissues	AR
b) symptomatic				
i) cystinosis	21,980 21,990 22,000	Same	Same	AR
ii) hereditary fructose intolerance	22,960	Same	Same	AR
iii) galactosemia(s)	23,040	Same	Same	AR
iv) hereditary tyrosinemia	27,670	Same	Same	AR
v) Wilson's disease	27,790	Same	Same	AR
vi) oculocerebro-renal syndrome	30,900	Same	Same	XL recessive
vii) vitamin-D dependency(s)	26,470	Same (particularly PO_4^{2-})	Kidney	AR
c) subforms				
glucoglycinuria	13,810	D-glucose, glycine	Kidney	AD
Luder-Sheldon syndrome	15,250	Many solutes and water	Kidney	AD
Busby's syndrome	26,850	Generalized aminoaciduria	Kidney, muscle(?) other tissues	AR
Absorbtive endocytosis hypercholesterolemia (Type II)	14,389	Increased LDL-cholesterol in serum	Nonparenchymal tissues (fibroblasts, lymphocytes, arterial smooth muscle)	AD

*Adapted largely from material in References 42, 45, and 46. McKusick catalogue numbers from Reference 8. Selected literature citations are given with each entry in the catalogues.
a Personal communication, Olli Simell (Helsinki).

the gene product to be altered quantitatively without total loss of activity. Another explanation postulates a regulatory gene mutation that alters the amount of gene product available for insertion into the membrane. And yet another postulates that the particular gene locus affected by the mutation specifies only a portion of the total membrane activity available to transport the substance.

MURINE COUNTERPARTS DELINEATING TRANSPORT PROCESSES

Among the murine mutants of medical interest only two, so far, delineate transport functions relevant to their human counterparts. However, both are of unusual interest because they bring insight to prevailing problems in human biology and disease.

Lessons From the PRO/Re Mouse

Hyperprolinemia in man was discovered some 20 years ago [24, 47, 48]. In its own modest way, it was an exciting discovery — not so much for the finding of yet another inborn error of amino acid metabolism, as for the delineation of a renal tubular transport system in man selective for the reabsorption of proline, hydroxyproline, glycine, and sarcosine [47–49]. Under normal conditions, proline and its fellow passengers are avidly reclaimed from glomerular filtrate by two selective transport mechanisms. The two carriers for proline have recently been identified in isolated renal brush border membranes in the rat [50].

In human hyperprolinemia, the hyperprolinuria is initiated at a concentration of proline in plasma somewhat below that which appears necessary to saturate the transport process in normal subjects. Until the discovery of the PRO/Re mouse, this finding was taken to be a simple aberration in the quantitative measurement of proline reabsorption and its saturability; it was assumed that hyperprolinuria in hyperprolinemia was merely the consequence of proline accumulation in filtrate, saturation of capacity in the reabsorption process (equivalent to Vmax), and "overflow" into bladder urine. We shall see that this assumption was wrong. We learned of the error in our ways from the PRO/Re mouse.

Blake and Russell [21] first reported on the hyperprolinemic PRO/Re strain of mouse eleven years after the description of human hyperprolinemia. Subsequent studies by Blake [22] and by Blake and colleagues [23] indicated that proline dehydrogenase activity in mitochondria involves two components. Component 1 — the enzyme missing in the PRO/Re homozygote — is under the control of the *pro-1* gene; the enzyme is present in the inner membrane of normal mitochondria, where it functions as a flavoprotein linked to the respiratory chain. Component 2 of the enzyme is not under the control of the *pro-1* locus, but is a constituent of the micochondrial proline dehydrogenase system, perhaps representing an embryonal form of the enzyme. The apparent Km of Component 2 for proline is about 100 times that of Component 1. In the normal mouse, Component 2 activity is a small fraction of total activity; it constitutes the significant residual activity observed in liver and kidney of PRO/Re homozygotes. These studies in the mouse have added a new dimension to our understanding of proline dehydrogenase activity in mammalian tissues. But whether the PRO/Re mouse is a homologue of the human disorder is still unknown, because no studies have been done yet in human subjects to delineate Component 1 or its absence in Type 1 hyperprolinemia. What then are we to make of the extraordinary hyperprolinuria in the PRO/Re mouse? Another line of investigation has explained that anomaly.

Comparative observation in PRO/Re and control mice, by our group [51], showed, that while plasma proline was eight times elevated in the former and renal cortex proline content was increased fourfold, urine proline was increased fiftyfold (Table III). We also observed that proline oxidation in normal mouse kidney is exceptionally vigorous, resulting in a lower intracellular content of proline and its metabolites than in other species. One can see the significance of this, when proline uptake and oxidation by renal cortex slices are compared in mouse [51], rat [52], and man [53]; it is also apparent in the very low plasma-proline concentration in the mouse, relative to rat and man. In the PRO/Re mouse, the deficiency of proline dehydrogenase (Component 1) causes proline to accumulate both in plasma and in tissues; in kidney, this accumulation occurs in the very cells responsible for proline reabsorption from filtrate. This event has no bearing

TABLE III. Characteristics of L-Proline Transport and Tissue Distribution in PRO/Re Mice*

	Control	PRO/Re[a]	P value
Plasma proline (mmole/L plasma water)	0.087	0.728	< 0.01
Renal cortex proline (mmole/L of cell water)	0.545	2.22	< 0.01
Urine proline (μmol/gm creatinine)	91	4960	< 0.001
Clearance of proline[b] (μl/min at 0.5 mM proline in plasma and normal glycine conc)	0.12	4.2	< 0.01
Clearance of glycine[b] (μl/min at 0.5 mM proline in plasma and normal glycine conc)	1.5 (approx)	1.5 (approx)	NS

*Data from Scriver et al. Reference 51; mean values given. Variation omitted for simplicity.
[a]PRO/Re mouse is deficient in mitochondrial proline oxidase, the first enzyme in the catabolic pathway for proline oxidation. Enzyme-specific activity is normally very high in mouse renal cortex.
[b]Data adapted from Reference 51, Figure 4.

on the transport runout of proline at the basal-lateral membrane; the fractional distribution of proline in vivo and in vitro between cellular water and extracellular fluid – including blood plasma – and the movement of proline on its carriers in that membrane do not differ from normal in the PRO/Re mouse. On the other hand, net reabsorption of proline across the luminal membrane in vivo clearly is not normal in the PRO/Re mouse (Table III). We can presume that luminal carriers for proline are intact, because normal interactions with glycine (and hydroxyproline) are observed during reabsorption. It follows that the most likely explanation for the striking hyperprolinuria in the PRO/Re strain is two-fold: an excess of proline in filtrate derived from plasma; and a deficit in metabolic runout of proline in absorbing epithelial cells. In the absence of efficient proline oxidation in renal cells, the concentration of proline must rise in the mitochondria and cytoplasm of renal epithelium. At this elevated concentration, proline gains access to its carrier in the luminal membrane and is carried back into the lumen of the nephron. However, because the column of urine in the lumen is moving distally in vivo, the system cannot come to equilibrium and, in the absence of distal nephron sites for its reclamation [44], proline must be lost in excess to bladder urine. From these observations in PRO/Re mice, we now realize that the appearance of prolinuria at a venous plasma threshold value below normal in the hyperprolinemic human subject is probably a reflection of the analogous mechanisms for prolinuria unmasked in the mouse.

Additional studies [Simell O, Mohyuddin F, Scriver CR, unpublished] have shown that proline oxidation is greatest in the very region of kidney where its reabsorption is greatest – the cortical aspect of the proximal nephron; the capsular region, the medulla, and the papilla also possess proline dehydrogenase activity, but observations in the PRO/Re mutant suggest that proline oxidation in these regions may comprise a larger fraction of Component 2 activity than in cortex.

Thus, the PRO/Re mutant has revealed that membrane transport and intracellular metabolism are, indeed, independent functions. For students of inborn errors of transport, in general, and of renal physiology, in particular, the PRO/Re mouse also offered these two specific lessons: Intracellular metabolism influences the maximal rate of tubular reabsorption (the so-called Tm, analogous to Vmax); bidirectional movement of solute takes place at the luminal membrane during net reabsorption. Neither of these revelations seem so startling now, but at one time they bordered on heresy. PRO/Re, teaching by analogy if not by homology, was an early apostle of now accepted doctrine.

Lessons From the *Hyp* Mouse

X-linked hypophosphatemia (XLH) in man was the first form of rickets to be classified as "resistant" to the action of vitamin D; this occurred in 1937 [54]. It was also the first to be recognized as both mendelian and X linked; this finding was reported in 1958 [55]. Males inheriting the allele are uniformly affected; females carrying it have a variable phenotype, with regard to bone disease, but always have hypophosphatemia. A long-standing tradition held that the disease was a disorder of vitamin-D metabolism; it was not until 1972 [56] that sufficient evidence had been accumulated to sustain an alternate hypothesis — that XLH is an inborn error of phosphate transport. However, argument and counterargument were to continue [26], and the significance of the proposed transport defect, in relation to phenotype, was to remain obscured until the discovery of the *Hyp* mouse [25].

The *Hyp* gene confers hypophosphatemia on the mouse. It is linked to the X chromosome and maps at the distal end. Expression of the mutant allele is dominant and apparently additive: Males (*Hyp*/Y) are uniformly more hypophosphatemic and rachitic than heterozygous females (Hyp/+). Ohno's "law" [20] on conservation of the X chromosome and its gene products in mammalian evolution implies that similar mechanisms are awry in human XLH and murine hypophosphatemia; that is, XLH and *Hyp* are homologous at the level of the mutant gene product. Accordingly, careful study of the *Hyp* mutant could divulge the basic defect in XLH.

Hyp/Y mice have increased fractional renal excretion of phosphate anion [57–59] (see Table IV); removal of the inhibitory action of parathyroid hormone on renal phosphate reabsorption by parathyroidectomy does not ablate the transport deficit in vivo [59], implying that it is an intrinsic cellular problem. Micropuncture studies [58, 59] reveal that the loss of transport is confined to the proximal convoluted tubule and to a less proximal site, probably in the pars recta of the nephron, but clearly not in the distal convoluted tubule. Since renal conservation of phosphate is critical for phosphate homeostasis in mammals, it is likely that the renal transport defect is an important determinant of the mutant phenotype in the *Hyp* mouse and, by homology, in human XLH.

In vitro investigation of the renal defect in phosphate transport is possible in the *Hyp* mouse; this is not the case in the human disease. We found renal cortex slices from *Hyp* mouse to have no abnormality of phosphate uptake, of incorporation into organic and inorganic pools, or of efflux across exposed basal lateral membranes [57]. *Hyp*/Y and +/Y littermates, in fact, maintain the same phosphate content in renal cortex (Table IV). In the aggregate, these findings imply that the transport defect must be confined to the renal brush-border membrane; other modes of uptake are apparently capable of maintaining the renal cellular phosphate pool.

Brush-border membranes have been isolated from *Hyp* and normal kidney, and phosphate uptake into membrane vesicles measured [57, 60–62]. Sodium-dependent, concentrative transport of phosphate is impaired in the vesicles obtained from *Hyp*/Y and severely affected *Hyp*/+ mice. However, loss of this mediated phosphate transport is incomplete; about half of the normal activity

TABLE IV. Characteristics of Renal Transport and Tissue Distribution of Phosphate (Phosphorus) in *Hyp*/Y Mice*

	+/Y[a]	*Hyp*/Y	P value
Serum phosphorus (mg/dl)	5.68	3.46	< 0.001
Fractional excretion of phosphate anion	0.21	0.35	< 0.01
Renal cortex phosphorus (matoms/gm protein)	46.6	46.6	NS
Distribution ratio of ^{32}P in cortex slice in:[b] organic pool	3.08	2.80	NS
total pool	5.74	5.31	NS
Efflux of ^{32}P (time course) from cortex slices in: slope	−0.0296	−0.0296	NS
Y intercept	4.15	4.13	NS
Uptake of ^{32}P by renal brush-border membrane vesicles (pmole P/pmole D-glucose at 30 sec)[c]	9.05	4.2	< 0.01

*Data summarized from References 57 and 60; variation (SD and SEM) omitted for simplicity.
[a]+/Y animals are littermate controls for *Hyp*/Y mice.
[b]Ratio is cpm/L cell water: cpm/L medium after 60-minute incubation at 30° pH 1.9 matom/l phosphate solution; inorganic phosphate and total (inorganic and organic) phosphate extracted and measured by method of Vestergaard-Bogind).
[c]Preparation and incubation of vesicles described in References 57 and 60.

is retained in mutant hemizygotes, yet the apparent Km for phosphate is not different from normal. We believe this finding can be interpreted quite simply from the following observations. First, Evers et al [63] have shown that in vivo inhibition of phosphate reabsorption by parathyroid hormone is associated with a fall in Vmax without change in the apparent Km of phosphate transport measured in vitro in isolated renal brush-border membrane vesicles. Second, the hemizygous XLH phenotype in man is accompanied by loss of the inhibitory response in renal phosphate reabsorption to parathyroid hormone [56]. Third, an autosomal-dominant phenocopy of XLH in man has been identified [64], in which comparable hypophosphatemia results from a defect in renal reabsorption and in which the residual renal phosphate transport remains sensitive to the inhibitory effect of parathyroid hormone. Together, these findings indicate the presence of at least two forms of phosphate transport in kidney, apparently in renal brush-border membranes: one sensitive to parathyroid hormone; the other less sensitive. The former is presumed to be deficient in the *Hyp* mouse and its human counterpart. Studies are in progress to investigate this proposal further.

Recently, we have shown that both the *Hyp* and the normal mouse adapt to severe dietary deprivation of phosphorus by reducing drastically the fractional excretion of phosphate (measured in vivo) and by augmenting brush-border membrane transport activity for phosphate (measured in vitro) [61, 62]; the latter response involves an increase in Vmax without change in the apparent Km for phosphate anion. This line of investigation has been informative in several ways: First, it showed that the difference between *Hyp* and + mice (wild-type) in renal phosphate transport persisted, both in vivo and in vitro, in the adapted state; second, it revealed that renal adaptation to the signal of phosphorus deprivation was intact in the *Hyp* mouse; third, it implied that hypophosphatemia in the *Hyp* mouse was not itself a sufficient signal to initiate the renal adaptation; fourth, it revealed that membrane adaptation involved a process apparently not under the control of the X-linked gene.

Many other facets of the X-linked phenotype are accessible to investigation in the *Hyp* mouse, for example: 1) Both the distribution of X-chromosome inactivation (in whole nephrons or individual cells along the nephron) and the evidence for additive inheritance of the X-linked gene affecting renal phosphate conservation could be studied in the *Hyp* mouse; 2) Transmural transport of phosphate in the intestine is reported to be abnormal in *Hyp* [65, 66], yet the defect is not expressed in the brush-border membrane of enterocytes [D Fast, HS Tennenhouse, CR Scriver; also B Sacktor et al, unpublished observations]. Where then is the intestinal defect? The mouse will tell – eventually; 3) Impairment of phosphate uptake by epiphyseal and woven bone may account for bone lesions that are more severe in the X-linked phenotype than in the autosomal form of hypophosphatemia [64], in which case, one can hypothesise that phosphate homeostasis in bone may be regulated by the X-linked gene. The hypothesis can be investigated better in the *Hyp* mouse than in man; 4) Placental transport of phosphate is poorly understood, and the *Hyp* mutation is a unique probe to delineate how placental transport of phosphate resembles or differs from the process in renal (and other) epithelium. The X-linked mutation has already been used in man to show that different gene products serve phosphate transport in membranes of the erythrocyte and renal epithelium [67]; 5) Hormonal responsiveness of renal epithelium in vitro, both to parathyroid hormone and to calcitonin, is modified in the *Hyp* mouse [68; Sacktor et al., unpublished data], and responsiveness to parathyroid hormone is attenuated in human XLH. The problem could be investigated definitely in the animal model. 6) The relative importance of phosphate replacement of vitamin-D analogues – in particular, $1,25$-$(OH)_2$ vitamin D – in treatment of the hypophosphatemic phenotype can be studied rapidly and specifically in the mouse model. One can think of other uses for the mouse homologue of a classic human medical puzzle. Enough has been said to indicate that the murine counterpart reveals XLH in man to be an inborn error of phosphate transport and homeostasis and to suggest that the many persisting lacunae in our understanding of XLH, in particular, and phosphate metabolism, in general, may yield to further study of the *Hyp* mouse.

CONCLUSION

Garrod's vision of human biochemical individuality has now been upheld and refined many times. His postulates of inborn errors of metabolism have taught us to appreciate that prediction of risk, through genetic screening and diagnosis, helps us to prevent the consequences of harmful interaction between genotype and environment [7, 69]. We have also learned that mutation sometimes permits useful interaction with the environment, for example, when the genes coding for sickle-cell heterozygosity or homozygous Duffy blood-group deficiency are expressed in populations living in malarial regions. We are beginning to realize that human biochemical genetics and population genetics are but two sides of the same coin [70]. When disease that originates in genotype and is related to environment comes to account for the major portion of the total disease burden of society, we will need our knowledge of genetics to reduce such a burden, and the context in which we will use it is already poignant. Human longevity truly has increased since Garrod's day, because we have learned to control the infectious and nutritional scourges of earlier times. Yet, at the same time, the heritability of most common diseases has increased and, pari passu, we most certainly have failed to achieve the goal of healthy longevity. Unhealthy longevity is the modern equivalent of an ancient plague: It consumes our resources; it has macroeconomic consequences [71]. Such circumstances call up our combativeness: We announce wars on disease; and we used the language of war to describe our view of the campaigns. In so doing, we forget who the enemy is; we do not recognize that the enemy is ourselves, that the heterozygotes among us fill the terminuses of our unhealthy longevity — our hospital beds.

In this setting, we need all the help we can get to understand disease. Animal "models" become appealing. All the more reason, then, that we recognize when the model is a homologue and when it is only an analogue. Society will not cherish investment in a false model if it has been sold as the genuine article. This symposium is an occasion to voice such a cautionary note.

We, the participants in the symposium, have been invited to make some predictions about events that might be of relevance to the Jackson Laboratory in the future. Well, if the inborn error of membrane transport went unpredicted by Garrod who possessed such profound wisdom and foresight, I am not likely to satisfy anyone's desire for great insight on things to come. Yet, I will venture some thoughts:

We might try to do better at learning about the evolutionary origins of membranes — so little is known about this important problem [72].

I anticipate an increased interest among biologists in membrane events because these events are the stuff of cellular organization and compartmentation; they dictate cellular hegomony and functional order. Disorder, as we know it in malignancy, must have something to do with membranes. Knowledge of membrane structure and function is advancing rapidly and it will continue to do so. Witness the understanding we have gained, in only a few years, of receptor-mediated endocytic processes [42].

As geneticists, we will probably want to delve further into particular membrane events such as the control of polio virus infection, diphtheria toxin binding, and invasion of the vivax malaria parasite, since these phenomena involve a membrane receptor, binding site, or blood group, respectively, each under the control of a specific gene locus in man [8].

I think greater success with cultivated epithelial cells from kidney, for example [73], would be a boon to students of the inborn errors of transport and of the normal asymmetry of cellular functions.

But I am voicing particular preferences. I return to my larger theme: The study of counterparts in man and mouse to gain a better understanding of health and disease. I began this essay somewhat irreverently, with reference to Amos. I will finish it by quoting in awe, a Maine writer's homage to our relationships with animals:

> "Remote from universal nature and living by complicated artifice, man in civilization surveys the creature through the glass of his knowledge and sees thereby a feather magnified and the whole image in distortion. We patronize them for their incompleteness, for their tragic fate of having taken form so far below ourselves. And therein we err, and greatly err. For the animal shall not be measured by man. In a world older and more complete than ours, they move finished and complete, gifted with extensions of the senses we have lost or never attained, living by voices we shall never hear. They are not brethren, they are not underlings; they are other nations, caught with ourselves in the net of life and time, fellow prisoners of the splendour and travail of the earth."
>
> (Henry Beston, "The Outermost House," Penguin, p 25)

Beston wrote on animals from his personal laboratory — the Fo'castle — the year before the Jackson Laboratory was founded. His reverence for animals — the mouse included — still commands our admiration at the mid-passage of your laboratory.

ACKNOWLEDGMENTS

It is a great pleasure to thank Elizabeth Russell and Eva Eicher for their contributions to my education and research. I also thank many of their colleagues at the Jackson Laboratory who have contributed, in various ways, to my appreciation of mouse genetics. F. Clarke Fraser, Francis Glorieux, Rod McInnes, Fazl Mohyuddin, and Harriet Tenenhouse have played key roles in the specific studies described in this paper. The original work was supported by the Medical Research Council of Canada, the Quebec Network of Genetic Medicine, and the Research Institute of the Montreal Children's Hospital.

REFERENCES

1. Garrod AE: "Inborn Errors of Metabolism," ed 2. London: Oxford Medical Publications, Frowde and Stoughton, 1923.
2. Childs B: Sir Archibald Garrod's conception of chemical individuality: A modern appreciation. N Engl J Med 282:71, 1970.
3. Garrod AE: The incidence of alkaptonuria. A study in chemical individuality. Lancet 2:1616, 1902.
4. Harris H, Hopkinson DA: Average heterozygosity per locus in man: An estimate based on the incidence of enzyme polymorphisms. Ann Hum Genet 36:9, 1972.
5. Childs B, der Kaloustian VM: Genetic heterogeneity. N Engl J Med 279:1205, 1968.
6. Childs B: Persistent echoes of the nature-nurture argument. Am J Hum Genet 29:1, 1977.
7. Scriver CR, Laberge C, Clow CL, Fraser FC: Genetics and medicine; an evolving relationship. Science 200:946, 1978.
8. McKusick VA: "Mendelian Inheritance in Man. Catalogs of Autosomal Dominant, Autosomal Recessive and X-linked Phenotypes," ed 5. Baltimore: Johns Hopkins University Press, 1978.
9. Stanbury JB, Wyngaarden JB, Fredrickson DS (eds): "The Metabolic Basis of Inherited Disease," ed 4. New York: McGraw-Hill, 1978, 1862 pp.
10. Altman PL, Katz DD: Inbred and genetically delivered strains of laboratory animals. Part 1. Mouse and rat. Bethesda: Biological Handbooks III, FASEB, 1979, Table 19, pp 77–95.
11. Kacser H, Bulfied G, Wallace ME: Histidinaemic mutant in the mouse. Nature 244:77, 1973.
12. Bulfield G, Kacser H: Histidinaemia in mouse and man. Arch Dis Child 49:545, 1974.
13. Kacser H, Mya KM, Duncker M, Wright A, Bulfield G, McLaren A, Lyon MF: Maternal histidine metabolism and its effect on foetal development in the mouse. Nature 265:262, 1977.
14. La Du BN: Histidinemia. In Stanbury JB, Wyngaarden JB, Frederickson DS (eds): "The Metabolic Basis of Inherited Disease," ed 4. New York: McGraw-Hill, 1978, p 317.
15. Lyon ICT, Gardner RJM, Veale AMO: Maternal histidinemia. Arch Dis Child 49:581, 1974.
16. Neville BGR, Harris RF, Stern DJ, Stern J: Maternal histidinaemia. Arch Dis Child 46:119, 1971.
17. Komrower GM, Sardharwalla IB, Coutts JMJ, Ingham D: Management of maternal phenylketonuria: An emerging clinical problem. Br Med J 1:1383, 1979.
18a. DeMars R, LeVan SL, Trend BL, Russell LB: Abnormal ornithine carbamyltransferase in mice having the sparse-fur mutation. Proc Natl Acad Sci USA 73:1693, 1976.
18b. Qureshi IA, Letarte J, Ouellet R: Ornithine transcarbamylase deficiency in mutant mice I. Studies on the characterization of enzyme defect and suitability as animal model of human disease. Pediatr Res 13:807, 1979.
19. Shih VE: Urea cycle disorders and other congenital hyperammonemic syndromes. In Stanbury JB, Wyngaarden JB, Frederickson DS (eds): "The Metabolic Basis of Inherited Disease," ed 4. New York: McGraw-Hill, 1978, p 362.

20. Ohno S: "Sex chromosomes and sex linked genes." New York: Springer-Verlag, 1967, p 46.
21. Blake RL, Russell ES: Hyperprolinemia and prolinuria in a new inbred strain of mice, PRO/Re. Science 176:809, 1972.
22. Blake RL: Animal model for hyperprolinaemia: Deficiency of mouse proline oxidase activity. Biochem J 129:987, 1972.
23. Blake RL, Hall JG, Russell ES: Mitochondrial proline dehydrogenase deficiency in hyperprolinemic PRO/Re mice: Genetic and enzymatic analyses. Biochem Genet 14:739, 1976.
24. Scriver CR: Disorders of proline and hydroxyproline metabolism. In Stanbury JB, Wyngaarden, JB, Frederickson DS (eds): "The Metabolic Basis of Inherited Disease," ed 4. New York: McGraw-Hill, 1977, p 336.
25. Eicher EM, Southard JL, Scriver CR, Glorieux FH: Hypophosphatemia: Mouse model for human familial hypophosphatemic (vitamin D-resistant) rickets. Proc Natl Acad Sci USA 73:4667, 1976.
26. Rasmussen M, Anast C: Familial hypophosphatemic (vitamin D-resistant) rickets and vitamin D-dependent rickets. In Stanbury JB, Wyngaarden JB, Frederickson DS (eds): "The Metabolic Basis of Inherited Disease," ed 4. New York: McGraw-Hill, 1978, p 1537.
27. Dent CE, Rose GA: Amino acid metabolism in cystinuria. Q J Med 20: 205, 1951.
28. Thier SO, Segal S: Cystinuria. In Stanbury JB, Wyngaarden JB, Frederickson DS (eds): "The Metabolic Basis of Inherited Disease," ed 4. New York: McGraw-Hill, 1978, p 1578.
29. Krane SM: Renal glucosuria. In Stanbury JB, Wyngaarden JB, Frederickson DS (eds): "The Metabolic Basis of Inherited Disease," ed 4. New York: McGraw-Hill, 1978, p 1607.
30. Scriver CR: The William Allan Memorial Award Address: On Phosphate Transport and Genetic Screening. Understanding Backward – Living Forward in Human Genetics. Am J Hum Genet 31:243, 1979.
31. Tanford C: The hydrophobic effect and the organization of living matter. Science 200:1012, 1978.
32. Silverman M: Specificity of membrane transport. Receptors and recognition. (Ser A) 3:133, 1977.
33. Rothstein A, Grinstein S, Ship S, Knauf PA: Asymmetry of functional sites of the erythrocyte anion transport protein. TIBS 3:126, 1978.
34. Kenny AJ, Booth AG: Microvilli: Their ultrastructure, enzymology and molecular organization. Essays Biochem 14:1, 1978.
35. Diamond JM: Tight and leaky junctions of epithelia: A perspective on kisses in the dark. Fed Proc 33:2220, 1974.
36. Scriver CR, Stacey TE, Tenenhouse HS, MacDonald WA: Transepithelial transport of phosphate anion in kidney. Potential mechanisms for hypophosphatemia. Adv Exp Med Biol 81:55, 1977.
37. Oxender DL, Christensen HN: Transcellular concentration as a consequence of intracellular accumulation. J Biol Chem 234:2321, 1959.
38. Crane RK: The gradient hypothesis and other models of carrier-mediated active transport. Rev Physiol Biochem Pharmacol 78:99, 1977.
39. Scriver CR: The human biochemical genetics of amino acid transport. Pediatrics 44:348, 1969.
40. Christensen HN: Some special kinetic problems of transport. Adv Enzymol 32:1, 1969.
41. Christensen HN, Oxender DL, Ronquist G: "Biological Transport," ed 2. Reading: WA Benjamin Inc, 1975.
42. Goldstein JL, Anderson RGW, Brown MS: Coated pits, coated vesicles, and receptor-mediated endocytosis. Nature 279:679, 1979.
43. Scriver CR, Hechtman P: Human genetics of membrane transport with emphasis on amino acids. Adv in Hum Genet 1:211, 1970.

44. Scriver CR, Bergeron M: Amino acid transport in kidney. The use of mutation to dissect membrane and transepithelial transport. In Nyhan WL (ed): "Heritable Disorders of Amino Acid Metabolism." New York: John Wiley & Son, 1974, p 515.
45. Scriver CR, Chesney RW, McInnes RR: Genetic aspects of renal tubular transport. Diversity and topology of carriers. Kidney Int 9:149, 1976.
46. Scriver CR: Hereditary and acquired aminoacidopathies, Entry #75. In Altman PL, Katz DD (eds): "Biological Handbooks II." Bethesda: Human Health and Disease, FASEB, 1977.
47. Scriver CR, Schafer RA, Efron ML: New renal tubular amino-acid transport system and a new hereditary disorder of amino-acid metabolism. Nature 192:672, 1961.
48. Scriver CR, Efron ML, Schafer IA: Renal tubular transport of proline, hydroxy-proline and glycine in health and in familial hyperprolinemia. J Clin Invest 43:374, 1964.
49. Scriver CR: Familial iminoglycinuria. In Stanbury JB, Wyngaarden JB, Frederickson DS (eds): "The Metabolic Basis of Inherited Disease," ed 4. New York: McGraw-Hill, 1978, p 1593.
50. McNamara PD, Ozegovic B, Pep LM, Segal S: Proline and glycine uptake by renal brush border membrane vesicles. Proc Natl Acad Sci USA 73:4521, 1976.
51. Scriver CR, McInnes RR, Mohyuddin F: Role of epithelial architecture and intracellular metabolism in proline uptake and transtubular reclamation in PRO/Re mouse kidney. Proc Natl Acad Sci USA 72:1431, 1975.
52. Baerlocher KE, Scriver CR, Mohyuddin F: The ontogeny of amino acid transport in rat kidney, I. Effect on distribution of proline and glycine. Biochim Biophys Acta 249:353, 1971.
53. Holtzapple P, Genel M, Rea C, Segal S: Metabolism and uptake of L-proline by human kidney cortex. Pediatr Res 7:818, 1973.
54. Albright F, Butler AM, Bloomberg E: Rickets resistant to vitamin D therapy. Am J Dis Child 54:529, 1937.
55. Winters RW, Graham JB, Williams TF, McFalls VW, Burnett CM: A genetic study of familial hypophosphatemia and vitamin D resistant rickets with a review of the literature. Medicine 37:97, 1958.
56. Glorieux F, Scriver CR: Loss of a parathyroid hormone-sensitive component of phosphate transport in X-linked hypophosphatemia. Science 175:997, 1972.
57. Tenenhouse HS, Scriver CR, McInnes RR, Glorieux FH: Renal handling of phosphate in vivo and in vitro by the X-linked hypophosphatemic male mouse: Evidence for a defect in the brush border membrane. Kidney Int 14:236, 1978.
58. Giasson SD, Brunette MG, Danan G, Vigneault N, Carriere S: Micropuncture study of renal phosphorus transport in hypophosphatemic vitamin D resistant rickets mice. Pfluegers Archiv 371:33, 1977.
59. Cowgill LD, Goldfarb S, Goldberg M, Slatopolsky E, Agus ZS: Demonstration of an intrinsic renal tubular defect in mice with familial hypophosphatemic rickets. J Clin Invest 63:1203, 1979.
60. Tenenhouse HS, Scriver CR: The defect in transcellular transport of phosphate in the nephron is located in brush border membranes in X-linked hypophosphatemia (*Hyp* mouse model). Can J Biochem 56:640, 1978.
61. Tenenhouse HS, Scriver CR: Renal adaptation to phosphate deprivation in the *Hyp* mouse with X-linked hypophosphatemia. Can J Biochem 57:938, 1979.
62. Evers CA, Murer H, Kinne R: Effect of parathyrin on the transport properties of isolated renal brush border vesicles. Biochem J 172:49, 1978.
63. Evers CA, Murer H, Kinne R: Effect of parathyrin on the transport properties of isolated renal brush border vesicles. Biochem J 172:49, 1978.
64. Scriver CR, MacDonald W, Reade T, Glorieux FH, Nogrady B: Hypophosphatemic non-

rachitic bone disease: An entity distinct from X-linked hypophosphatemia in the renal defect, bone involvement, and inheritance. Am J Med Genet 1:101, 1977.
65. O'Doherty PJA, Deluca HF, Eicher EM: Lack of effect of vitamin D and its metabolites on intestinal phosphate transport in familial hypophosphatemia of mice. Endocrinology 101:1325, 1977.
67. Tenenhouse HS, Scriver CR: Orthophosphate transport in the erythrocyte of normal subjects and of patients with X-linked hypophosphatemia. J Clin Invest 55:644, 1975.
68. Brunette MG, Chabardes D, Imbert-Teboul M, Clique A, Montegut M, Morel F: Hormone-sensitive adenylate cyclase along the nephron of genetically hypophosphatemic mice. Kidney Int 15:357, 1979.
69. Scriver CR: Human biochemical genetics. A view on individuality. Excerpta Medica ICS 411. Human Genetics. "Proc 5th International Congress on Human Genetics." Amsterdam: Excerpta Medica, 1977, p 142.
70. Harris H: Genetic heterogeneity in inherited disease. J Clin Pathol 8:(suppl 27):32, 1974.
71. Gori GB, Ritcher BJ: Macroeconomics of disease. Prevention in the United States. Science 200:1124, 1978.
72. Fox SW, Dose K: "Molecular Evolution and the Origin of Life." San Francisco: WH Freeman and Co, 1972.
73. Bergeron M, Fast DK, Scriver CR, Simell O: On the use of cell and organ culture methods to study renal birth defects. In Danes BS, Cox R (eds): "Study of Human Birth Defects Using Cultured Epithelilium." New York: National Foundation—March of Dimes Birth Defects, Original Article Series 16(2):305–314, 1980.

SESSION IV. IMMUNOGENETICS

DOROTHEA BENNETT, Chairman

Introduction to Session IV

In recent years the extent, diversity, and complexity of immunogenetic research have increased enormously, so that it seems inevitable that the three up-to-the-minute papers presented in this session deal with very different aspects of the field.

George Klein, as a tumor immunologist, reminds us that the field of immunogenetics started as the genetics of tumor transplantation, and then was extended to general tissue transplantation and immune responses after Snell's discovery of H-2, the complex major histocompatibility locus of the mouse. Today H-2 effects are of greater interest in relation to tumorigenesis than to tumor transplants. Klein reviews historical developments from a modern perspective, including helpful summaries of tumor induction, enhancement, and immunostimulation. Especially instructive are his own studies of hybrid resistance and "natural killer" cells with their probable role in tumor surveillance. Klein effectively crosses over into the field of Session V with his implication that viruses are not the sole causes of either mammary tumors or leukemias. In leukemogenesis, he considers viruses and physical and chemical carcinogens to be initial agents, with additional changes required for overt leukemogenesis.

One of the most fascinating developments in immunogenetics, and one of closest links between the genetics of man and mouse, has been the growing evidence that all mammalian species tested have one major histocompatibility locus (plus numerous minor loci), and that this locus, or compact chromosomal region, is complex, highly polymorphic, and almost certainly homologous between species. Although the pathway of discovery necessarily differed markedly between man and mouse, much parallel information exists about the human HLA system and the mouse H-2 system.

Walter Bodmer, who combines creative insight with skills of the population geneticist and of the immunologist, has been responsible for much of our understanding of the HLA system. His masterly presentation of this information provides an excellent comparison of HLA and H-2. Certainly genetic components responsible for very similar (but in each case highly polymorphic) products are expressed in both systems, and detailed information on the arrangement of these components in the mouse H-2 complex has provided a very useful model for presumptive arrangement of HLA components. Comparable molecular structures have been found in the two species. Bodmer's closing report of characteristics of the major histocompatibility complexes of certain higher primates and Old World monkeys suggests patterns followed in the diversification and evolution of this complex locus.

George Snell, who has had so much to do with the development of mouse histocompatibility genetics, discusses present and future potentialities in immunogenetic research and application. Readers relatively new to this field will find two especially useful features early in his paper. One, which will facilitate understanding of all three papers in this session, is an excellent glossary of symbols and abbreviations widely used in immunogenetics. The second organizes the field by depicting graphically the recent rapid growth of knowledge of three distinct types of antigen-determining loci in the mouse; loci associated with H-2; a growing number of non-H-2 histocompatibility loci; and an even more rapidly increasing number of serologically detected loci. Most alloantigens are located in cell surfaces, and much current research dealing with the highly complex H-2 region is concerned with understanding the chemistry, functions, and shifting distribution of many different cell membrane components — at present largely in diverse kinds of lymphocytes. Snell puts forth the suggestion that the K, D, L antigens of H-2 may, as organizers of the cell surface, regulate cell interactions and hence differentiation. He also makes the excellent point that while immunologists, cell physiologists, and cell chemists are all deeply engrossed in membrane research, progress would be facilitated by integration of their differing approaches. With considerable specific attention to "how," he discusses potential roles both for extensive polymorphism and for specific cell surface antigens in initiation of appropriate immune responses against a diverse group of virus infections and possibly also against autoimmune diseases. His section on medical implications provides highly pertinent stimulus for the clinician.

Modern immunogenetics, a diverse but highly productive field, is pulled together neatly by the excellent summarizing comments of the chairman of this session, Dorothea Bennett, whose own developmental research has made great use of the immunogenetic approach.

Mouse Histocompatibility Genetics and Tumor Immunology

George Klein

If tumor transplantation strangely and unexpectedly fathered transplantation immunology, the latter first appeared to kill, but later actually revived, tumor immunology. This is a strange family relationship indeed, conceived, fostered, and nurtured at Bar Harbor, still continuing, and far from concluded.

Tumor transplantation was started with the emphasis on the *first,* not the second word. The inbred strains were established to study the biology and the genetics of *cancer*, not of transplantation. What was first discovered, however, were the now classical laws of transplantation, as they were first formulated at Bar Harbor.

We may recall the expression "histocompatibility gene requirements." The idea that tumors required the presence of certain genes in the host in order to take and grow, was completely novel. It stemmed from experiments with F_2 and backcross mice. The fact that the tumor takes fell into neat parcels of $(1/2)^n$ and $(3/4)^n$ ratios, in the corresponding backcross and F_2 tests suggested that a certain number of genes had to be shared by tumor and host — a number that could be different for different tumors [1]. Although the phenomenology was quite clear, it was not easy to understand its meaning. It clearly reflected the multiplicity of what were later called the histocompatibility loci, however, and suggested that tumors differed in their ability to transgress the barriers set by them. After the discovery of *H-2*, it turned out that most tumors were rejected across this "major" barrier and that although transgressions could occur, they were relatively rare.

It is perhaps unfortunate that this problem has not been reinvestigated since the arrival of the congenic resistant mice. One would like to know whether there is a hierarchy between the different histocompatibility barriers (*H*-genes), affecting different tumors according to a more or less fixed "pecking order"? Alternatively, tumors might differ in their sensitivity to the effector cell population generated by a given *H*-gene difference, just as they are known to differ in their vulnerability to different types of killer cells in more thoroughly investigated model systems [2].

A curious, interesting, almost forgotten, but still-to-be-explained episode was the Barrett-Deringer effect [3]. It will be recalled that Barrett and Deringer found a decrease in the "histocompatibility gene requirement," ie, an increased take incidence in F_2 and backcross hybrids, following a single passage through the genetically corresponding F_1 hybrid. We have analyzed this phenomenon to some extent [4] and found, to our surprise, that it was not due to the selection of a variant minority (as we had expected on the basis of our previous work on ascites conversion and drug resistance) but to some kind of adaptive modulation induced in the majority of the tumor cells.

Looking at our present ideas, it is intriguing to seek parallels between the phenomena of "hybrid resistance" (also discovered at Bar Harbor), "histocompatibility gene requirements," and the Barrett-Deringer effect.

Snell and Stevens have shown that certain F_1 hybrid mice were relatively resistant to tumors derived from one parental strain, compared to the homozygous strain of origin [5]. This was confirmed by later reports [6, 7]. Recently, hybrid resistance against some lymphomas was found to be mediated by natural killer (NK) cells. The genetics of NK-mediated cytotoxicity in vitro and hybrid resistance in vivo were remarkably similar, at least as far as the YAC lymphoma (Moloney virus-induced, of strain A origin) was concerned [8, 9]. There were, however, major differences in the genetic resistance pattern against different tumors.

With the exceptions of one NK-cell—resistant lymphoma line, all lymphomas, sarcomas, and carcinomas tested showed characteristic hybrid resistance patterns. Backcross tests showed that at least one major resistance factor was *H-2* linked, as far as the lymphomas and leukemias were concerned, whereas there was no detectable *H-2*—linked resistance factor in relation to any of the relatively few sarcomas and carcinomas that were tested [10].

Within the lymphoma/leukemia group, different *H-2*—linked reactivity patterns could be identified for tumors of different genotypes and, to some extent against different tumors of the same genotype as well. At present it is not clear whether this reflects the action of different alleles of the same genetic locus or different loci within a complex pseudoallelic system, although preliminary evidence favors the latter possibility.

The potential importance of this complex, polymorphic system lies in the fact that its effectors are capable of rejecting small numbers of tumor cells of certain types, *without* previous immunization. It is also noteworthy that NK-mediated lysis can be boosted by a variety of agents [11—13]. Interferon and interferon inducers can have a dramatic effect, particularly in some genetically or physiologically (age-dependent) low reactive host—tumor cell combinations. Small numbers of tumor cells can also boost the response.

In conclusion, it is highly likely that Snell and Stevens' unexpected discovery of what first appeared as a minor and somewhat irksome exception from the transplantation laws will lead to the unraveling of a powerful surveillance mechanism capable of eliminating aberrant cells of certain types, including both virus-infected cells and cancer cells.

Viewed against this background, the Barrett-Deringer phenomenon may represent the adaptive modification of tumor cells, after F_1 passage, leading to resistance against a certain (genetically) specific effector. Preliminary findings suggest that such modulations do, in fact, occur [14].

THE BIRTH OF *H-2*, ITS GRADUATION INTO MHC, AND ITS IMPERIAL GROWTH INTO A SUPERGENE SYSTEM

Recognized first as a mouse blood group, *H-2* has traveled a long way from its humble beginnings. Antigen II, as it was originally called, was identified by Gorer in 1936 from a rabbit anti-mouse strain A antiserum [15]. The antiserum was absorbed with CBA red cells to remove species-specific antibodies. The absorbed serum did not react with CBA and C57BL but still reacted with strain A erythrocytes. Segregation analysis showed a clear relationship between the blood group gene that was identified and the hypothetical genes that governed susceptibility or resistance to tumor homografts. Cross-absorption experiments made it clear that normal and neoplastic tissues shared the same II antigens, or *H-2* antigens, as they were soon to be called. Rejection of *H-2* incompatible grafts was accompanied by production of alloantibodies. This led to the isoantigen concept of transplantation, which was fully developed by 1938.

There was a time, not many years ago, when George Snell could count the number of people who understood the serology and genetics of *H-2* without using all his fingers. The recent rapid growth of the topic into a highly sophisticated field, with generally recognized relevance (not only for transplantation and immunogenetics but also for immunology and cell differentiation in their widest aspects), came after the monumental development of the congenic mice by George Snell, the identification of recombinants, the definition of SD vs LD loci, the linking of serology, cell-mediated in vitro reactions, and in vivo rejection — a saga that is still far from being finished. The continuously growing number of functions localized within or near what is now known as the MHC region, or "supergene," is staggering. Most, but not all, known sites code for surface molecules. Some of them, like the serologically defined *H-2* antigens, are present on all nucleated cells; others, like the Ia products, represent relatively broadly distributed differentiation products; still others are restricted to only one or a few cellular subclasses of a given differentiation lineage (eg, the TL antigens or the products of the *I-J* locus). Some of the inserted genes within the MHC code for other ingredients of the immune system — eg, complement components — and some specify serum proteins that have no apparent relationship to surfaces or to immunological phenomena (eg, the *Ss* protein).

In all likelihood, the MHC region arose by gene duplication. Speculations as to how this happened and why the whole was conserved with such tenacity in so many mammalian (if not vertebrate) species is a favorite pastime of transplantation geneticists. While the basic mechanism, gene duplication, with subsequent functional differentiation is very likely indeed, the selective forces responsible for the preservation of the region are more difficult to understand, mainly because of uncertainty about the biological functions of the MHC products. Tumor surveillance, protection against the growth of implanted fetal cells in the mother, and prevention against the spread of certain leukemia viruses are some of the functions that have been considered in trying to explain the remarkable polymorphism of the system. It could not be excluded, however, that a high mutation rate and random genetic drift are responsible for the latter.

The recent discovery of "syngeneic restriction," ie, the requirement for at least partial *H-2* compatibility for T-cell mediated killing, B-T cell cooperation, delayed-type hypersensitivity, and a number of other fundamental immunological phenomena [16, 17] has strongly increased the suspicion that *H-2* polymorphism is not merely the result of random genetic drift but has a profound biological significance. It would be outside the scope of this paper to discuss the various theories that have been proposed to explain these wholly unexpected findings, particularly since none of them explains all the facts. It is quite clear, however, that MHC products must play a crucial role in most, if not all, immune recognition and cooperation phenomena that involve T-cells.

Recently, we have encountered an interesting experimental situation that supports the importance of syngeneic restriction in vivo for tumor rejection by specifically preimmunized, syngeneic hosts. With Tina Dalianis, we have selected variants from the Moloney virus–induced YAC lymphoma by repeated passage of antibody-coated tumor cells through anti-Ig columns [18]. The variants had a reduced antibody-binding capacity for the corresponding antigen. Depending on the coating antiserum, selection was either directed against the Moloney virus–determined cell surface antigen (MCSA) or against H-2^a. In both cases, variant sublines were obtained with a reduced antigen expression. The anti-MCSA selected subline showed only a slightly increased incidence of takes in YAC-preimmunized semisyngeneic mice. The anti-H-2^a selected sublines showed an increased allotransplantability, as expected. Surprisingly, however, they were *also* completely resistant to the rejection of the YAC-preimmunized semisyngeneic mice (Table I).

These results could be explained both by the "altered self" or the "dual recognition" model of syngeneic restriction. We failed to find any co-capping between *H-2* and MCSA in the same tumor system, however. Moreover, somatic hybridization and segregation studies on another (MC-induced) tumor with a well-defined TSTA gave no support to the idea that TSTA represents modified *H-2* antigens [19, 20]. Therefore, these findings speak for some type of dual recognition.

TABLE I. Antigen Expression and Rejectability of Column-Passaged Selected YAC Sublines

Cell line	Antigen expression		Allotransplantability	Rejectability in pre-immunized syngeneic host
	H-2^a	MCSA		
YAC	High	High	No	High
YAC-anti-MCSA 12	High	Reduced	No	Slightly reduced
YAC_1-anti A7	Reduced	Reduced	Very high	Abolished
YAC_2-anti A10	Reduced	Reduced	Very high	Abolished

From Dalianis et al [18].

MHC-PRODUCTS AS MARKERS IN CELLULAR GENETICS

In the early 1950s, it was already quite clear that most tumors evolve by a stepwise process, or, in the terms of Peyton Rous, "go from bad to worse." This process was termed "tumor progression" and was analyzed in great detail by Leslie Foulds [21]. Progression is essentially a *focal* process: whenever visible, each new progressional step was seen to arise in a small area of the previous lesion. We were interested to explore whether progression was due to variation and selection or to adaptive, essentially epigenetic (differentiation-type), modulations. A variety of experiments on relatively well-defined laboratory models such as the development of drug resistance or ascites conversion pointed to variation and selection as one, if not *the* most, important process [22]. At the same time, more or less reversible modulation phenomena (including the Barrett-Deringer effect, already discussed), also showed that the Darwinian mechanism was not exclusive. This is not surprising, since tumor cells have a dual nature. On the one hand, they are asexually reproducing microorganism-like populations subject to genetic change. On the other, they are abnormally differentiated somatic cells more or less responsive to differentiation-inducing stimuli.

As one approach toward the understanding of progression, it was thus necessary to study variation and selection. This required a cellular marker with a selective value. The obvious choice was the *H-2* antigens. They could be readily demonstrated on the surface of all nucleated cells. But would they also have the required selective value? This obviously depended on the selectivity of the homograft reaction. Would a small minority of *H-2*–compatible tumor cells grow out from among a large excess of *H-2*–incompatible cells, undergoing graft rejection? We have tested this in model experiments involving artificial mixtures of compatible and incompatible cells [23]. We found that as few as one in a million compatible cells survived and actually grew better in the presence of a violent homograft reaction provoked by the allogeneic cells. The desired selective system was thus clearly available. In addition, however, this experiment also had some important implications for the understanding of the homograft reaction itself.

Some of these implications are still not widely recognized. Clearly, rejection must be due to direct cell-mediated lysis, with killer T-cells as the most likely effectors, rather than to the various soluble cytotoxic factors that can be demonstrated in vitro. The latter must be either in vitro artifacts or, if they are produced in vivo, they must be rapidly inactivated. In view of the considerable importance of this conclusion, the experiments were repeated more recently and on a large series of tumor–host combinations, and involved preimmunized hosts as well [24]. The conclusions, however, were identical, showing the exquisite specificity of the homograft reaction and the lack of any demonstrable "innocent bystander" killing.

After the first success of the model experiments, we have performed a large series of isoantigenic variant selection experiments on *H-2* heterozygous tumors [25–27]. The material included carcinomas, sarcomas, and lymphomas induced in F_1 hybrids derived from the cross of two *H-2*-congenic lines. The main conclusions can be summarized as follows:

1) Phenotypically hemizygous, *H-2* haplotype loss variants can be selected from most, if not all, *H-2* heterozygous tumors by passage in one parental strain.

2) Parent-selected variants have irreversibly lost the entire *H-2* haplotype derived from the opposite parental strain.

3) The frequency of variant formation and the degree of asymmetry between the two complementary variants differed from tumor to tumor, even in cases where the tumors were induced by the same agent and in the same host genotype.

4) Extensive attempts to select variants that have lost *both* haplotypes were unsuccessful. Some tumors yielded "nonspecific" immunoresistant variants with a reduced *H-2* antigen concentration. They could grow indiscriminately across *H-2* barriers.

5) Separate selection of one haplotype brought interesting and unexpected results [28]. Selection against D-end antigens led to the loss of D and the corresponding K, located in the *cis* position. Selection against K has either led to the loss of K only or to the loss of both D and K in the *cis* position. In no case did selection affect the antigens controlled by the *trans* haplotype. Although the mechanism is still not understood, this finding is of considerable interest, since it suggests the existence of some mechanism that controls the expression of the functionally similar but widely separated D and K subloci within the large MHC-supergene. (For a detailed consideration of possible mechanisms see reference 29).

These findings have reemphasized the dynamic nature of genetic variation in tumor cell populations and, more important, they have demonstrated the individuality of different tumors. They also showed that the same end result, the loss of a complete haplotype, can be due to either a one-step change or to two consecutive steps.

TABLE II. Analogies Between Tumor Progression and H-2 Variant Formation

Progression	Variant formation
Independent in different tumors; independent reassortment of different cellular characters	Different tumors show individual occurrence and frequency of variants
Different unit characters progress independently	Different isoantigenic variants can be obtained from the same tumor
Progression can occur in successive, distinct steps	Same for variant formation
Progression is essentially a one-way process	Variant formation is irreversible

Table II shows some of the parallels between tumor progression and the formation of isoantigenic variants.

THE DEVELOPMENT OF TUMOR IMMUNOLOGY

When it turned out that the same genetic rules of histocompatibility govern the transplantability of normal and tumor tissues, the earlier hope of a rapid immunological solution of the cancer problem dwindled rapidly. This hope was originally based on the successful immunization of nonhistocompatible hosts against "transplantable" tumors. As so often happens in cancer research, the pendulum moved to the opposite extreme. It was now believed that all tumor resistance phenomena were transplantation artifacts. When they occurred within inbred strains, they were assumed to reflect the residual heterozygosis of the strain.

The tide turned in the late 1950s and the early 1960s, following a series of experimental surprises. The present director of the Jackson Laboratory was one of the main initiators of this development. It was first shown that some chemically induced tumors could induce rejection responses in critically histocompatible hosts [30–32]. These responses were relative rather than absolute, in contrast to the homograft rejection; ie, they could only protect the host against a certain limited number of tumor cells, although this number could exceed the threshold dose required for growth in untreated mice by 3–4 log units, in favorable cases.

One of the most remarkable features of this newly discovered system was the antigenic individuality of the chemically induced tumors shown for tumors induced by such widely different agents as aromatic hydrocarbons, azo dyes, and even some physical carcinogens, as in the case of the ultraviolet-induced skin carcinomas. The meaning of antigenic individuality has not been clarified. The number of possible antigenic alternatives has not been determined; there are practically no educated guesses. Conceivably, the antigenic changes may reflect a whole variety of more or less randomly afflicted modulations of a critical membrane structure.

A few years later, when the virus-induced mouse tumors acquired general use, the picture appeared to have become more diversified. The virally induced tumor-specific transplantation antigens (TSTA) were cross-reactive between all tumors induced by the same agent, in contrast to their chemically induced counterparts [33]. Virus-induced tumors did not cross-react with chemically induced tumors or with tumors induced by other viruses.

The cross-reactivity of the virally induced tumors and the antigenic individuality of the chemically induced tumors appeared first as a law. It is possible, however, that the picture may have been oversimplified to some extent. Evidence is increasing to suggest that virally induced tumors may also contain individually distinct antigens, over and above the virally induced group-specific antigens [34–36]. The difference between the chemically and the virally induced tumors may therefore not be as profound as it first appeared. It is possible that the virally induced antigens merely overshadow individually distinct antigens in conventional immunization-rejection experiments.

At first it was taken for granted that immune responses must be beneficial to the host. The development of immunology, with allergies, autoimmune diseases, graft vs host reactions, etc soon brought a different and more complex insight. As far as graft and tumor rejections were concerned, the first important memento came from the Jackson Laboratory. It was the phenomenon of immunological enhancement, as studied by Kaliss and Snell in relation to the *H-2* system [37]. Later, enhancement was also shown to operate in other systems, including tumor-specific or tumor-associated antigens. The mechanisms of enhancement are complex and still are not completely understood. Its experimentally demonstrated components include inhibition of antigen release by humoral antibody, antibody-mediated blocking of effective antigenic sites on the target cell, and according to the latest developments, probably suppressor cells as well. The possible relationship between enhancement and the immunostimulation phenomena discovered by Richmond Prehn is not yet clear; they may both express different facets of the same phenomenon, or they could be due to independent mechanisms.

In the classical Kaliss-Snell system, the administration of the antigen in the form of lyophilized tissue, instead of living cells, was of crucial importance to achieve enhancement, rather than rejection. The finding that such relatively simple changes in antigen administration can lead to such diametrically opposed results in the same antigen–host combination was of fundamental significance for the development of this area.

With the later advent of tolerance and suppression the picture appeared to become still more complicated. It is quite possible, however, that all these phenomena reflect different facets of the same basic system. As far as tumors are concerned, it is important to note that a given host–tumor combination can be geared to rejection or to its opposite – acceptance, or even enhancement – from the inception of the tumor. "Rejection geared" systems can be best illustrated by the potential tumor cells induced by ubiquitous oncogenic viruses in their natural hosts, like polyoma virus in mice, Herpesvirus saimiri in the squirrel monkey, or Epstein-Barr virus (EBV) in man [38]. Here, rejection is the rule. The antigenic target, often a relatively minor modification of the cell membrane (eg, the TSTA in SV40 or polyoma), is recognized with faultless precision, and rejection occurs with a watertight efficiency. This is clearly the end result of a prolonged selection process that has led not only to the fixation of the appropriate *Ir* genes required for the recognition of the common antigenic (TSTA) target, but, even more importantly, to the concerted action of the immune system, geared to avoid suppression and to bring about rejection. A very different situation prevails in spontaneous mouse mammary carcinomas, where enhancement is an equally or more common occurrence after immunization [39, 40].

A very instructive case is represented by the UV-induced skin carcinomas of mice [41]. In the normal syngeneic host rejection is the rule, even without any preimmunization. Rejection can be readily converted into acceptance by the usual immunosuppressive treatments. Most surprising, however, local UV-irradiation of the skin leads to acceptance as well in otherwise untreated syngeneic hosts. It has been shown that UV light acts by generating suppressor T-cells.

In view of the powerful mutagenic action of UV light, it is conceivable that the system has evolved suppressor cells in order to protect the host against the rejection of mutagenized and therefore antigenic skin. If so, the diminished possibility of rejecting the correspondingly antigenic tumor product would be an undesirable but unavoidable consequence.

DO VIRUSES INDUCE TUMORS DIRECTLY IN VIVO?

One of the most important tumor viruses, the mammary tumor agent, was discovered at Bar Harbor. This was the result of a coordinated, systematic search for the transmission routes of the "maternal influence," as it was then called — clearly an important contributor to the development of mammary cancer in certain high cancer strains [42]. All three conceivable routes of transmission were explored: the egg cytoplasm and the transplacental route could be ruled out, whereas milk transmission could be demonstrated by Bittner.

It may be recalled that Bittner first refrained from calling the milk agent a virus, due to the prevailing negative climate of opinion concerning tumor viruses in mammals. It may be asked, in retrospect, why the discovery of MTV, of monumental importance in itself, failed to change the climate of opinion, in contrast to what has happened since the discovery of the Gross virus a couple of decades later. At least in part, this may have been due to the relatively indirect involvement of MTV in the genesis of mouse mammary carcinoma in contrast to the dramatic result of the Gross experiment, suggesting a relatively direct viral induction of mouse leukemia. Viewed from our present horizon, the two systems are not so different as it would appear on first sight, however. Table III shows a schematic comparison of the two systems.

The virus is neither necessary nor sufficient for tumor induction in either sys-

TABLE III. Comparison Between the Involvement of MTV in Mammary Carcinoma Induction and of MuLV (Gross-Virus) in Murine Leukemia Induction

	MTV/mammary carcinoma	MuLV/leukemia
Is the virus absolutely necessary?	No	No
Is the virus alone sufficient?	No	Probably not
What conditioning factors?	Hormonal, genetic	Host genetics, somatic genetics
Does the virus induce pre-neoplastic cells?	Yes	Yes
What genetic mechanisms contribute?	MTV reproduction, hormonal environment, susceptibility, target level	Integr. virus (facultative), Fv-1 amplification, Rgv-1^S (imm. unresponsive), susceptibility at target level
Somatic genetics	Stepwise progression (focal)	Chromosomal change (trisomy 15)

tem. Malignization is dependent on the contribution of important additional factors. In mammary carcinoma, those factors are both hormonal and genetic; in mouse leukemia they are largely genetic. In the former, the known genetic mechanisms include factors that influence 1) replication of the virus, 2) the hormonal environment, and 3) the likelihood of the neoplastic change at the target cell level [43]. In the Gross virus–associated leukemia of the AKR strain, the integrated provirus acts as a genetic factor in itself. However, it is equally facultative as MTV, since the same type of leukemia can also arise in its absence. Other genes like Fv-1 contribute by amplifying the replication of the virus, and still others (exemplified by Rgv-1) modulate the immune responsiveness of the host to the leukemic cells. Still other genes act at the level of target transformability [44].

Perhaps the most important parallels are to be sought at the somatic level of cancer development per se. The mammary tumor develops by stepwise progression, an essentially focal process that can be postulated to involve several genetic changes in succession [21]. In mouse T-cell leukemia, trisomy of chromosome 15 is the outstanding feature. The recent development in this area can be summarized as follows:

Dofuko et al [45] reported that the cells involved in "spontaneous" T-cell leukemias of the AKR mouse frequently contain 41 chromosomes instead of 40, with trisomy of chromosome 15 as the most common change. We found a similar predominance of trisomy 15 in T-cell leukemias induced in C57BL mice by two different substrains of the radiation leukemia virus [46] and by the chemical carcinogen dimethylbenz(a)anthracene [47]. Trisomy 17 was the second most common anomaly, but was much less frequent than trisomy 15 and was never found without the latter. Trisomy 15 was also identified as the main cytogenetic change in x-ray-induced mouse lymphomas [48]. In contrast, lymphoreticular neoplasias of non-T-cell origin, induced by the Rauscher, Friend, Graffi, and Duplan viruses; some B lymphomas of spontaneous origin; and a series of mineral oil–induced plasmocytomas showed no trisomy 15 (F. Wiener, S. Ohno, N. Haran-Ghera, J. Spira, and G. Klein, unpublished data 1980). The question whether they have other types of distinctive chromosomal changes has not yet been answered.

It is sometimes postulated that all murine T-cell lymphomas are due to the activation of latent type-C viruses. Careful examination of the pathogenesis of these lymphomas makes this most unlikely, however [for review see reference 49]. It is more likely that x-rays and chemical and viral carcinogens can all play the role of initiating agents that can create long-lived preleukemic cells. The development of overt leukemia depends on additional changes that occur during the prolonged latency of the preleukemic cells in their host. It is very likely that the duplication of a certain gene(s), reflected by the trisomy 15, plays a key role in this process.

Is there a specific region on chromosome 15 that needs to be duplicated for the development of leukemia? Recently, we have examined the karyotype of dimethylbenz(a)anthracene-induced T-cell leukemias in CBAT6T6 mice [50]. The T6 marker has arisen by the breakage of chromosome 15, not far from the centromere, and translocation of the distal part of the long arm to chromosome 14. Six independently induced leukemias showed trisomy of the 14;15 translocation, whereas the small T6 marker was present in only two copies. This suggests the involvement of a specific region(s) in leukemogenesis, localized in the distal part of the long arm of chromosome 15. Additional translocations will be helpful in defining the region more precisely.

It might be objected that trisomy 15 is not causally involved in the genesis of T-cell leukemia but represents the only surviving cell type, from among a wide variety of chromosomal anomalies that arise at random in the overt leukemia clone. This was ruled out, however, by our recent study of T-cell leukemias, induced by DMBA and by Moloney virus in Robertsonian translocation mice [51]. We have examined mouse stocks where chromosome 15 has fused with chromosomes 1, 4, 5, and 6, respectively, at the centromere. In the leukemias, the entire Robertsonian translocation appeared in three copies. This shows that the duplication of some chromosome 15–associated gene is of critical importance for the development of overt leukemia, to the extent that it compels additional genetic elements, attached by translocation, to become "fellow travelers," even if they are located on the longest autosome (number 1).

Most recently, we have induced T-cell leukemias in F_1 hybrids between CBAT6T6 mice (where the leukemia-associated duplication occurs in the cytogenetically recognizable 14;15 marker) and three strains with normal 15 chromosomes – AKR, C3H, and C57BL. In the AKR × T6T6 hybrid, T-cell lymphomas showed a regular duplication of the AKR-derived chromosome 15, whereas in the other two combinations, the CBAT6T6 derived 14;15 element was duplicated. This suggests the existence of genetic variation at the chromosome 15–associated locus that influences leukemogenesis. Furthermore, it also implies that duplication of the chromosomes carrying chromosome 15–associated genes from one strain or the other exert an unequal influence on the likelihood that the trisomic cell will grow into an autonomous tumor. This may be a question of rather subtle differences in selective advantage influencing the latency period or the growth rate.

It is intriguing to speculate that the chromosome 15–associated gene may be akin to the gene(s) that was previously shown to influence T-cell leukemia development at the target cell level itself. In all likelihood, this area will provide another fruitful field for genetic studies on leukemia.

THE CONCEPT OF CONVERGENCE IN TUMOR EVOLUTION

There are three major tumor-associated, nonrandom chromosomal changes presently known (in addition to minor ones): 1) the Philadelphia chromosome in human CML; 2) the reciprocal 8;14 translocation in Burkitt lymphoma; 3) 15-trisomy in murine T-cell leukemia.

In all three conditions, a given, highly specific chromosomal change is associated with tumors that arise in the same target cell. In Burkitt lymphoma (BL), both EBV-carrying (and presumably EBV-initiated) and EBV-negative lymphomas were found to show the same change, together with the rare type of B-cell acute lymphocytic leukemia that is believed to originate from the same target cell type as BL [52]. In the murine T-lymphoma, discussed above, clearly the same chromosomal change is involved, no matter what the etiology of the tumor.

On the basis of these facts, we have recently suggested [53] that the development of at least some tumors may follow a "convergent" pathway from the genetic point of view. This concept is not new. In essence, it corresponds to one of the rules of tumor progression, as formulated by Foulds [21]. He stated that the "multiple reassortment of unit characteristics" that formed the basis of the progression concept "could follow one of several alternative pathways of development." Some aspects of this process can be stated here in a more specific way. They are as follows:

1) Like chemical or physical carcinogens, *viruses* play essentially the role of *initiators* in tumor progression. Their major effect is the establishment of *long-lived preneoplastic cells.*

2) *Specific genetic changes* are responsible for the transition of preneoplastic to frankly malignant cells. In some systems they are expressed as cytogenetically detectable chromosomal anomalies, characteristic for the majority of the tumors that originate from the same target cell. The changes may arise by random mechanisms. They are selectively fixed, owing to the increased growth advantage of

the clone that carries them. This advantage is based on a decreased responsiveness to growth-controlling or differentiation-inducing host signals. This selection process, rather than any specific induction mechanism, is responsible for the "cytogenetic convergence" of preneoplastic cell lineages, initiated ("caused") by widely diverse agents, toward the same nonrandom chromosomal change.

3) The cytogenetic changes act by shifting the balance between genes that favor progressive growth in vivo and genes that counteract it. Changes in effective gene dosage are brought about by nonrandom duplication of a whole chromosome, as in trisomy, or by reciprocal translocation that may affect gene expression on the donor or the recipient chromosome.

REFERENCES

1. Snell GD, Stimpfling JH: Genetics of tissue transplantation. In Green EL (ed): "Biology of the Laboratory Mouse, The Staff of the Jackson Laboratory," Ed 2. New York: McGraw-Hill, 1966, p 457.
2. Becker S, Klein E: Host response against oncornavirus induced lymphomas in mice. In Bentvelzen et al (eds): "Advances in Comparative Leukemia Research 1977." Amsterdam: Elsevier/North Holland, 1978, p 61.
3. Barrett MK, Deringer MK: Induced adaptation in a tumor: Permanence of the change. J Natl Cancer Inst 12:1011, 1952.
4. Klein G, Klein E: The evolution of independence from specific growth stimulation and inhibition in mammalian tumour cell populations. Symposia of the Society for Experimental Biology 11:305, 1957.
5. Snell GD: Histocompatibility genes of the mouse. II. Production and analysis of isogenic resistant lines. J Natl Cancer Inst 21:843, 1958.
6. Hellström KE: Differential behavior of transplanted mouse lymphoma lines in genetically compatible homozygous and F_1 hybrid mice. Nature 199:614, 1963.
7. Oth D, Donner M, Burg C: Measurement of the antitumoral immune reactions against a strain-specific chemically induced sarcoma in syngeneic and F_1 hybrid mice. Eur J Cancer 7:479, 1971.
8. Kiessling R, Klein E, Wigzell H: "Natural" killer cells in the mouse. I. Cytotoxic cells with specificity for mouse Moloney leukemia cells. Specificity and distribution according to genotype. Eur J Immunol 5:112, 1975.
9. Kiessling R, Klein E, Pross H, Wigzell H: "Natural" killer cells in the mouse. II. Cytotoxic cells with specificity for mouse Moloney leukemia cells. Characteristics of the killer cell. Eur J Immunol 5:117, 1975.
10. Klein G, Klein GO, Kärre K, Kiessling R: "Hybrid resistance" against parental tumors: One or several genetic patterns? Immunogenetics 7:391, 1978.
11. Kiessling R, Wigzell H: An analysis of the murine NK cell as to structure, function and biological relevance. Immunol Rev 44:165, 1979.
12. Gidlund M, Örn A, Wigzell H, Senik A, Greser I: Enhanced NK cell activity in mice injected with interferon and interferon inducers. Nature 273:759, 1978.
13. Trinchieri G, Santoli D, Knowles BB: Tumor cell lines induce interferon in human lymphocytes. Nature 270:611, 1977.
14. Hansson M, Kiessling R et al: (in preparation).
15. Gorer PA: The detection of a hereditary antigenic difference in the blood of mice by means of human group A serum. J Genet 32:17, 1936.

16. Doherty PC, Zinkernagel RM: T-cell mediated immunopathology in viral infections. Transplant Rev 19:89, 1974.
17. Benacerraf B, Katz DH: The histocompatibility linked immune response genes. Adv Cancer Res 21:121, 1975.
18. Dalianis T, Klein G, Andersson B: Column selection of antigenic variants from tumors. YAC (Moloney) lymphoma variants with reduced antigen expression. (in preparation).
19. Klein G, Klein E: Are methylcholanthrene-induced sarcoma-associated, rejection-inducing (TSTA) antigens, modified forms of H-2 or linked determinants? Int J Cancer 15:879, 1975.
20. Fenyö EM, Yefenof E, Klein E, Klein G: Immunization of mice with syngeneic Moloney lymphoma cells induces separate antibodies against virion envelope glycoprotein and virus-induced cell surface antigens. J Exp Med 146:1521, 1977.
21. Foulds LMA: The natural history of cancer. J Chronic Dis 8:2, 1958.
22. Klein G: Variation and selection in tumor cell populations. Can Cancer Conf 3:215, 1959.
23. Klein G, Klein E: Detection of an allelic difference at a single gene locus in a small fraction of a large tumor-cell population. Nature 178:1389, 1956.
24. Klein E, Klein G: Specificity of homograft rejection in vivo assessed by inoculation of artificially mixed compatible and incomparable tumor cells. Cell Immunol 5:201, 1972.
25. Klein G, Klein E: Histocompatibility changes in tumors. J Cell Comp Physiol 52:125, 1958.
26. Klein E: Parental variants. Transplant Proc 3:1167, 1971.
27. Pjaring B, Klein G: Antigenic characterization of heterozygous mouse lymphomas after immunoselection in vivo. J Natl Cancer Inst 41:1411, 1968.
28. Klein E, Klein G: Studies on the mechanism of ioșantigenic variant formation in heterozygous mouse tumors. III. Behavior of H-2 antigens D and K when located in the *trans* position. J Natl Cancer Inst 32:569, 1964.
29. Klein J: Biology of the Mouse Histocompatibility-2 Complex. Berlin: Springer-Verlag, 1975, p 1.
30. Foley EJ: Antigenic properties of methylcholanthrene-induced tumors in mice of the strain of origin. Cancer Res 13:835, 1953.
31. Prehn RT, Main JM: Immunity to methylcholanthrene-induced sarcomas. J Natl Cancer Inst 18:769, 1957.
32. Klein G, Sjögren HO, Klein E, Hellström KE: Demonstration of resistance against methylcholanthrene-induced sarcomas in the primary autochthonous host. Cancer Res 20:1561, 1960.
33. Klein G: Tumor antigens. Annu Rev Microbiol 20:223, 1966.
34. Witz IP, Lee N, Klein G: Serologically detectable specific and cross-reactive antigens on the membrane of a polyoma virus-induced murine tumor. Int J Cancer 18:243, 1976.
35. Morton DL, Miller GF, Wood DA: Demonstration of tumor-specific immunity against antigens unrelated to the mammary tumor virus in spontaneous mammary adenocarcinomas. J Natl Cancer Inst 42:289, 1969.
36. Weiss D: Immunological parameters of the host–parasite relationship in neoplasia. Ann NY Acad Sci 164:431, 1969.
37. Kaliss N: Immunological enhancement of tumor homografts in mice: A review. Cancer Res 18:992, 1978.
38. Klein G, Klein E: Immune surveillance against virus-induced tumors and non-rejectability of spontaneous tumors – Contrasting consequences of host versus tumor evolution. Proc Natl Acad Sci USA 74:2121, 1977.

39. Prehn RT: Sixteenth annual symposium fundamental cancer research. In Cumley RW (ed): "Conceptual Advances in Immunology and Oncology." Houston, Texas: Texas University Press, 1963, p 475.
40. Hewitt HB, Blake ER, Walder AS: A critique of the evidence for active host defence against cancer, based on personal studies of 27 murine tumours of spontaneous origin. Br J Cancer 33:241, 1976.
41. Fisher MS, Kripke ML: Systemic alteration induced in mice by ultraviolet light irradiation and its relationship to ultraviolet carcinogenesis. Proc Natl Acad Sci USA 74:1688, 1977.
42. Staff Jackson Memorial Laboratory, The existence of non-chromosomal influence of mammary tumors in mice. Science 78:465, 1933.
43. Nandi S, McGrath CM: Mammary neoplasia in mice. Adv Cancer Res 17:353, 1973.
44. Lilly F, Pincus T: Genetic control of murine viral leukemogenesis. Adv Cancer Res 17:231, 1973.
45. Dofuku R, Biedler JL, Spengler BA, Old LJ: Trisomy of chromosome 15 in spontaneous leukemia of AKR mice. Proc Natl Acad Sci USA 72:1515, 1975.
46. Wiener F, Ohno S, Spira J, Haran-Ghera N, Klein G: Chromosome changes (trisomies 15 and 17) associated with tumor progression in leukemias induced by radiation leukemia virus. J Natl Cancer Inst 61:227, 1978.
47. Wiener F, Spira J, Ohno S, Haran-Ghera N, Klein G: Chromosome changes (trisomy 15) in murine T-cell leukemia induced by 7,12-dimethylbenz(a)anthracene (DMBA). Int J Cancer 22:447, 1978.
48. Chang TD, Biedler JL, Stockert E, Old LJ: Trisomy of chromosome 15 in X-ray induced mouse leukemia. Proc Am Assoc Cancer Res (abstracts) 225, 1977.
49. Haran-Ghera N, Peled A: Induction of leukemia in mice by irradiation and radiation leukemia virus variants. Adv Cancer Res 30:45, 1979.
50. Wiener F, Ohno S, Spira J, Haran-Ghera N: Cytogenetic mapping of the trisomic segment of chromosome 15 in murine T-cell leukemia. Nature 275:658, 1978.
51. Spira J, Wiener F, Ohno S, Klein G: Is trisomy a cause or a consequence of murine T-cell leukemia development? Studies on Robertsonian translocation mice. Proc Natl Acad Sci USA 76:6619, 1979.
52. Mitelman F, Andersson-Anvret M, Brandt L, Catovsky D, Klein G, Manolov Y, Marc-Vendel E, Nilsson PG: Reciprocal 8;14 translocation in EBV-negative B-cell acute lymphocytic leukemia with Burkitt-type cells. Int J Cancer 24:27, 1979.
53. Klein G: Lymphoma development in mice and humans: Diversity of initiation is followed by convergent cytogenetic evolution. Proc Natl Acad Sci USA 76:2442, 1979.

The Major Histocompatibility Gene Clusters of Man and Mouse

Walter F. Bodmer

The development of the HLA system has naturally in many respects paralleled that of the mouse H-2 system and has, to a fair extent, used H-2 as a model. There are, however, some intriguing differences between the two systems, as well as many increasingly remarkable similarities. A major difference is that there are no inbred human lines, and so the development of the HLA system has had to follow the path of population studies using the human blood groups as a model. This is, perhaps, more analogous to Gorer's original description of the H-2 system than to its subsequent development by George Snell and others using inbred and congenic mouse strains (see eg, references 1 and 2 for brief historical reviews). Another major difference has been that, at least until quite recently, it has not been possible routinely to produce HLA typing sera by planned immunization. This has led to the use of natural sources of sera, particularly (following the independent discovery by Van Rood and Rose Payne in 1958 that leukocyte agglutinins arose by fetal maternal stimulation) sera produced by multiparous women. These sera do not have the strength and specificity of those that can be produced experimentally. Statistical approaches have thus played a major role in the development of the HLA system following especially Van Rood's original description in 1962 of the system that he called Group 4, which was the precursor of what is now called the *HLA-B* locus. A series of international collaborative workshops initiated by Amos in 1964 and organised successively by Van Rood, Ceppellini, Terasaki, Dausset, Kissmeyer-Nielsen, and most recently in Oxford by W. F. and J. G. Bodmer together with their colleagues Batchelor, Morris, and Festenstein have been integral to the development of knowledge of the HLA system, and their proceedings provide the best documentation of its development [3–8].

THE HLA SYSTEM AND ITS HOMOLOGY WITH H-2

A complete list of the recognised HLA specificities is given in Table I. The expression "HLA" refers to the whole region, whereas the letters A, B, C, and D(R) refer to the constituent loci thus far defined. The HLA-A, B, and C antigens correspond to H-2K, D, and L and have a broad tissue distribution, being present on most nucleated cells, though not on red blood cells. The HLA-D determinants are defined by the mixed lymphocyte culture reaction and the use of homozygous typing cells as typing reagents, whereas the DR determinants are identified serologically on cells of B lymphocyte origin either in the peripheral blood or, often, using the B lymphocyte—derived EB (Epstein-Barr) virus transformed lymphoblastoid cell lines. These antigens are the counterpart of at least some of the mouse Ia antigens controlled by either the I-A or the I-E/C regions of H-2.

The HLA-DR antigens were defined independently of any knowledge of their relationship to other antigens, though their relationship to the HLA-D types identified by the mixed lymphocyte culture reaction was suspected. This was because the antisera used to define the DR determinants were known to block the mixed lymphocyte culture reaction. The very close association subsequently established between the DRW and DW types, especially in families, where antigen assignment can be done with greater confidence than in unrelated individuals, is shown in Table II. Many of the discrepancies between the serological and cellular typing can still be explained by technical problems, though some, especially those between DRW 4 and DW 4, simply reflect the present incomplete state of definition of both sets of determinants and the fact that they are both bound to be further refined and subdivided. The close correlation between these two sets of determinants naturally leads to a numbering which is such that they correspond to each other and, together with other evidence, supports the supposition that they occur on the same gene products.

A schematic comparison of the genetic maps of the mouse H-2 and human HLA regions is shown in Figure 1. The recombination fractions between the constituent loci are comparable in magnitude. The main difference in the organisation is that in the mouse the I region and its neighbouring loci are between *H-2K* and *D*, whereas in man the *D(R)* locus, whose neighbourhood is presumed to correspond to the H-2 I region, is found outside the *HLA-A* to *B* interval. It has usually been assumed that the *HLA-B* locus corresponds to *H-2K* and the *HLA-A* locus to *H-2D*, mainly because of the fact that the I region is adjacent to *H-2K* in the mouse and the *D(R)* locus is adjacent to *HLA-B* in man. However, now that it is known that the *Ss*-determined protein corresponds to C4 and that the *C4* locus most probably lies in the interval between *HLA-D* and *HLA-B*, a more appropriate analogy might, as pointed out by Barnstable et al [9], be to assume that *HLA-B* corresponds to *H-2D*. In this case the I region and *HLA-DR*, C4 and *Ss*, and *HLA-B* and *H-2D* would be correspondingly aligned, as illustrated in Figure 1. Another piece of evidence in favour of this arrangement is that the recently described *H-2L* locus in the mouse, which seems perhaps to correspond to *HLA-C*, lies near *H-2D* [10], again emphasising the possible homology between *H-2D* and *HLA-B*. If this alignment of the two systems is correct, then it seems

MOUSE H-2

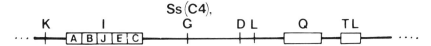

HUMAN HLA

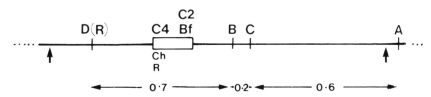

Fig. 1. Schematic comparison of the genetic maps of the mouse H-2 and human HLA regions. They are aligned so that I corresponds to *DR*, *Ss* to C4, C2, and Bf, and *K, D, L* to *A, B,* and *C,* respectively. It is suggested that the interval between *A* and *C* contains the human equivalents of *Q* and *TL* regions and that, therefore, *HLA-A* corresponds to *H-2K*. The numbers underneath the line for HLA are the approximate recombination fractions, in percent. The vertical arrows to the left of *A* and *DR* indicate the positions of break points for an inversion that would convert the HLA into the H-2 sequence.

possible that the human equivalents of the loci in the Q and TL regions will lie between *HLA-A* and *C,* and so fill in this genetic gap. The mouse H-2 and human HLA regions could then be related by an inversion whose two break points lie first between *HLA-A* and *C* or, in the mouse, on the distal side, of the Q and TL regions, beyond the *H-2D,* and second to the "left" of *HLA-DR* (ie, outside the presently defined genetic region) and, in the mouse, between *H-2K* and the I region. It is interesting to note that it has recently been shown that the *H-2G* blood group locus is probably C4 attached to red cells (C. David, personal communication), exactly as in man the Chido and Rodgers blood groups have been shown to be C4 [11]. Moreover, it has also recently been suggested that in man the C4 locus is duplicated, and perhaps these duplicates correspond to *Ss* and *Slp* in the mouse [12, 13]. As will be discussed later, there is now evidence in man for possible homologous products to those of loci in the TL and perhaps Q regions, though as far as I am aware, there is not any evidence for genes for C2 and factor B in the mouse H-2 region.

TABLE I. Recognized HLA Specificities (previous names for some newer specificities are given in parentheses)

HLA-A	HLA-B (continued)	HLA-C
A1[a]	B15	Cw1 (T1)
A2	Bw16	Cw2 (T2)
A3	B17	Cw3 (T3)
A9	B18	Cw4 (T4)
A10	Bw21	Cw5 (T5)
A11	Bw22	Cw6 (T7)
Aw19	B27	
Aw23	Bw35 (w5)	HLA-D
Aw24	B37 (TY)	Dw1
A25	Bw38 (w 16.1)	Dw2
A26	Bw39 (w 16.2)	Dw3
A28	B40 (w 10)	Dw4
A29	Bw41 (Sabell)	Dw5
Aw30	Bw42[b] (MWA)	Dw6
Aw31	Bw44 (B12 not TT*)	Dw7 (LD107)
Aw32	Bw45 (TT*)	Dw8 (LD108)
Aw33 (w 19.6)	Bw46[c] (HS, SIN2)	Dw9 (TB9)
Aw34 (Malay 2)	Bw47 (407*)	Dw10 (LD16)
Aw36[b] (MO*)	Bw48 (KSO)	Dw11 (LD17)
Aw43[b] (BK)	Bw49 (Bw21.1)	
	Bw50 (Bw21.2)	HLA-DR
HLA-B	Bw51 (B5.1)	DRw1
B5	Bw52 (B5.2)	DRw2
B7	Bw53 (HR)	DRw3
B8	Bw54[c] (Bw22.1)	DRw4
B12	Bw4 (4a)	DRw5
B13	Bw6 (4b)	DRw6
B14		DRw7

The following is a list of those specificities that have arisen as clear-cut splits of other specificities:

> A9 into Aw23 and Aw24
> A10 into A25 and A26
> B5 into Bw51 and Bw52
> B12 into Bw44 and Bw45
> Bw16 into Bw38 and Bw39
> Bw21 into Bw49 and Bw50

Historically, Aw19 has included Aw29, Aw30, Aw31, Aw32, and Aw33.

Aw before the number indicates that the specificity may still need further confirmation or, sometimes, that only limited amounts of the key defining sera are available.

[a]Formal usage is, eg, HLA-A1. Where such terms occur frequently in a text, the HLA is customarily omitted.
[b]Found so far only in Black populations.
[c]Found so far only in Mongoloid populations.
Data from reference 8.

HLA Population Distributions and Disease Associations

The HLA system is extremely polymorphic, more so than any other known human polymorphism. The relatively large number of different determinants of the various loci can generate between them nearly one billion genetically different individuals and about 100 million distinguishable combinations of antigens. The distribution of the gene frequencies in various populations is relatively even, with few having frequencies over 10% and many with frequencies at most up to 5%. This leads to an extraordinarily high level of heterozygosity, which is close to 90% for the *HLA-A, B,* and *DR* loci and little less for the *C* locus in European Caucasoids, with comparable figures in other major racial groups (see Table III).

TABLE II. Association Between HLA-DRw and Dw Types in 189 Haplotypes From Healthy Caucasoids

DRw	Dw	++	+−	−+	−−	r[a]
1	1	22	1	2	164	0.92
2	2	39	3	0	147	0.95
3	3	28	4	0	157	0.92
4	4	9	10	0	170	0.66
5	5	9	2	0	178	0.89
6	6	8	2	1	178	0.83
7	7	18	3	4	164	0.81

[a]Correlation coefficient measure of association between the DRw and Dw types. All values are highly significant.
++, +−, etc: positive for DRw and Dw, positive for DRw and negative for Dw, etc.
Data from reference 8.

TABLE III. Average HLA Heterozygosity (in percentage)

	HLA-A	HLA-B	HLA-C	HLC-DR
Number of alleles	18	27	6	8
European Caucasoids (228[a])	86	92	73	87
African Blacks (102)	89	89	68	75
Japanese (195)	77	87	63	73

998,530,800 genotypes
96,087,112 distinguishable phenotypes
32,076 haplotypes

[a]Numbers tested based on data from "Histocompatibility Testing," 1977 [8].

These levels are much higher than the overall average frequency of heterozygotes per locus in man, based on studies of blood group and enzyme polymorphisms, which is about 15% [14].

Population associations between the products of linked genes, such as the antigens from the HLA loci, arise if there is a tendency for alleles of the different loci to occur together more often on the same chromosome, or haplotype, than is expected by chance. This tendency has been called "linkage disequilibrium," and its extent is measured by the difference between the observed haplotype frequency and the expectation in the absence of an association, which is the product of the separate gene frequencies. For example, in the case of the *A1, B8* haplotype, its frequency is 0.088 in a typical European population, whereas the frequencies of the *A1* and *B8* alleles are 0.17 and 0.11, respectively, so that the measure of linkage disequilibrium is

$$D = 0.088 - 0.17 \times 0.11 = 0.069$$

Thus, in this case, the contribution to the total haplotype frequency, 0.088, due to linkage disequilibrium is 0.069, or nearly 80% of the total frequency. Linkage disequilibrium is a relatively common phenomenon in the HLA system and there are many well-known associations such as (*A1, B8*), (*A3, B7*), (*DW2, B7*), and (*DW3, B8*) in typical European Caucasoid populations. Associations between *A* and *B*, and between *B* and *D* tend to be different for each major population group, whereas those between the *HLA, B,* and *C* loci are, mostly, common to all the major racial groups. This may reflect the fact that the *B* and *C* loci are closer together than the other pairs, and that, as a result, there simply has not been enough time for the expected equilibrium, where linkage disequilibrium is zero, to have been reached. The rate at which linkage disequilibrium approaches zero in a random mating population, and in the absence of selection, is $1 - r$ per generation, where r is the recombination fraction between two loci, and so clearly the closer together the loci are, the longer it takes for linkage disequilibrium to be dissipated. It seems likely that the persistent linkage disequilibria observed between alleles of the *HLA-A* and *B* or *HLA-B* and *D* loci are due to natural selection. Fisher in 1938 [15] first suggested how certain types of natural selection favouring particular pairwise combinations of alleles at different loci could provide an explanation for persistent linkage disequilibrium. The selective interactions needed to maintain persistent linkage disequilibrium between alleles at pairs of loci suggest a functional relationship between the loci of a sort that may be expected within a complex genetic system such as HLA. In other words, the existence of linkage disequilibrium between alleles at pairs of closely linked loci within the HLA system helps to define the loci that constitute the system [see eg, references 1 and 16].

The evidence for the existence of immune response genes in man within the
HLA system is mainly indirect, and comes particularly from HLA and disease
association studies. A number of striking associations have been observed
between certain HLA determinants and a variety of diseases, the most significant
being that between B27 and ankylosing spondylitis; more than 90% of people
with the disease have the antigen. Many of the HLA-associated diseases have a
presumptive or suspected immune etiology, for example, coeliac disease, active
chronic hepatitis, myasthenia gravis, juvenile-onset diabetes, and rheumatoid
arthritis, and these diseases are particularly associated with the HLA-D(R) determinants D(R)W3 and D(R)W4. The main explanation for these associations, as
suggested by McDevitt and Bodmer [17] and others, is that they are due to immune response genes in the HLA region near to the *HLA-D* locus, or perhaps involve the *HLA-D* locus determinants themselves through their role in T–B lymphocyte and macrophage–T lymphocyte interactions. These data, which are of great
potential clinical interest, support the view that the I region of the H-2 system
corresponds to the neighbourhood around the *HLA-D* locus.

STRUCTURE AND FUNCTION OF HLA PRODUCTS

The molecular structures of the four known classes of gene products in the
HLA region — namely, those controlled by the *HLA-A, B,* and *C* loci, by the
HLA-D(R) locus, C2 and Bf, and C4 — are at least to some extent known (see
Table IV). The molecules recognised by antisera to HLA-A, B, and C specificities
are composed of a 43,000 molecular weight glycosylated polypeptide, which
carries the polymorphic specificities and which is noncovalently associated with
a nonglycosylated polypeptide of molecular weight 12,000, which is beta-2
microglobulin (β2m). The products of these three loci are structurally very similar to one another and are presumed to have a common origin by duplication.
They are found on most nucleated cells of the body, excluding, in particular,
sperm and trophoblast [18]. Their absence from human red cells, in contrast to
the homologous mouse products H-2K, D, and L, which are present on red cells,
may simply reflect a difference in maturation rate or rate of turnover of the red
cell, as it is known that in man reticulocytes carry the HLA-A, B specificities [19].
Presumably *HLA-A, B,* and *C* products are lost gradually during the maturation
of the red cells in the absence of any new synthesis. The HLA region lies on the
short arm of chromosome 6, and the gene for β2m is on chromosome 15, and
so not in the HLA region (see eg, reference 9 for a more detailed review).

Molecules recognized by antisera to the *HLA-DR* products also have a two-chain structure. These antigens do not contain β2m, but comprise two noncovalently associated glycosylated polypeptides of molecular weights 33,000 and
28,000, one of which may be coded for by the *HLA-DR* locus while the other,

TABLE IV. Known Loci, Products, and Immune Functions of the HLA and H-2 Regions

Loci	Products	Function
HLA-A,-B,-C H-2K, D, L	43,000 mol wt; present on most nucleated cells, associated with β^2-microglobulin (12,000 mol wt)	T cell recognition
HLA-D(R) H-2 I-A, E/C	33,000 or 28,000 mol wt; present mainly on B lymphocyte, monocytes, or macrophages, and some epithelial cells	Mixed lymphocyte culture reaction, T_H-B and T_H-Mϕ interactions
H-2 I-J and others	Not known	T_S interactions (immune response)
C2, Bf	100,000 mol wt	Classical 2nd and alternate factor B complement components
C4 (Ss also Chido and Rodgers human red blood cell groups, and H-2G)	200,000 mol wt (three constituent chains – 30,000, 80,000, and 90,000 – probably synthesized as a single unit)	Classical 4th complement component

T_H = helper T lymphocyte.
T_S = suppressor T lymphocyte.
Mϕ = macrophage.
B = B lymphocyte.

as in the case of β2m, may be coded for by a gene elsewhere [20, 21]. This problem, and the homology of these products with their mouse counterparts, will be further discussed later. Factor B of the alternate complement pathway and C2 of the classical complement pathway have similar, apparently homologous, structures, each being a single polypeptide chain with a molecular weight of about 100,000. C4, the fourth component of the classical complement pathway, has a molecular weight of about 200,000 and is composed of three polypeptide chains of molecular weights about 90,000, 80,000, and 30,000, which are thought to be synthesised from a single polypeptide, presumably coded for in the HLA region [22]. Recent data, as already mentioned, suggest that there are two C4 genes corresponding to the C4F and C4S variants [12]. If, therefore, one accepts the possibility suggested by the mouse H-2I region that there will probably be two or more similar loci in the neighbourhood of HLA-D(R), then each of the four classes of gene products of the HLA region occurs as a cluster of two or more duplicated loci.

The structural studies show that the genes of the HLA region code for at least four distinct classes of polypeptide chains. This poses a problem for the simple suggestion that all the genes in the HLA region may be derived from a common ancestor by duplication. It is possible, of course, that a diverse collection of genes has been trapped in the HLA region and that these have subsequently been duplicated to different extents. However, it is much more appealing to try to explain the origin of the system in terms of duplication of a single, perhaps admittedly complex, nucleotide sequence. A possible model for this taking into account recent dramatic advances in our understanding of genetic organisation in higher organisms at the DNA level will be discussed later.

The complement components of both the classical and alternate pathways play an obvious role in protection against infectious disease, as effectors of pathogen destruction and removal following an immune response. The products of the *HLA-A, B,* and *C* loci were identified using serological tests because of genetic differences between individuals, which of course are the basis for histocompatibility differences, but these provide no direct clue as to their functions. A major clue to the function of the H-2K, D, and L products, and therefore also their presumed human counterparts HLA-A, B, and C, comes from the work of Zinkernagel and Doherty [23]. They have shown that these products are required for immune T cell recognition of specific target antigens, using as their main model immune cellular killing of virus-infected cells expressing virally determined surface antigens. They have therefore emphasised the importance of these products for protection against viral infection, a role that parallels, to some extent, the role of the complement components in protection against other types of infection. The H-2I-A and I-E/C products of the mouse seem to be involved in an analogous way in controlling interactions between B and T lymphocytes and also between B lymphocytes and macrophages in the immune response. Other H-2I region products appear to control interactions involving suppressor T cells, and no doubt similar determinants are likely eventually to be found controlled by genes in the neighbourhood of the *HLA-D* locus in man.

Many explanations have been put forward for these cellular interactions, including the so-called dual recognition and altered self models, as extensively discussed in a recent review by George Snell [24]. Another explanation, based on an analogy with complement activity, was suggested by Barnstable et al [25], and I should like to expand on this possibility here. The suggestion is that the HLA-A, B, and C products, and their mouse counterparts, behave like complement molecules on the cell surface in relation to the T cell "receptor" or, following Snell, "recogniser," and the surface antigen it recognises. An antibody, when complexed with antigen, undergoes a conformational change that enables fixation of the first component of complement and subsequent activation of the complement pathway to take place. Similarly, it may be that when the T cell recogniser complexes with its specific antigen on another cell, it undergoes a conformational change that allows the HLA-A, B, and C molecules on the target cell (or perhaps sometimes on the cytotoxic T cell) to interact with a specific region of the T cell recogniser that is analogous to the Fc region of the immunoglobulin molecule. This interaction may then trigger "effectuation" of whatever activity is involved in T cell recognition, such as specific killing of the target cell.

This model for the role of these products in T cell recognition avoids such problems as the formation of specific complexes between any surface antigen and the HLA-A, B, and C molecules, and also the need for a separate specific recognition system for the *HLA-A, B,* and *C* polymorphic determinants. To accommodate the data on the restriction phenomenon by which the role of these molecules in T cell recognition was identified, namely that H-2K and D polymorphic differences interfere with specific immune killing, it must be supposed that the association of the HLA-A, B, and C products with the T cell recogniser may interfere with antigen recognition by the T cell recogniser, or vice versa. This would mean that only appropriate variable regions of the T cell recogniser could associate properly with the HLA-A, B, and C products following antigen recognition. It is as if polymorphic variation for complement components interfered, in some cases, with the binding of complement to antigen–antibody complexes. The role of the thymus during the maturation of T lymphocytes can then be interpreted as being to eliminate T cells bearing recognisers whose variable regions would interfere with the association with HLA-A, B, and C products. In the case of the antibody molecule, the fact that the complement binding Fc region is on a separate, more or less independent domain of the molecule circumvents this problem. Perhaps the T cell recognition system is an evolutionarily more primitive precursor of antibody recognition for which the independence between specific recognition and effectuation has not yet evolved. T cell recognition as we now see it may, as Snell [2], Bodmer [26], and others have suggested, be an evolutionary development of the processes required for specific cell–cell recognition during differentiation and development. Such recognition would, of course, involve only specific, well-defined differentiation antigens and not, as in the immune system, arbitrary, "foreign" determinants recognised by one of a pool of recognisers generated in somatic cells during development. The original mechanism for cell–cell recognition presumably could have evolved in parallel with the differentiation antigens, so that these did not interfere with the complement-like binding of the equivalents of the HLA-A, B, and C products required for effectuation. Interference between the antigen recognised and the effectuation step, leading to the restriction phenomenon discovered by Doherty and Zinkernagel, would on this basis only have arisen with the evolution of T cell immunity.

The mechanism suggested here for the role of the HLA-A, B, and C products in T cell recognition, and more generally perhaps in cell–cell recognition during differentiation, provides a natural rationale for associating the determinants of complement components and the HLA-A, B, C, and D products in the same genetic region. Perhaps one should look for esterase-like activity in these products, by analogy with that of the first and third components of complement. Undoubtedly, the real challenge is to understand these structures and their functions at the detailed molecular level.

MONOCLONAL ANTIBODIES FOR ANALYSIS OF HLA DETERMINANTS

The sera that come from multiparous women and that have been used to define the antigens of the HLA-A, B, C, and DR series are generally comparatively weak, serologically complex, and available in only limited quantities. Large-scale screening programmes are required to find usable sera, especially for the rarer specificities. Antisera to HLA determinants have also been produced by planned immunization of human volunteers and by heterospecific immunization, especially using purified antigens, but both approaches have their limitations. The heteroantisera usually require extensive absorption to reveal polymorphic specificities and have not so far proved generally useful as routine tissue typing reagents. The development of techniques by Kohler and Milstein [27, 28] for producing somatic-cell hybrid lines secreting monoclonal antibodies with defined specificities now provides a way of circumventing these problems and is being used to produce monoclonal antibodies against HLA region antigenic determinants. Here I shall describe briefly some results obtained in our laboratory with studies using such monoclonal antibodies.

The principle of the technique is to rescue immune lymphocytes from a hyperimmunized mouse by fusing them with myeloma cells of the appropriate line. The resulting somatic cell hybrid has the properties of a myeloma cell, and so secretes the specific antibody produced by the parental immune lymphocyte at much the same level as the parental myeloma cell secretes its particular protein. Hybrids can generally be selected using standard techniques, involving a drug-resistance marker in the myeloma parent line. The supernatant culture medium from individual hybrids can then be tested against appropriate sources of antigen to detect and characterize the specificity of the antibodies being produced. An indirect radioactive binding assay, using ^{125}I-labeled rabbit–anti-mouse immunoglobulin [29], is commonly used for the detection and characterization of mouse monoclonal antibodies. Hybrids are initially grown in plates containing 24 × 2 ml capacity wells, and the supernatants from these wells can be checked for their activity before the hybrid is grown further and eventually cloned.

Whole cells, cell membranes, and purified antigen have been used as immunogens, though antibody characterization has mostly been carried out, at least initially, on whole cells. The yield of antibodies against HLA products, and especially against the polymorphic determinants, is naturally greatest when using purified antigen for immunization.

The first antibody produced against an HLA determinant, a monomorphic antibody reacting with each of the products of the *HLA-A, B,* and *C* loci (in all members of the species), was initially recognised by the fact that it failed to react with the lymphoblastoid cell line Daudi, which lacks HLA-A, B, C, and β2m, while reacting with all of a wide variety of other cell lines tested [30]. The antibody was then shown to react with human–mouse somatic cell hybrids that contained chromosome 6 but not chromosome 15, so ruling out $\beta 2m$ as its specificity. The antibody specificity was subsequently confirmed by inhibition studies with purified HLA-A, B, C, and $\beta 2m$, as well as by specific immunoprecipitation. Antibodies to $\beta 2m$ can similarly be recognised by their lack of reaction with Daudi, but in this case reaction with somatic cell hybrids containing human chromosome 15, but not 6. These monomorphic-reacting antibodies have proved useful for confirming the tissue distribution of these determinants, as well as for purification of the antigens and for immunochemical studies. The monomorphic anti-HLA-A, B, C antibody, for example, does not react with sperm or with human trophoblasts, confirming the absence of these antigens from these two cell types [18].

A further application of the use of these antibodies for characterizing tissue distribution is illustrated in Figure 2. This shows the binding of a monomorphic anti-$\beta 2m$ and a monomorphic anti-HLA-A, B, C antibody to different B and T cell sources. The level of binding is shown for different dilutions of antibody; the plateau at the higher concentrations (reflecting the saturating level at which antibody is bound) is a reasonable measure of the amount of antigen per cell. Thus, for the B cell-derived line, Bristol 8, and the T cell line, HSB2, both antibodies plateau at about the same level, as is the case with nearly all cell types tested, indicating approximately equivalent amounts of $\beta 2m$ and HLA-A, B, C on the surfaces of these cells. This is exactly as expected if $\beta 2m$ occurred only on these cells in association with the HLA-A, B, and C products. The results for thymocytes and the thymocyte-like line, Molt 4, are, however, quite different. In this case, at saturating levels, the anti-$\beta 2m$ antibody binds much more strongly than the anti-HLA-A, B, C, indicating, for these cells, a marked excess of $\beta 2m$ over HLA-A, B, C product on their surface. This result parallels data obtained by Tada and his colleagues [31] and is most easily interpreted by assuming that thymocytes have a significant proportion of their surface $\beta 2m$ associated with a non HLA-A, B, C product, which by analogy with the mouse may well be the human equivalent of a TL region product.

Polymorphic reacting antibodies have, so far, only been produced by immunization with purified antigen. Thus, using the papain digestion purified HLA-A, B, C product we have produced antibodies whose main specificity is HLA-A2 and B7, as shown in Figure 3. Antibodies PA2.1 and BB7.2 appear to have identical specificities. They react with all individuals who are HLA-A2 and, at first, were thought not to react with any individuals carrying the highly cross-reactive determinant, A28. Subsequently, however, these antibodies have been found to react with the cells of one out of six A28 individuals tested, but with no individuals of any other HLA types. The antibodies, in addition to their A2 specificity, therefore apparently reveal an unexpected split in the HLA-A28 specificity, since, using our conventional antisera, there is nothing to distinguish the individual who reacts with these antibodies from other individuals who are A28. The third antibody illustrated in Figure 3, BB7.1, is a straightforward anti HLA-B7, which does not react with the cross-reactive determinant B40.

The patterns of reactions of two anti HLA-DR antibodies against a series of lymphoid cell lines are shown in Figure 4. The first of these antibodies, DA2, was produced by immunization with a membrane extract of the line LKT. It reacts with all the cell lines except Molt 4, which is the only T cell line in this panel,

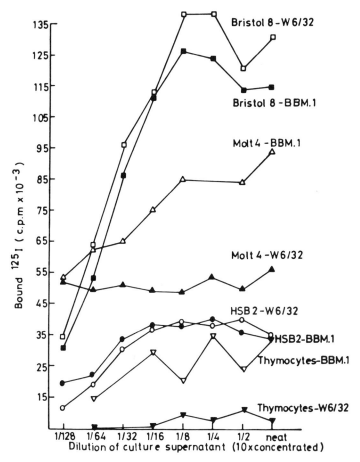

Fig. 2. Binding of monoclonal antibodies to $\beta 2m$ (BBM1, filled symbols) and to the HLA-A,B,C determinants (W6/32, open symbols) to different cell types. The assay is done under saturating conditions of the labeled rabbit anti-mouse antibody (25 μg/ml with 3 ×10^6 cells in 50 μl). The graphs show reactivity in terms of numbers of counts bound at different dilutions of culture supernatant as a source of antibody. Bristol 8 is a B cell line, Molt 4 is a thymocyte-like T cell line, HSB2 is a T cell line with no thymocyte-like properties (from reference 18).

all the rest being EBV (Epstein Barr virus) transformed B cell lines. Antibody DA2 is a monomorphic anti HLA-DR, which has proved useful, for example, in separating out DR-positive cells from peripheral blood (Kretzer, personal communication). The antibody Genox 3.53, which was produced by immunization with a partially purified glycoprotein fraction containing the HLA-DR products, has a polymorphic specificity directed against the antigens HLA-DRW1, DRW2, and DRW6. This is a common pattern of cross-reaction seen with the antisera used

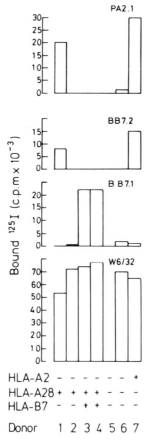

Fig. 3. Binding of various monoclonal anti-HLA-A,B,C antibodies to lymphocytes of different HLA-A and B specificities. Assays were done under trace binding conditions on 5×10^5 cells. Each bar is a mean of duplicates, and background binding is subtracted. PA2.1 and BB7.2 have similar specificities and bind only to HLA-A2 lymphocytes and lymphocytes from donor 1, who is A28. BB7.1 binds only to HLA-B7 lymphocytes. W6/32 reacts with all HLA-A, B, and C determinants and is included as a control (from reference 18).

to define these determinants [32]. Both these antibodies show exactly the tissue distribution expected for the HLA-DR products and are inhibited by the purified antigen.

A monoclonal mouse antibody to rat Ia antigens has been produced by McMaster and Williams [33] which cross-reacts with human HLA-DRW determinants [34]. This antibody, though it has a lower affinity for the human antigens than for the rat antigens, shows the same pattern of cross-reactivity —

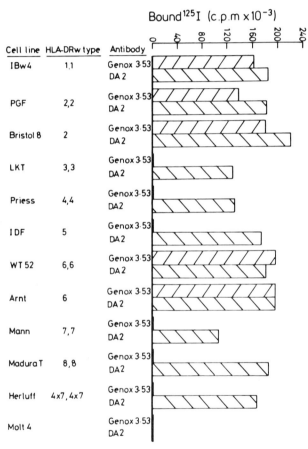

Fig. 4. Binding of monoclonal anti-HLA-DR antibodies Genox 3.53 and DA2 to cell lines with different HLA-DRW specificities. The assay was done under saturating conditions using 5×10^5 cells. Each bar is a mean of duplicates, and background binding is subtracted. The T cell line, Molt 4, is included as a control. All the remaining lines are EBV-transformed lymphoblastoid cell lines. DA2 binds uniformly to all B cells tested and is a monomorphic anti-HLA-DRW antibody. Genox 3.53 binds only to those lines expressing HLA-DRW 1, 2, or 6 (from reference 18).

namely, reacting with DRW1, 2 and 6, as expressed by the mouse anti-human antibody, 3.53. The anti-rat Ia antibody, as might be expected, reacts equally well with mouse cells and with rat cells, and by its strain distribution it appears to segregate with I-A. The most remarkable feature of this anti-rat Ia antibody is that it detects polymorphic differences, which if not identical are at least closely related, in such widely differing species as man, rat, or mouse. This is analogous to the existence of antisera produced in one species which, in a number of cases, have been shown to react with polymorphic determinants in widely different species. This phenomenon led to the suggestion that some of the apparent allelism in complex genetic systems, such as HLA, may really reflect polymorphism for the control of the expression of two or more tandemly duplicated genes [35]. The possibility that the same, or even similar, allelic polymorphic differences now found within widely separated species predate their evolutionary separation is not easily accommodated by present concepts in population genetics or by data on protein evolution. The observation is easily explained, however, if the apparent allelic products are really different gene products, related by duplication and subsequent divergence, such as, for example, the β and δ chains of haemoglobin. The analogous situation for haemoglobins would be that there was polymorphism with respect to the expression of these two chains, so that some individuals expressed only the haemoglobin β chain and others expressed only the δ chain. I shall return to these ideas after a discussion of data on the species distributions of reactivity of these and other monoclonal antibodies and their implications.

Earlier data from our laboratory [25] suggested that the 33,000 molecular weight glycopolypeptide constituent of the DRW antigens was coded for in the HLA region, whereas that for the other chain was coded for elsewhere and was probably not polymorphic. This issue is not yet clearly resolved in man, nor indeed, in the mouse. While some have supported our original suggestion [36], others [37, 38] have suggested that it is the 28,000 molecular weight chain that carries the polymorphic determinants. Preliminary data obtained from peptide mapping of the products from different homozygous lymphoid lines indicate that the 28,000 chain shows more variability than the 33,000 chain, but the latter may vary, at least to some extent (M. C. Crumpton, personal communication). Similar data have also been obtained by Strominger and colleagues (personal communication [39]). In the mouse, while the I-A and I-E/C products appear superficially to be quite similar, there is now good evidence that their two constituent polypeptide chains, which are analagous to the 33,000 and 28,000 chains of the HLA-DRW products, differ quite substantially from each other [40–42]. The data of these authors, together with those of Jones et al [43], suggest that while both of the I-A chains may be controlled by the H-2I-A genetic region, the smaller molecular weight, or β chain of the I-E/C product is also controlled by the I-A region, leaving only the larger molecular weight chain, α, of the I-E/C product, apparently controlled by the I-E/C genetic region. This is unusual since, as pointed out by Barnstable et al [25], in other cases where polymeric proteins are made up of different subunits, the genes for these are unlinked. Exceptions to this empirical rule occur when, as in the case of insulin and C4, a single polypeptide is processed to form two or more separable subunits in the final molecule. There is, at least so far, no evidence of this for the Ia antigens, and so the distinct possibility remains that the I-E/C region is not coding for the α polypeptide chain but is controlling its expression, possibly from a genetic region quite remote from H-2. Chemical comparison of the human DR and mouse Ia antigens clearly indicates that it is the I-E/C product of the mouse that corresponds to the human HLA-DRW products so far identified and analysed [39, 44]. In this respect, therefore, the data on the anti-rat Ia antibody, which identify a polymorphic DRW determinant but maps in the mouse to the I-A region, are very suggestive. Taken together with the above information, it indicates that the human polymorphic DRW chain corresponds to the mouse I-E/C β chain coded for by the mouse I-A region. This would suggest, therefore, in contrast to Barnstable et al [25], that it is the 28,000 molecular weight chain that controls the polymorphism and is coded for by *HLA-DR*, and it leaves open the possibility that the other chain is coded for by a gene outside the HLA region. An added complication is that Tosi et al [38] have suggested the existence of another product, which they call DC-1, which is different from the presently identified DRW products but is controlled by a gene closely linked to it. In our laboratory we have obtained some preliminary evidence for the existence of an extra product on certain cell lines by looking at the ratio of reactivities of the monomorphic anti-DRW antibody, DA2, and the polymorphic antibody, 3.53 [Brodsky et al, in preparation]. These products may have a relationship to the HLA-DRW products analogous to that between the HLA-C and HLA-B products. Clearly, further chemical and genetic characterization of these products, especially using appropriate somatic cell hybrids, should resolve these issues and eventually, no doubt, will identify the human equivalents of the other mouse I region products.

SPECIES DISTRIBUTION OF MONOCLONAL ANTIBODY REACTIONS

Major histocompatibility systems homologous to HLA and H-2 have been found in a wide variety of other species [45]. The patterns of reaction of monoclonal antibodies in different species are beginning to help unravel the relationships between their major histocompatibility systems. These patterns have also contributed to the further definition of the determinants recognised by the monoclonal antibodies and have, as in the case of the anti-rat Ia antibody and others now to be discussed, posed some intriguing problems about the nature of the HLA region.

The monoclonal anti-$\beta 2$m antibodies so far react only with the African higher primates — namely, chimpanzee and gorilla — in addition to human cells, and do not react with cells from orangutan or gibbon or any more distantly related species. This suggests that the mouse antibodies are picking up a difference between human and mouse $\beta 2$m that has arisen quite recently, in evolutionary terms. The monomorphic anti-HLA-A, B, C antibody, W6/32, however, shows a more widely distributed pattern of reactions. It reacts with all higher primate and Old World monkey cells tested, but, among other species more distantly related, it reacts only with a particular subset of the New World owl monkeys defined by their karyotypes [46]. Thus, antibody W6/32, which is monomorphic in humans, and probably in all higher primates and Old World species of monkeys, is nevertheless polymorphic in the owl monkey. This may, perhaps, be expected for a determinant that arose near to the time at which the Old and New World monkeys diverged, placing the owl monkey near the dividing line between those species that do and those that do not react with this antibody. A somewhat different but related pattern is shown by another monomorphic anti-HLA-A, B, C antibody (PA2.5), which reacts with all primates, including higher primates and New and Old World monkeys but is polymorphic in the New World owl and spider monkeys [46]. An example of another pattern of reaction with a monoclonal antibody that is predominantly monomorphic against HLA-A, and B, is shown in Table V. When tested on a panel of some 20 human cell lines, the antibody failed to react with only one line, which was homozygous *AW32, B13* but, strangely, showed no evidence of any other specificity. The antibody is more effectively inhibited by papain-solubilized *HLA-A* than *HLA-B* locus antigens, suggesting that it detects a determinant that once was common to both loci but has changed somewhat in the *B* locus product and appears to be absent specifically from AW32 and B13. It is always possible, of course, that a difference in re-

TABLE V. Reactions of Primate Lymphocytes (5×10^5) With PA2.2 Antibody in a Trace Binding Assay

Species	Number of animals	Positive	Negative
Chimpanzee	8	8	0
Baboon	2	1	1
Rhesus monkey	2	1	1
Owl monkey	24	0	24
Spider monkey	12	6	6
Cotton top marmoset	6	6	0
Tamarin	5	5	0
Cebus	Pooled cells for 9 animals		
Squirrel monkey	Pooled cells for 20 animals		

From reference 18.

activity such as that shown by the PA2.2 antibody for *A* and *B* locus products is due not so much to a change in the amino acid sequence controlling the determinant recognised by the antibody as to changes elsewhere in the molecule. These could, for instance, affect conformation in such a way as to lead to weaker binding to the specific region of the molecule, which itself is still determined by the same amino acid sequence. The three-dimensional conformation of a protein may lead to complex relationships between amino acid variations in noncontiguous regions of a molecule and the specification of an antigenic determinant. The antibodies, although they are very valuable as preliminary probes for protein structure and evolutionary divergence, are not substitutes for a knowledge of the complete amino acid sequence and three-dimensional structure of a protein. The pattern of reactions of antibody PA2.2 on lymphocytes from a variety of primates differs markedly from the other antibodies so far discussed. It is polymorphic in some Old World monkey species (even though the numbers tested are small), and in the New World spider monkeys. It is positive with all cells

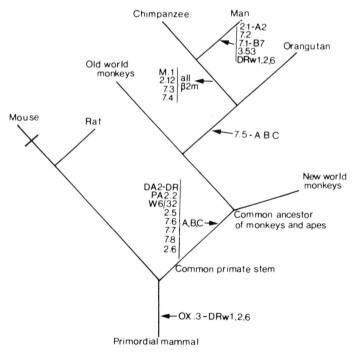

Fig. 5. Appearance of antigenic determinants defined by monoclonal antibodies during primate evolution. 2.1, 2.5, 2.6, and 2.12 are PA2.1, 5.6, and 12, respectively; M1 is BBM1, and 7.1, 2, 3, 4, 6, 7, 8 are the same as BB7.1 etc (based on data from reference 18).

tested from some New World monkey species and negative with others. In some respects, the antibody is similar to the anti-rat Ia antibody discussed earlier, in that it appears to detect polymorphisms in relatively widely separated species. The fact that polymorphic variation is fixed in some species and lost in others is not particularly surprising.

A summary of the patterns of reaction of a variety of monoclonal antibodies, including those already referred to, in the form of a much simplified evolutionary tree, is shown in Figure 5. The antibodies to the human polymorphic determinants, which do not react with any other species, presumably identify differences that have arisen comparatively recently, after man diverged from the higher primates, and such patterns of reaction are generally expected. The patterns of reactivity of the anti-$\beta 2m$ antibodies, and the interpretation of antibodies such as OX3 and perhaps PA2.2, which detect polymorphisms in widely differing species, have already been discussed. There is one further feature of the evolutionary pattern of reactions that at first sight is particularly puzzling. This is the fact that antibodies such as W6/32 and 7.5, which have been shown to react with the products of all three of the *HLA-A, B,* and *C* loci, only react with a limited range of primary species. This is particularly striking for 7.5, which does not react even with Old World monkeys. Such antibodies must recognise determinants that have arisen comparatively recently in evolutionary terms. If one assumes a simple gene duplication model for the origin of the *HLA-A, B,* and *C* loci, which clearly from all the biochemical data produce very similar products, then the results with antibodies such as 7.5 would imply that duplication (or triplication) of the loci occurred, at the earliest, after the separation of the line leading from the Old World monkeys to the higher primates and man. Extensive data on the major histocompatibility systems of a variety of mammalian species suggest, however, that for all those sufficiently studied, there exists at least a similar duplicated genetic organization to the mouse *H-2K* and *D*, and human *HLA-A* and *B* loci. In other words, it seems most likely that these duplications are comparatively old in mammalian evolutionary terms [45]. If the assumption is correct that the triplication of the *HLA-A, B,* and *C* loci, and their equivalents in other species, occurred early in mammalian evolution, if not predating the mammals themselves, then this must have happened well before the evolutionary divergence of the higher primates. In this case, sites such as that recognised by antibody 7.5, once having arisen in one of the three genes, would have to have been "spread" to the appropriate sequences for the other gene products, which seems an exceedingly unlikely event. A simple alternative explanation that can now be envisaged in the light of recent advances in our knowledge of gene structure at the DNA level (see Leder, this volume [47]), is that the three gene products have a common region coded for by the same nucleotide sequence. The enigmatic mouse polymorphic determinant H-2.28 and its partner, H-2.1, raise an exactly analogous

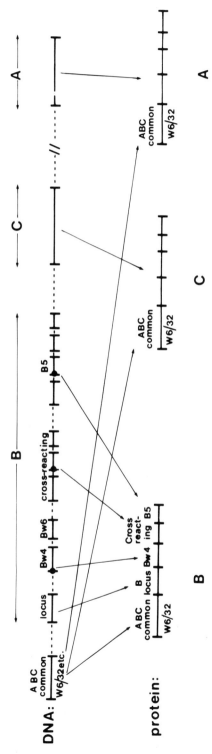

Fig. 6. Scheme for the production of HLA-A,B,C products from different regions corresponding to different domains. Continuous lines in the DNA sequence correspond to coding regions, which may, however, contain intervening sequences. The dotted lines are flanking regions or intervening sequences. The large dot at the left of a region, such as Bw4, indicates that this is the one, of a series of alternatives, that is expressed. While the details are not shown, the derivation of the C and A protein products is assumed to be similar to that for B. Arrows from DNA to protein indicate which DNA region is coding for a given protein domain. The double slash between A and C indicates the fact that other regions, probably corresponding to Q and TL, lie between A and C.

question for the three products, H-2K, D, and L; namely, how can a given polymorphic specificity be found on three different gene products. A similar explanation was given by Demant et al [10] to this problem as to the evolutionary one raised by antibodies such as 7.5 — namely, in terms of a region common to the three gene products. Indeed, in the owl monkey, where antibody W6/32 is polymorphic, it gives rise to an exactly analogous situation to H-2.28 and H-2.1 in the mouse. So far at least, there seem to be no examples in man of polymorphic variants common to the three loci, *HLA-A, B,* and *C*, and this has certainly simplified their serological description.

THE HLA REGION, DOMAINS AND GENES IN PIECES

The recent advances in our understanding of gene structure at the DNA level have completely changed our view of the relationship between DNA sequence and protein product. No longer is it possible to think in terms of a simple one-to-one relationship between a given amino acid sequence and its corresponding nucleotide sequence in the DNA. Now we know that coding sequences are interrupted by intervening noncoding sequences, that separate coding sequences may correspond to different domains of a protein, and that a given coding region may be associated with different noncontiguous coding regions and so contribute to two or more gene products. We have, therefore, as Blake [48] put it, to think in terms of "genes in pieces" corresponding to "proteins in pieces." The following are some speculative suggestions as to how the HLA region, or at least part of it, may look, taking into account these notions.

Previously known serological patterns of cross-reactivity between HLA determinants, the new patterns of complexity revealed by monoclonal antibodies, and data on the chemical structure of the HLA-A, B, C determinants, as obtained particularly by Strominger et al [39], all suggest that these products can be subdivided into molecular regions corresponding to distinguishable domains, analogous to the domains of other complex proteins such as the immunoglobulins. In the absence of any detailed understanding of the relationship between antigenic sites and protein structure for these particular products, a minimum set of domains can, perhaps, most easily be defined in terms of patterns of serological reactivity. Thus, antibodies such as W6/32 and 7.5 identify "class" specific determinants common to the HLA-A, B, and C products. Antibodies such as PA2.2 suggest the existence of locus-specific determinants. Patterns of cross-reaction among the presumed allelic products within any one series, such as HLA-A or HLA-B, are well known [9], and these define another class of determinants, to which may be added the Bw4 and Bw6 (4a and 4b) antigens associated with the *HLA-B* locus. These two antigens constituted the first genetic system described by van Rood in 1962 and are part of the HLA-B series in the sense that each *B* locus product carries with it either the Bw4 or the Bw6 determinant. Lastly, there is, of course, the class of allele-specific determinants by which the various alleles are primarily identified. Cross-blocking studies such as those carried out by Legrand and Dausset [49] suggest, as discussed by Brodsky et al [18], that these antigenic determinants may be on clearly separable regions of the protein product, supporting the notion that they may lie in different molecular domains of the polypeptide chain.

Assuming that each domain may correspond to a separate coding region (no doubt perhaps with its own intervening sequences), that such regions can be put together in different combinations and can sometimes be used to make more than one protein product, and that polymorphism for control should now be interpreted in terms of regions corresponding to a domain of a protein rather than whole genes, we are led to the highly schematic view of a part of the HLA region shown in Figure 6. The "A, B, C, common" region is presumed to code for the antigenic sites corresponding to antibodies such as W6/32 and 7.5 and contributes to the formation of each of the A, B, and C products. An expanded version of the B region suggests separate subregions for the locus, the Bw4 and Bw6, the cross-reacting, and the allelic specificities, with some duplications, of which only one (in this case Bw4, with its particular cross-reacting determinant and B5) is expressed. The picture is obviously highly speculative and poses a number of questions. If the loci in the *Q* and *TL* regions lie between *A* and *C*, will they have the same common region as A, B, and C? Is it correct to group the regions for each "locus"? Should the grouping rather, for example, be by domains, putting together say the A, B, and C locus-specific regions, then the various cross-reacting regions, and finally the allele-specific regions? How does the splicing of the A, B, C, common region with the locus-specific regions take place over what appears to be, in terms of nucleotides, rather a long distance, and at what level does the splicing take place? Though association at the DNA level, analogous to that involved in the V-C joining of immunoglobulins, cannot be ruled out, it seems more likely that most of the splicing takes place at the RNA level, as has been observed in other systems. In this case, given that the recombination fraction between *A* and *B* is of the order of 0.08%, and so represents a distance that may be as much as a million nucleotides, it seems unlikely that the whole region is transcribed into a single giant initial transcript that is subsequently processed, because there is no evidence at this time for transcripts of this size. Is it possible to read off smaller transcripts that include the A, B, C common region and other regions far away from it by looping out a major part of the intervening nucleotide sequence? What is the nature of the signal that determines which of a set of duplicates is chosen to be expressed by any given chromosome?

A similar picture may presumably hold for HLA-DR and their associated products, as well as for the complement components coded for in the HLA region. In particular, if it is assumed that the DRW-1,2,6 cross-reacting site corresponds to a separate domain of the DRW product with its own coding region, and that this associates in different individuals with a region and domain corresponding to the allelic specificities DRW1, 2, and 6, then it is easy to explain how the antihuman DR antibody, 3.53, reacts only with humans while the anti-rat antibody OX3, with apparently the same specificity, reacts with cells from mouse, rat, and man. The 3.53 antibody presumably reacts with a site created by a comparatively recent mutation in the DRW-1,2,6 cross-reacting region, whereas the anti-rat antibody reacts with a determinant that distinguishes DRW-1,2,6 from other crossreacting domains, and corresponds to a difference within the region that may have originated very early during mammalian evolution.

Can the origin of the HLA system be explained in terms of duplication of a single, perhaps admittedly complex, nucleotide sequence? One possible answer to this question has been suggested by Bodmer [50]. When a complex region, such as that depicted in Figure 6 for the HLA-A, B, and C determinants, itself duplicates, there are many ways in which completely different protein products could be formed from the duplicated sequence. Regions could be inverted and will diverge by the normal evolutionary processes. Regions could be read in different reading frames, and unrelated "insertion sequences" could be introduced into a duplicate that did not exist in the original sequence. Allowing for all these possibilities, a gene cluster such as HLA could become quite complex and yield a number of apparently unrelated protein products.

The final answer to all these intriguing questions may only come when we have established the nucleotide sequence of the HLA region using DNA cloning techniques and matched this to the various known gene products. The challenge is in some ways comparable to that of sequencing the whole of the genome of E coli, since a rough estimate of the size of the HLA region would be about three million nucleotide pairs. Duplication, however, renders the complexity of the HLA region less than that of E coli, and the major histocompatibility system gene cluster may well be at the limit of the complexity to be expected for gene clusters of eukaryotes. Because of my interests in the HLA system and human work, the emphasis of this discussion has been on the HLA rather than the H-2 system. The similarities between the two systems and presumably between all the mammalian major histocompatibility systems, mean however that models based on any one system should be applicable to others. When we understand how the complex genetic major histocompatibility region works from the level of DNA, through RNA, to protein, the cell, the individual, and the population, we shall have solved many problems at different levels of biological complexity, and no doubt we will have found many more surprises on the way.

ACKNOWLEDGMENTS

I am particularly grateful to Frances Brodsky, Peter Parham, and Colin Barnstable for their data on the monoclonal antibodies and for many helpful discussions. The original work reported in this paper was supported, in part, by a grant from the UK Medical Research Council.

REFERENCES

1. Bodmer WF: HLA:A super supergene. The Harvey Lectures 72:91–138, 1978.
2. Snell GD: The major histocompatibility complex: Its evolution and involvement in cellular immunity. The Harvey Lectures, Vol. 74, pp 49–80, 1980.
3. Van Rood JJ (ed): "Histocompatibility Testing." Copenhagen: Munksgaard, 1965.
4. Curtoni ES, Mattiuz PL, Tosi RM (eds): "Histocompatibility Testing." Copenhagen: Munksgaard, 1967.
5. Terasaki PI (ed): "Histocompatibility Testing." Copenhagen: Munksgaard, 1970.
6. Dausset J, Colombani J (eds): "Histocompatibility Testing." Copenhagen: Munksgaard, 1972.
7. Kissmeyer-Nielsen F (ed): "Histocompatibility Testing." Copenhagen: Munksgaard, 1975.
8. Bodmer WF, Batchelor JR, Bodmer JG, Festenstein H, Morris PJ (eds): "Histocompatibility Testing." Copenhagen: Munksgaard, 1977.
9. Barnstable CJ, Jones EA, Bodmer WF: "Genetic Structure of Major Histocompatibility Regions." In Lennox ES (ed): International Review of Biochemistry, "Defence and Recognition. IIA. Vol 22, Cellular Aspects." Baltimore: University Park Press, 1979, p 151.
10. Demant P, Ivanyi D, Neauport-Sautes C, Snoek M: H-2.228, an alloantigenic marker allelic to H-2.1, is expressed on all three known types of H-2 molecules. Proc Natl Acad Sci USA 75:4441–4445, 1978.
11. O'Neill GJ, Young-Yang S, Tegoli J, Berger R, Dupont B: Chido and Rodgers blood groups are distinct antigenic components of human complement. Nature 273:668–670, 1978.
12. O'Neill GJ, Young-Yang S, Dupont B: Two HLA-linked loci controlling the fourth component of human complement. Proc Natl Acad Sci USA 75:5165–5169, 1978.
13. Roos MH, Atkinson JP, Shreffler DC: Molecular characterization of the Ss and Slp(C4) proteins of the mouse H-2 complex: Subunit composition, chain size polymorphism, and an intracellular (PRO-Ss) precursor. J Immunol 121:1106–1115, 1978.
14. Harris H: "The Principles of Human Biochemical Genetics," Ed 2. Amsterdam: North-Holland, 1975.
15. Fisher RA: "The Genetical Theory of Natural Selection." London: Oxford University Press, 1930.
16. Bodmer WF, Bodmer JG: Evolution and function of the HLA system. Br Med Bull 34: 309–316, 1978.
17. McDevitt HO, Bodmer WF: HL-A, immunoresponse genes, and disease. Lancet 1:1269, 1974.
18. Brodsky FM, Parham P, Barnstable CJ, Crumpton MJ, Bodmer WF: "Monoclonal Antibodies for Analysis of the HLA System." Immunological Rev 47:3–61. Copenhagen: Munksgaard, 1979.
19. Harris R, Zervas JD: Reticulocyte HLA antigens. Nature 221:1062, 1969.
20. Snary D, Barnstable C, Bodmer WF, Goodfellow P, Crumpton MJ: Human Ia Antigens – Purification and molecular structure. Cold Spring Harbor Symp Quant Biol 41:379–386, 1976.
21. Springer TA, Kaufman JF, Siddoway LA, Giphart M, Mann DL, Terhorst C, Strominger JL: Chemical and immunological characterization of HL-A-linked B-lymphocyte alloantigens. Cold Spring Harbor Symp Quant Biol 41:387–396, 1976.

22. Porter RR: Complement. Int. Review of Biochemistry. "Defense and recognition. IIB." Baltimore: University Park Press, p 23, 1979.
23. Zinkernagel RM, Doherty PC: Restriction of in vitro T cell-mediated cytotoxicity in lymphocytic choriomeningitis within a syngeneic or allogeneic system. Nature 248:701, 1974.
24. Snell GD: T cells, T cell recognition structures, and the major histocompatibility complex. Immunol Rev 38:3–69, 1978.
25. Barnstable CJ, Bodmer WF, Bodmer JG, Arce-Gomez B, Snary D, Crumpton MJ: Genetics and serology of the HL-A linked human Ia antigens. Cold Spring Harbor Symp Quant Biol 41:443, 1977.
26. Bodmer WF: Evolutionary significance of the HL-A system. Nature 237:139, 1972.
27. Kohler G, Milstein C: Continuous cultures of fused cells secreting antibody of predefined specificity. Nature 256:495, 1975.
28. Kohler G, Milstein C: Derivation of specific antibody-producing tissue culture and tumor lines by cell fusion. Eur J Immunol 6:511, 1976.
29. Williams AF: Differentiation antigens of the lymphocyte cell surface. Contemp Topics Mol Immunol 6:83, 1977.
30. Barnstable CJ, Bodmer WF, Brown G, Galfré G, Milstein C, Williams AF, Zeigler A: Production of monoclonal antibodies to Group A erythrocytes HLA and other-human cell surface antigens – new tools for genetic analysis. Cell 14:9, 1978.
31. Tada N, Tanagaki N, Pressman D: Human cell membrane components bound to β^2-microglobulin in T cell-type cell lines. J Immunol 120:513, 1978.
32. Bodmer JG: Ia antigens. Definition of the HLA-DRw specificities. Br Med Bull 34:233, 1978.
33. McMaster WR, Williams AF: Identification of Ia glycoproteins in rat thymus and purification from rat spleen. Eur J Immunol 9:426–433, 1979.
34. McMaster WR, Winearls BC, Parham P: A monoclonal mouse anti-rat Ia antibody which cross reacts with a human HLA-DRw determinant. Tissue Antigens 14:5, 453–458, 1979.
35. Bodmer AF: A new genetic model for allelism at histocompatibility and other complex loci: Polymorphism for control of gene expression. Transplant Proc 5:1471, 1973.
36. Klareskog L, Rask L, Fohlman J, Peterson PA: Heavy HLA-DR (Ia) antigen chain is controlled by the MHC region. Nature 275:762–764, 1978.
37. Silver J, Ferrone S: Structural polymorphism of human DR antigens. Nature 279:436–437, 1979.
38. Tosi R, Tanigaki N, Centis D, Ferrara GB, Pressman D: Immunological dissection of human Ia molecules. J Exp Med 148:1592, 1978.
39. Strominger JL, Engelhard VH, Fuks A, Guild BC, Hyafil F, Kaufman JF, Korman AJ, Kostyk TG, Krangel MS, Lancet D, Lopez de Castro JA, Mann DL, Orr HT, Parham PR, Parker KC, Ploegh HL, Pober JS, Robb RJ, Shackelford DA: Structure of MHC products. In Benacerraf B, Dorf ME (eds): "The Role of the Major Histocompatibility Complex in Immunobiology." New York: Garland Press Publishing, Inc., 1980.
40. Cook RG, Uhr JW, Capra JD, Vitetta ES: Structural studies on the murine Ia alloantigens. J Immunol 121:2205–2212, 1978.
41. McMillan M, Cecka JM, Hood L: Peptide map analyses of murine Ia antigens of the I-E subregion using HPLC. Nature 277:663–665, 1979.
42. Silver J, Russell WA: Structural polymorphism of I-E subregion antigens determined by a gene in the H-2K to I-B genetic interval. Nature 279:437–439, 1979.
43. Jones PP, Murphy DB, McDevitt HO: Two-gene control of the expression of a murine Ia antigen. J Exp Med 148:925–939, 1978.

44. Allison JP, Walker LE, Russell WA, Pellegrino MA, Ferrone S, Reisfeld RA, Frelinger JA, Silver J: Murine Ia and human DR antigens: Homology of amino-terminal sequences. Proc Natl Acad Sci USA 75:3953–3956, 1978.
45. Gotze D (ed): "The Major Histocompatibility System in Man and Animals." Berlin: Springer-Verlag, 1977.
46. Parham P, Sehgal PK, Brodsky FM: Anti-HLA-A,B,C monoclonal antibodies with no alloantigenic specificity in humans define polymorphisms in other primate species. Nature 279:639, 1979.
47. Crick F: Split genes and RNA splicing. Science 204:264, 1979.
48. Blake CCF: Do genes-in-pieces imply proteins-in-pieces? Nature 273:267, 1978.
49. Legrand L, Dausset J: Serological evidence of the existence of several antigenic determinants (or factors) on the HL-A gene products. In Dausset J, Colombani J (eds): "Histocompatibility Testing." Copenhagen: Munksgaard, 1972, p 441.
50. Bodmer WF: Gene clusters and the HLA system. Ciba Foundation Symposium 66:205–229, 1979.

Mammalian Genetics and Cancer: The Jackson Laboratory Fiftieth
Anniversary Symposium, pages 241–272
© 1981 Alan R. Liss, Inc., 150 Fifth Avenue, New York, NY 10011

The Future of Immunogenetics

George D. Snell

INTRODUCTION

It is a pleasure and a privilege to be a participant in this Fiftieth Anniversary Symposium of The Jackson Laboratory. The Laboratory has provided many people an opportunity to seek answers to the riddles of mammalian genetics and cancer in ways of their own choosing. It is perhaps unique in the congeniality and openness of its environment. All of us who have worked here owe it and Dr. Clarence Little an enormous debt of gratitude. I also want to express my personal gratitude to the many wonderful people I have been privileged to work with over the years, whose major contributions and friendship have meant so much.

When Dr. Russell asked me to speak on the future of immunogenetics, I thought this would be a stimulating and challenging job. It has been. I wish we could all board a Wellsian time machine and get a glimpse of what immunogenetics really will be like in 2029. That route being closed to us, I turn to the more prosaic method of examining the past and extrapolating from it.

In 1929 I was just finishing my graduate studies at the Bussey Institution

Abbreviations and symbols: CMAD loci = cell membrane alloantigen determining loci; H loci = histocompatibility loci; HLA = the human major histocompatability complex; HLA-A, HLA-B = two of the major histocompatibility loci within HLA; HLA-DR = antigens of human cells corresponding to the mouse Ia antigens; H-2 complex = the major histocompatibility complex of the mouse; H-$2K$, H-$2D$, H-$2L$ = the histocompatibility loci that define the H-2 complex, sometimes abbreviated K,D,L; I region = a region of the H-2 complex between H-$2K$ and H-$2D$ that contains the Ia loci and has a major influence on the immune response; I-A, I-J = subregions of the I region; Ia antigens = antigens determined by the I region of H-2 and found on lymphocytes, macrophages, and a few other cell types; K,D,L = an abbreviation for the three major histocompatibility antigens, H-2K, H-2D, H-2L, determined by the H-2 complex; MHC = major histocompatibility complex; NK cells = natural killer cells (lymphocytes); Qa–Tla = a region of murine chromosome 17 adjacent to H-2 that contains the Qa, Qat, and Tla loci; SD loci = serologically defined loci; T cells = lymphocytes of thymic origin; Thy-1 = a murine locus whose end product occurs on T cells and some brain cells.

under Dr. Castle. Immunogenetics had been born four years earlier, when Bernstein [1] showed that the ABO blood groups are determined by three alleles at one locus. A fellow student at the Bussey, Bob Irwin, would christen the infant science seven years later when he and Cole [2] coined the term *immunogenetics*. In the same year, Gorer [3] would discover *H-2*.

At least until 1949, most vertebrate immunogenetics dealt with blood groups, with humans, cattle, and dove–pigeon hybrids being the favored species [4]. The fundamental studies of Little and co-workers on transplantation genetics [5, 6] constituted a major exception with, however, the qualification that they did not identify individual genes and that, since antisera were not employed, theirs has been referred to as histogenetic rather than immunogenetic research.

Gradually, the pace of immunogenetic research picked up. Two important turning points were the discovery of the thymus leukemia (TL or Tla) antigen, the first known lymphocyte antigen not determined by *H-2*, by Boyse et al [7], and the identification of McDevitt and Chinitz [8] of an immune response region within the *H-2* complex. Now the volume of research has become such that at least eight books have been written about a bit of the 17th chromosome of the mouse that measures about two centimorgans in length [9–17].

The pace of the past ten years shows no signs of abating. Barring a major cut in research funding, it should continue at the present level or even accelerate well into the twenty-first century. The resulting output of knowledge will be enormous.

How much of this can we hope to foresee? I see five areas in which the forecaster has a reasonable chance of success and one in which he has very little.

To take the negative case first, I see almost no prospect of predicting the real breakthroughs. By their very nature, these are unexpected. As Philip Handler [18] put it, "The only certainty [about what we shall learn in the field of biology] seems to be that we shall be surprised." This obviously leaves a huge gap in our forecasting.

I turn now to the positive. First, we can take existing trends and project them into the future. Second, we can look for important questions raised by recent discoveries and predict that they will be answered. Third, if we want to stick our necks out, we can guess what the answers will be. We can, fourth, look for areas of interaction that should generate fruitful interdisciplinary studies, and perhaps get a few glimpses of where they will lead. And finally, in the light of recent developments in mammalian immunogenetics, we can point out possibilities for important medical applications.

DISCOVERY OF CMAD GENES

At the center of mammalian immunogenetics stand those loci which determine cell membrane alloantigens (CMAD loci). Aside from studies of human blood groups and certain aspects of the HLA complex, work on the CMAD loci of mice has far outstripped that done on other mammals. I shall confine my projections to the mouse.

The murine CMAD loci can be divided into three categories: 1) histocompatibility loci, 2) loci identified by serological methods, and 3) loci that are part of or clearly associated with *H-2*, the murine major histocompatibility complex. Group 3 includes both histocompatibility loci and loci that are serologically identified, so that it overlaps groups 1 and 2, but because of its unique nature it is appropriate to consider it separately. Group 2 can be divided into 1) blood group loci, 2) loci whose end product is demonstrated by lymphocyte cytotoxicity, and 3) loci whose end product is identified on cells other than red blood cells and lymphocytes.

Figures 1 and 2 show the annual growth in the number of known loci in these different categories. Data for 1979, since incomplete, are not included. In some cases the curve is projected to 1980.

Histocompatibility (*H*) loci, which have been the major interest of Dr. Bailey and me, show both the longest history and the highest return to date in terms of the number of loci discovered (Fig. 1). Fifty loci are now known, exclusive of those in the *H-2* complex. Graff and Brown [20], using the classical method of transplants to segrating generations, but with the added factor of preimmuniza-

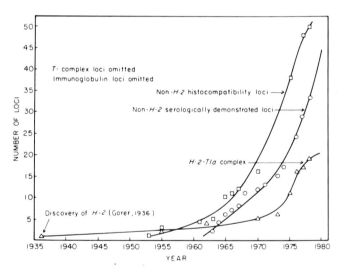

Fig. 1. Growth in the number of known alloantigen-determining loci in the mouse. *T*-complex loci, part of the *H-2–Tla* complex family, and the immunoglobulin loci, a large but numerically poorly defined portion of the non-*H-2* serologically demonstrated group, are omitted. Data are taken from many sources. Pertinent references are reviewed in Snell et al [16] and Snell [19].

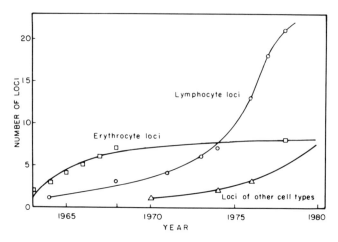

Fig. 2. Growth in the number of known non-*H-2* serologically demonstrated loci.

tion against the skin donor, found that strains B10.D2 and DBA/2, both *H-2^d*, differ at 88 ± 5.7 non-*H-2 H* loci. There are probably additional loci at which B10.D2 and DBA/2 share the same *H* allele but at which allelic differences occur in other strain pairs. Perhaps it is a reasonable estimate that the number of CMAD loci, exclusive of the *H-2* group, which will be demonstrated by skin grafting techniques is about 100. There is, in fact, a suggestion in Figure 1 that the rate of discovery of *H* loci has begun to slow down. Many of the future discoveries are likely to come from the analysis of complexes of closely linked genes [21].

Figure 1 also suggests that the rate of discovery of loci in the *H-2* complex is slowing. The number now stands at 21. I should note that this figure includes two loci, *Ss* and *Slp,* whose end products are immunologically demonstrated serum proteins rather than cell membrane alloantigens [22]. Such loci constitute a small but important part of immunogenetically identified loci. Certainly the number of known loci in the *H-2–Tla* region will increase further, but major emphasis has already shifted to an understanding of the chemistry and functions of the *H-2–Tla* end products.

One curve in Figure 1 unquestionably is still on the way up. This is the curve showing the serologically demonstrated (SD) non-*H-2* loci. The number of known loci in this category now stands at 33. This figure does not include the three clusters of serologically demonstrated immunoglobulin loci. There is a substantial number of these, but many of them are poorly defined [23]. I should note that in all the curves there is some uncertainty as to the validity of the evidence for, or concerning the distinctness of, a few of the loci included, and that this uncertainty is particularly great in the case of the SD non-*H-2* loci. Thus, I have treated *Ly-6, Ly-8,* and *Ala-1* as distinct loci, though there are no known crossovers between them and their properties are very similar [24, 25]. Both the *H-2* complex and the *Lyt-2, Lyt-3* locus pair [26] provide precedents, however, for linked, but distinct, loci with related end products. Such clusters are likely to prove of increasing interest.

Extrapolation from the known growth rate of SD loci will be assisted if we break the loci into subcategories. This is done in Figure 2. As the figure shows, the blood group curve has already tapered off. Intensive research could doubtless double the number of known loci, but on the basis of the thoroughly analyzed human blood groups, we can hardly expect much more than this. The lymphocyte alloantigen curve is still moving sharply up, with 21 loci already known, not counting the immunoglobulin loci. The curve now shows a hint of leveling, and I suspect that this may be real. The third curve is derived from a mere three loci, which is a poor basis for extrapolation. I think, however, that we can argue from the generous harvest reaped from the study of red cells and lymphocytes that future immunogenetic research on the dozens or hundreds of other cell types in the body will prove enormously fruitful. The obvious technical limitation that slows up progress is the problem of obtaining suspensions of free cells. I gather from talking to Charity Waymouth that tissue culture techniques appropriate to solving this problem are already far advanced. An example is a recent description of a serum-free medium that permits the culture of a kidney epithelium cell line in the absence of fibroblast overgrowth [27].

Either the epithelium cell or the macrophage is a likely candidate as the next cell type (perhaps more properly, family of cell types) to yield up its surface secrets. One macrophage alloantigen has already been described [28]. Because of its role in immune processes, I think that the macrophage is the cell that I would pick to work on if I were getting back into immunogenetic research.

How many CMAD loci, exclusive of *H* loci and of those loci defined by lymphocyte cytotoxicity, will have been found by 2029? If we can assume that virtually every cell type has its own differentiation antigen or antigens, a reasonable assumption in view of the demonstrated complexity of the differentiation antigens of lymphocytes, the number certainly will be substantial. In 1974, in a bit of crystal gazing, I predicted a total of 500 CMAD loci, 80 of these being *H* loci [29]. This implies at least 350 non-erythroid, non-lymphoid, serologically defined loci. This figure, if anything, now seems conservative.

An important factor in the future discovery of SD CMAD loci certainly will be improvement in relevant techniques. I have already mentioned progress in culture methods for providing single-cell suspensions. With respect to serological techniques, there are two areas in which we can anticipate significant developments. The first concerns the production of monoclonal antibodies through the generation of antibody-secreting hybrid cells. The hybrids are produced by the fusion of an antibody-secreting plasma cell tumor (myeloma) and a spleen cell from an immunized mouse. Appropriate selection of derivative cells after chromosome deletion can produce a hybrid line with the sustained antibody-producing capacity of the myeloma and the specificity of the normal spleen cell. Thus Lemke et al [30] have established three hybridomas from anti-H-2 spleen cells that produce seemingly pure anti-H-2.5, H-2.11, and H-2.25. Work in this area is expanding rapidly and will certainly play an important role in future immunogenetic studies.

A second area in which I foresee improvement in serological techniques is in the production of otherwise elusive antisera through the use of *coimmunization*. Whereas immune response genes (which also offer potential for methodological manipulation) act through the *recipient,* coimmunization acts through the *donor.* It is often found that a donor with an X + Y alloantigenic disparity from the recipient will yield an anti-X, whereas a donor with an X disparity alone or an X + Z disparity will not. One of the first observations suggesting the utility of coimmunization was made by Schierman and McBride [31] in poultry. These authors obtained a good antibody against a weak erythrocyte alloantigen by adding a disparity at a strong erythrocyte locus. Coimmunization has been used for a number of years by Cherry and her co-workers at The Jackson Laboratory. Recent studies by those investigators [32, 33] show that its effect can be very specific, one particular combination working when many others do not. A report by Wernet et al [34] also points to this conclusion. At present, the selection of working combinations is achieved only by trial and error; we know virtually nothing about the principles involved. Advances in this area should have both substantial theoretical and practical importance.

A technological trend likely to continue for a long time is the growing use of recombinant inbred strains for mapping loci or for determining the identity or nonidentity of loci with similar phenotypes [35, 36]. The beauty of recombinant inbred strains is that, like a rare violin, their value increases with use and age. The more genes they have mapped, the easier it is to map new genes. If this value could be converted to cash, Don and Ben would have the perfect hedge against inflation!

Another area of immunogenetic research that we can expect to see continue and probably expand is the study, so brilliantly pioneered by Bailey and Kohn [37] and by Egorov [38], of histocompatibility, and especially *H-2* mutations. One of the many beauties of *H* gene mutations is that they can give truly coisogenic lines; any functional difference between members of a congenic pair can be attributed with assurance to the defining locus.

IMMUNOGENETICS AND DIFFERENTIATION

An aspect of immunogenetic research that is still too undeveloped to call a trend, but that almost certainly will become one, is the study of the role of membrane alloantigens in differentiation. The antigens concerned probably will turn out primarily to be those demonstrated by serological methods. These, except for the H-2K, H-2D, and H-2L products, usually show a restricted tissue distribution. Because of this, Boyse and Old [39], anticipating the trend to which I refer, called them *differentiation antigens*. Firm evidence of their role in differentiation is still lacking, but some interesting clues have come from the study of Thy-1. This antigen, as originally shown by Reif and Allen [40], is found on mouse thymocytes and in brain. It has become, in the mouse, the standard marker for T-lymphocytes [41]. Thy-1 has been found in mouse epidermis and mammary gland and in mouse and rat fibroblasts [42–44], but it is not otherwise widely distributed. While rats, like mice, possess a Thy-1 antigen, the interesting point recently developed is that the tissue distributions shows significant differences. In the mouse Thy-1 is present in all T-lymphocytes, whereas in the rat it appears to be lacking in most peripheral T cells but is present in a substantial number of bone marrow cells [45, 46]. Since the mouse and the rat are closely related species that must follow similar differentiation patterns, this raises interesting questions about the differentiation antigen concept.

Yet there is evidence that Thy-1 is indeed concerned with cell interactions and differentiation. Thus, Acton et al [47] find that, in mouse brains, Thy-1 is associated with synaptosomes and synaptic junctional complexes. This suggests a role in the formation and/or maintenance of synaptic connections. Dulbecco et al [48] have drawn comparable conclusions from a study of Thy-1 in a rat mammary cell line. This line, when cultured in appropriate media, undergoes differentiation into domes or ridges or projections. Thy-1 was found to be localized at the borders between cells in the process of dome formation. When anti-Thy-1 was applied, domes were not formed, and existing domes disappeared. Other antisera reactive with the test cells did not have these effects.

A possible explanation of how a membrane alloantigen can be a differentiation antigen and yet show major interspecies differences in tissue distribution is provided by a chemical study of Barclay et al [49]. These authors found that, whereas the protein portions of the rat Thy-1 from thymus and brain are probably identical, the carbohydrate fractions, which make up about 30%, are very different. This suggests that, while the alloantigenic properties reside in the protein, the differentiation functions are determined by the carbohydrate. Lengerova et al [50] and Pouyssegur et al [51] have made similar suggestions.

I will not venture to guess in just what ways membrane alloantigens will be found to function in differentiation. This is the sort of situation where totally unexpected breakthroughs can be expected. I will predict, however, that this will become an increasingly important area of research, and that immunogenetics will make substantial contributions.

IMMUNOGENETICS AND THE CELL MEMBRANE

Another trend that we can detect and project into the future is the growing contribution of immunogenetics to cell membrane research. It is generally agreed that the great majority of the alloantigens demonstrated by immunogenetic and histogenetic methods are located in or on the membrane. The cell most intensively studied by these methods is the lymphocyte, and the number of known separate molecules that we can now attribute to the membrane of even a single category of lymphocytes is impressive. The problem is complicated by the large number of lymphocyte categories and the imperfections of our knowledge about each of them. We can say safely that every T, or thymus-derived, lymphoid cell carries on its surface the H-2K, H-2D, and H-2L (K,D,L) major histocompatibility antigens [19], something in the order of 100 minor histocompatibility antigens [this paper], and the Thy-1 antigen [52]. There are other antigens that probably are not found on all T cells, but that have been reported on some fraction or some subdivision of the T family. Lyt-1 (originally called Ly-1 or mu), Lyt-2, and Lyt-3 [53–55] undoubtedly will turn out to be markers for T groupings. Besides these three, the list includes Ia antigens determined by the *I-A, I-J,* and perhaps other *I* subregions [15, 56, 57], Ly-4 [58–60], Ly-5 [61, 62], Ly-6 [16, 59, 62–65], Ly-7 [16, 59], Ly-8 (which may be identical with Ly-6) [24, 66], Ala-1 (possibly identical with Ly-6 and Ly-8) [24, 25, 67], several alloantigens determined by X-linked loci [68], Tla [7], Qa-1 [69, 70], Qa-2, and Qa-3 [71], and Qat-4 and Qat-5 [72]. All T lymphocytes also possess an antigen recognition structure probably composed of a typical immunoglobulin variable region and a product of one of the Ia loci [review in 73].

The T cell, as thus conceived, is diagramed in the top half of Figure 3. The bottom half of the figure shows, probably inadequately though in a manner sufficient for our purposes, a typical cell as viewed by the cell physiologist and the cell chemist.

All three viewpoints reveal a complex membrane structure. The immunogeneticist perhaps has most thoroughly documented the diversity of the proteins in the cell membrane; the chemist has spelled out in much the greatest detail the complexity of the lipid bilayer. As to function, physiological studies have told us about receptors and transport proteins, the chemical studies about changes in and the behavior of the lipid bilayer [see, eg, 91] and various enzymes loosely or firmly attached to this bilayer, and the immunogenetical studies have indicated the important role that MHC products play in the immune response. Chemistry reveals the presence of both glycolipids and glycoproteins, with the carbohydrate component apparently always being on the outer surface of the membrane [88]. A number of the proteins revealed by immunogenetics are known to be glycoproteins. Electron microscopy, not specifically included in Figure 3, reveals a complex of structures internal to, but interacting with, the plasma membrane.

The major conclusion to be drawn from Figure 3, it seems to me, is the pressing need for integrating the membrane pictures revealed by the three disciplines we have been considering. Many questions suggest themselves. What is the functional role of the minor histocompatibility antigens? Are some of them receptors or transport proteins or enzymes? Have any of them arisen from viral genes incorporated into the murine genome? Do any of the differentiation antigens have

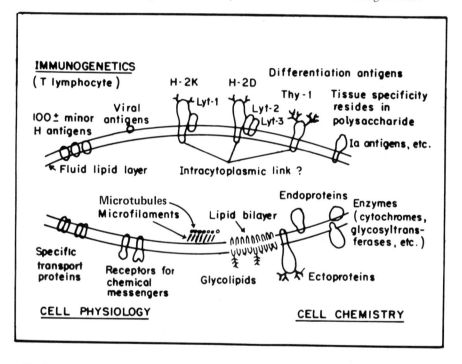

Fig. 3. The cell membrane as seen by workers in three different disciplines. The upper half of the figure is the T-lymphocyte as seen by immunogeneticists. Details are taken from Bailey [21, 74], Dalchau and Fabre [75], Morse et al [76], Edidin and Weiss [77], Graff [78], Koch and Smith [79], Rolink et al [80], Snell [19], Snell et al [16]. There are a number of important omissions, especially the H-2L molecule [81] and the antigen-recognition structures, which appear to be formed from typical immunoglobulin variable regions combined with an *Ia* locus product. The division of the H-2K, H-2D, and Ia antigens into heavy and light chains is also not shown. The lower half of the figure shows the cell membrane as seen by the cell physiologists, major sources being Pardee [82], Ramasamy et al [83], and Rowzkowski et al [84], and by the cell chemist, major sources being Edelman [85], Gulik-Drzywicki [86], Loor [87], Rothman and Lenard [88], Shur and Roth [89], and Singer and Nicolson [90].

enzymatic capability? Are the membrane enzymes of the chemist detectable by immunogenetic methods? Does immunogenetics pick up any of the chemist's endoproteins or any of the many external polysaccharides? (The human ABO blood group substances are, of course, one family of polysaccharides thus detected.) The next few decades should see answers to many of these questions.

There is one area of immunogenetic research still in its infancy, which may add a whole new dimension to our view of membrane structure. This concerns specific groupings of glycoproteins on the cell surface, with the H-2 antigens possibly playing an organizing role. In 1968 Boyse et al [92], using the blocking effect between antibodies applied sequentially to mouse thymocytes, concluded that Lyt-1 is associated with H-2K, Lyt-2 with H-2D, and that Thy-1 is part of both these membrane islands. Flaherty and Zimmerman [93] have confirmed the observations of Boyse et al and have added the finding that, in cells fixed in paraformaldehyde, the Thy-1 antigen is absent from the groupings. The implication is that this particular association occurs only under the impulse of certain external stimuli. These concepts are indicated in our T-lymphocyte membrane diagram (Fig. 3).

There are other antibody-blocking experiments that suggest further complexities in the groupings of proteins on the cell surface. Thus, Nakayama et al [94] found that either anti-Lyt-2 or anti-Lyt-3, but not anti-Lyt-1, directed against the effector T cell in an allogeneic cell-mediated lysis (CML) test, was able to block reactivity. This implies that Lyt-2, Lyt-3, and the T effector cell recognition structure form a grouping on the cell surface. From the study of Boyse et al, cited above, we would infer that H-2D must be in this grouping also, and indeed Nakayama et al [94] found that anti-H-2 also blocked. Freund et al [95] conducted a similar study using, however, the mixed lymphocyte reaction (MLR), rather than CML, and antibody against the target, rather than the effector. This test dealt primarily with the Lyb-4 antigen of B cells. This appeared to be associated with an Ia antigen, the typical MLR target, and also, though perhaps more weakly, with H-2D and Mls, both also MLR targets. A surprising finding was that antigen associations occurred only when the corresponding genes were derived from the same parent.

There also have been reports of an association between the Fc receptor and Ia antigens in B cells, but this subject is in a confusing state. Results seem to depend on the nature of the test used [96, 97].

These results suggest that there are, within the cell membrane, organized islands, each consisting of several proteins or glycoproteins bound together. Although the methods used are not conclusive, they all point in the same direction and are certainly strengthened in the case of the Flaherty and Zimmerman [93] study by the use of fixed cells. Another point of interest is that all groupings so far demonstrated contain an MHC product. On the two cell types studied, however, the associated non-MHC antigens appeared to be different.

This subject is in its infancy; it certainly will receive much more attention. My futurology here will take the form of speculations about some of the conclusions to which future studies will lead. I suggest, first, that the results I have been summarizing point to a role of the K,D,L antigens as organizers of the cell surface and hence, probably, as regulators of differentiation. Ohno's suggestion [98] that H-2 is an "anchorage site of organogenesis-directing proteins" has similar implications. The hypothesis is strengthened by evidence, from several quite distinct types of experiments, that H-2 plays a role in cell interactions [99, 100; review of 101]. The appearance of H-2 early in development points in the same direction. If *K,D,L* products were primarily concerned in some way with immune reactions, it would seem unlikely that they would appear so early. In mammals early appearance of K,D,L has the disadvantage that K,D,L must be a prime target of anti-fetal reactions. Another indirect indication that K,D,L may be concerned with cell interactions is their evident homology with the *Qa–Tla* locus products, which in the sense of Boyse and Old [39], are differentiation antigens.

Of interest in this connection is a consideration of the few cell types that appear to lack or at least fail to express, H-2. One of these is the trophoblast cell [102]. The lack in this instance undoubtedly is a device for protection of the fetus. Because of H-2 restriction (discussed in a later section), this would help to protect the fetus not only from anti-H-2 but also anti-non-H-2 effector T cells. There are also reports that H-2 has a very low representation on hepatocytes and is lacking from several types of epithelial cells – eg, pancreatic B-cells [103, 104]. The meaning of this is not clear, but one possibility is that H-2 is deleted from cells particularly subject to autoimmune attack. But this raises the question: if K,D,L are so essential in all interactions, how can normal cells do without them? Perhaps in some fully differentiated cells K,D,L are not indispensible. An alternative but highly speculative possibility that I find attractive is that the *H-2* complex is the source of a substantial but undiscovered group of differentiation antigens, analogous to the *Qa–Tla* products or perhaps to the Ia antigens, but occurring on non-lymphoid cells. As I have already noted, the differentiation antigens of non-lymphoid and nonerythroid cells are an almost totally unexplored area. The currently empty *G* region of the *H-2* complex would be a possible location for some of the determining loci.

THE MAJOR HISTOCOMPATIBILITY COMPLEX AND CELLULAR IMMUNITY

One of the surprising and delightful findings about *H-2* is the number and diversity of the areas of research for which it seems to have significant implications. The work I have just described implies that it has a role in cell interactions and probably in development. Although the *H-2* product is widespread rather than restricted in its tissue distribution, it may function as a differentiation antigen because it can react selectively with membrane components that are specific for individual tissues. But intensive research over the past decade also points to a major role for *H-2* in immunological processes, especially in the cellular defenses, which seem to be particularly important in viral infections.

This is a vast subject. I have reviewed it elsewhere [16, 19, 73, 101]. The only clear trends are the ever-growing volume of research and the increasing synthesis of *H-2* research with immunology and virology. These trends certainly will continue, but the most important new discoveries will be the sort of breakthroughs that cannot be foreseen. What I propose to do here is to list some of the background facts, citing references only where they are not included in my earlier reviews, and then to offer some speculative interpretations of the facts. I will not repeat the speculations I have already offered in a Harvey Lecture [101].

The link of *H-2* to immunology was first clearly established when McDevitt and Chinitz [8], just a decade ago, reported that there are immune response (Ir) genes associated with the *H-2* complex. This marked a major turning point in *H-2* research. Subsequent studies established the facts discussed in the paragraphs that follow.

The *I* region of the *H-2* complex, where the original *Ir* loci are located, also determines serologically demonstrable Ia (*I*-associated) membrane components. These *H-2* products show a highly specific tissue distribution and play an important role in regulating the interactions of lymphocytes with macrophages and of different categories of lymphocytes one with another, thereby acting as regulators of the immune response. *Ia* products also are part of antigen-specific helper and suppressor factors released by lymphocytes.

Ir genes were originally supposed to be confined to the *I* region, but there is now evidence that the *H-2K* and *H-2D* (and *H-2L?*) loci play an immune response role quite comparable to that played by the original *Ir* loci [19, 105–108]. It may be, however, that *H-2K* and *H-2D* function only in the T effector cell cytolytic response, whereas the *I* region loci function through helper cells concerned with both cell-mediated and humoral responses.

One of the most intriguing links between *H-2* and immunology is the phenomenon of *H-2* restriction. Briefly stated, *H-2* restriction is the inability of T-lymphocytes to react to non-H-2 antigens of whatever sort unless they react to an *H-2* product at the same time. Furthermore, the *H-2* product is usually an allelic form to which the lymphocytes were exposed during their development in the thymus, and which they therefore see as "self." The restricting antigen of helper cells is usually an *I* region product; that of effector cells, a *K,D,L* product. This is in keeping with the immune response links of these cell types noted in the preceding paragraph.

H-2 restriction has been demonstrated with a variety of antigens, including non-*H-2* histocompatibility antigens, but there is reason to believe that the normal target of T effector cells is cells carrying a viral infection. The effectors thereby curb the multiplication and spread of viruses of all types.

An extension of the immune response capability of the *K,D,L* products noted above has been reported by Meruelo and co-workers [109, 110], though confined, in results reported so far, to the *H-2D* antigen and a viral target. The virus used by the authors was obtained from an x-ray-induced lymphoma of C57BL mice. Resistance to this virus was found to be favored by the *H-2D^d* allele. *H-2* genotype influenced primarily the later stages of viral infection, but it was found that, in resistant strains, the *H-2D^d* antigen increased markedly on the thymocyte surface almost immediately after virus inoculation. Moreover, the concentration of this antigen showed an inverse relationship to the concentration of viral antigens; also the high *H-2D^d*, low virus antigen cells were particularly effective in stimulating T effector cells. In cell cultures transformed by the virus, and hence potentially leukemic, *H-2* antigens were undetectable.

Another link between *K,D,L* and viruses is a physical association sometimes seen between antigens of both derivations on the cell surface [review in 73]. Again, there is evidence that unpredicted variants of the *K,D,L* products appear on the surface of virally infected cells [reviews in 19, 111]. This contention has been disputed, but I find the sum total of the evidence rather convincing.

An important and widely known tie of MHC to immunity is the strong association of particular HLA haplotypes with various human diseases possibly autoimmune in nature. Multiple sclerosis, juvenile diabetes, and ankylosing spondylitis are three diseases in this category. There is increasing evidence that viruses, and perhaps other infectious agents, play a role in this tie, since specific prior infections as well as the possession of a particular HLA type apparently predispose to these diseases [112–114].

Although the primary link of *H-2* to immunology appears to occur through the T lymphocyte (including the interactions of T helper cells with B cells in generating antibody production), *H-2* is also linked to one other important category of lymphocytes. The natural killer, or NK, cell, which is neither thymus-processed nor a typical B cell, is able to attack, in a matter of hours and without prior multiplication, RNA virus-infected cells and many tumor cells. This suggests that immune surveillance against potentially cancerous cells is one of its major functions, and while this assumption is still debatable, evidence supporting it seems to be mounting [115–117]. The important point here is that while NK activity is under multiple gene control, *H-2* plays an important role [118].

SOME SPECULATIONS

In consideration of these fascinating observations concerning the interactions of *H-2* with immune processes, the crystal ball gazing I shall allow myself is the addition to already numerous speculations of three suggested interpretations of some of the data.

I have mentioned that there is extensive, though still disputed, evidence that virus-infected cells, including many tumor cells, develop unexpected *H-2* specificities. This is a very puzzling phenomenon. I suggest that the explanation is to be found in a report by S.M. Phillips et al [119] that one of the most effective activitors of the RNA Moloney leukemia virus endogenous to T lymphocytes is "allogenic stimulation, either *in vivo* or *in vitro.*" The drug iododioxyuridine was also noted as an effective activator, though nonspecific mitogens such as PHA and concanavalin A were not. If there is selective value to the virus in replication, it would thus not be surprising if the virus had acquired the ability to induce the appearance on the surface of the infected cell of real or simulated MHC alloantigenic products. And replication probably is advantageous, since it would be a prerequisite for lateral transmission. Such transmission of RNA virus has been reported [120, 121] and would be an effective route to survival. In brief, then, I suggest that the unexpected *H-2* specificities detected on infected cells are engineered by the virus to serve its own evolutionary ends.

There are several ways in which a virus might induce an apparently allogeneic form of H-2, one of the more likely being the production of *altered self*. One of the explanations of *H-2* restriction that has been proposed is that T effector cells responding to a virus or other non-H-2 antigens are actually reacting to an altered H-2. It is not easy to see how a virus could so modify H-2K, H-2D, or H-2L that they would be seen by T cells as allogeneic. If, however, the virus gained from so doing, altered self-modification becomes plausible. I may add that I think the major alternative to altered self, the so-called dual recognition hypothesis, is also valid in many circumstances, including some mouse anti-viral responses [73, 101].

A second property of the MHC where I think understanding may be on the way concerns its extraordinary polymorphism, at least in some species [122]. The polymorphism is most striking in the case of the major histocompatibility (K,D,L) loci in the mouse. I think it would be correct to say that no other locus in mammals approaches these three (especially K and D) for the known number of alleles. The HLA-A, B, and C loci of man are not far behind, and so far as we can judge from available evidence, the same is true of the MHC of other primates [12]. The *Ia*-type loci of mouse and man are also polymorphic, but somewhat less so [123]. The reason for this polymorphism is one of the puzzles of MHC research.

The genetic findings in the mouse are, at first sight, made even more puzzling by recent evidence that the MHC of the rat, a closely related species, is far less polymorphic [19, 124, 125]. Even more surprising is a report by Phillips et al [126] concerning the Syrian hamster. It has been known for some time that it is difficult to get anti-MHC antibodies in this species, but it has been possible to explain this as due to the origin of all domestic lines from a very few ancestral animals. J.T. Phillips et al, using wild as well as domestic hamsters, obtained an alloantibody, but this reacted with molecules corresponding in molecular weight to mouse Ia rather than K,D,L antigens. By using a xenogenic anti-B_2-microgulbulin, which should precipitate the K,D,L homologs, they obtained molecules of the expected size, but they found no serological evidence of allelic diversity. It does appear, then, that there is an extraordinary *lack* of MHC polymorphism in this species.

We can thus arrange these species, with respect to their MHC polymorphism, as follows:

$$\text{mouse} \geqslant \text{man (and other primates?)} > \text{rat} > \text{hamster}$$

Is there any explanation for these curious results?

A clue, I think, can be found in evidence suggesting that these species can be similarly arranged with respect to their load of endogenous RNA viruses. The evidence is clearest with respect to the two species at the extreme ends of the gradient, the mouse and the hamster. Endogenous RNA viruses are common in the mouse [127]. The many leukemia viruses — eg, the Tennant [128] virus of strain BALB/c — are examples. Hamsters, however, have a very low load of endogenous viruses and a low incidence of spontaneous tumors, but at the same time are susceptible to viruses from virtually any other species and to carcinogenic agents [129]. The information on rats is not extensive, but though RNA viruses are present, the load of ecotropic tumor-causing varieties, laterally transferable between strains, appears to be substantially lower than in mice [130]. This gives them a middle position with respect to viral load as well as polymorphism. The position of man in the viral scale is the least clear because of the impossibility of carrying out the sort of study that has provided the most information with mice. Tests for endogenous viruses have to be performed in vitro rather than in vivo. One relevant study has been reported [131] based on the hybridization of nucleotide sequences from human neoplastic cells with known mammalian RNA tumor viruses. The results indicated that RNA viruses are involved in several types of human cancer, including sarcoma, leukemia, prostatic adenocarcinoma, and breast adenocarcinoma. We can also say that endogenous RNA viruses appear to be widely distributed in man's simian relatives [132].

Thus, while the order of the mouse, man, rat, and hamster on a scale of viral load is not entirely clear, there is some indication of a parallel to the polymorphism scale. I may add that poultry also seems to conform, fitting near the high end on both scales. I should also note that, while I have been speaking entirely of RNA viruses, it may be that the correlation of the threat from viruses with the degree of polymorphism will be found to be more easily documented by reference to some other specific category of virus or to virus in general.

If this correlation is indeed real, what can it mean? I suggest that it indicates that MHC polymorphism strengthens the defenses against virus infection and/or virus replication and thereby has developed in species which, for whatever reason, carry a threatening viral load. This leaves us with another question: why is it that some species do and some do not possess dangerous (endogenous?) viral levels? I shall not try to answer this question in detail; perhaps the answer is to be found in such factors as population structure, certain components of the environment such as amount of humidity and ultraviolet radiation, and the anti-viral effectiveness of other components of the immune system.

We have yet another question to consider: how can MHC polymorphisms strengthen anti-viral defenses? I believe that there are several ways in which this might, and perhaps does, happen, but I will examine only one. The critical link, I suggest, is provided by the Jerne mechanism [133, 134], which explains the somatic generation of antibody diversity in terms of the interaction of anti-self-MHC clones with MHC antigens in the thymus. MHC heterozygosity would double, or at least substantially increase, the number of initial clones from which the generation process could start. It should thus lead to a larger antibody vocabulary and hence, presumably, a broader defense against infection. Klein [122] and Snell [101] make the same point. The presence of three major histocompatibility loci (the mouse *K,D,L,* the human *A,B,C*) would be advantageous for the same reason. In the hamster, which lacks MHC polymorphism, a reduced antibody diversity would be expected, and indeed it has been reported that adult hamsters contain Ig-bearing lymphoid cells with a heterogeneity comparable to neonatal, rather than mature, murine B cells [129].

For this concept to be relevant in the present context, it has to apply to the generation of T effector cell clones. This actually fits nicely, since these clones appear to be slanted toward recognition of the K,D,L-type antigens, and it is these antigens that, in the mouse and man, show the extraordinary polymorphism. It is not surprising, however, that the *I*-region genes also are polymorphic, since this would contribute to the generation of clones of those T helper cells that appear to be necessary for effector cell activation [134, 135].

In seeking explanations for polymorphism, it must be kept in mind that the most likely mechanism for the maintenance of multiple alleles in a population is a selective advantage for the heterozygote [136]. It is thus important that the von Boehmer et al [134] hypothesis does predict a selective advantage for the heterozygote, and that *I*-region immune response loci generally show dominance of responsiveness [137, and many others]. Recessive behavior, however, has been reported for the *K*- and *D*-associated immune response effects seen in certain viral infections in the mouse [135]. There is also direct evidence that the MHC heterozygote possesses a selective advantage. Thus, in poultry, MHC heterozygosity has been found both to be specifically advantageous and to persist longer than expected during inbreeding [138, 139]. In both poultry and the rat, there is evidence that MHC heterozygosity is favorable to early survival [140, 141], and in man, HLA-B heterozygosity seems to increase with age, suggesting a greater viability for the heterozygotes [142]. Curiously, in the mouse, although dominance of the responder status is the rule, evidence for *H-2* selective advantage is ambiguous. Taylor (personal communication) failed to find persistance of *H-2* heterozygosity beyond theoretical expectations during the production of a large group of recombinant inbred strains, and Tennant and Snell [143], in a study of the dominance relations of *H-2*-related resistance to viral leukemogenesis, found complex patterns with no clear trend. Possibly one factor in the lack of selective advantage of MHC heterozygosity in the mouse is the high degree of sanitation maintained in most present-day mouse colonies. The advantage may be manifest only when there is a high risk of infection.

Doubtless the situation is complex in all species, but if the mouse does, in fact, show less MHC-related heterozygous advantage than other species, it perhaps has other mechanisms for maintaining MHC polymorphism. Snell [144] pointed out that the T locus in the mouse helps to maintain chromosome 17 heterozygosity. The combination of numerous lethal alleles, segregation ratio distortion, and crossover suppression between T and H-2 should have just this effect. There may well be T homologs in other species, but if there are, they probably lack these curious effects. Only a species like the mouse, which has large litters and a generally high reproductive capacity, could afford to evolve a mechanism that would perpetuate numerous lethal alleles.

Another bit of crystal ball gazing that I shall allow myself concerns the Meruelo phenomenon [109, 110]. How can an increase in the H-$2D$ product on virally infected thymocytes increase resistance to the infection? The clue, perhaps, can be found in the additional observation that resistance to the particular infection studied in H-$2D$-restricted. In terms of the von Bochmer et al theory [134, 101], this means that the T effector cells responding to this particular agent were generated by interaction in the thymus with the H-2D antigen. The responder status is then explained as the ability of the responding H-$2D$ allele to generate clones responsive to the given agent. However, besides their virally directed recognizer, the effector cells also carry an anti-D recognizer. To become activated, both receptors have to become bound to their respective targets.

I suggest, then, that the value of H-2D increase on the target cell surface can be explained by applying the law of mass action to this system. If D is increased, the effector cell can achieve the necessary double triggering at *lower* levels of virus. Infected cells will be eliminated at an earlier stage in the infective cycle, and the spread of the virus will be restricted.

The intimate and complex relationships between the MHC and immune defense mechanisms, especially defenses against viral agents, which we have been discussing, have been studied primarily in mammals, but in their major aspect, they apply to all vertebrates. It is quite possible that they have prevertebrate antecedents [101]. They thus have a long evolutionary history. The driving force in this evolutionary process has been the age-old struggle between host and parasite, with viruses being the inciting agent of primary importance in the evolution of cellular immunity. The host has developed complex defenses, illustrated by the numerous, recently discovered categories of lymphocytes, each with its highly specific functions, and the parasite has evolved ways to evade or subvert these defenses.

Bloom [145] has pointed out ways in which parasites other than viruses evade immune surveillance. Evidence concerning the stratagems of viruses is more recent, but it is likely to grow rapidly. One example is the report that in strains low in natural killer cell activity, Friend virus is able to induce suppressor cells that restrict anti-viral immunity [146] (these authors refer to the natural killer cell as the M cell). In a somewhat comparable situation, some tumors, both native and transplanted, protect themselves from host defenses by shedding blocking factors [147–149], and perhaps by secreting a layer composed of hyaluronic acid and other polysaccharides [150].

MEDICAL IMPLICATIONS

If viruses and tumors can manipulate the immune response for their own benefit, certainly man can do so too. It would be an insult to think otherwise. Of course, we have been doing it for more than a century by means of vaccination. What is new is a recognition of cellular immunity as a separate compartment of the immune mechanism and the achievement of a sufficient understanding of this compartment so that we can begin to think intelligently about controlling it. I take an optimistic view of the prospects. Whereas viruses and tumors have achieved their capability largely through an evolutionary process of trial and error, man must and will do it primarily, though not entirely, through intelligent exploitation of a knowledge of immune processes. Our knowledge is extensive and growing rapidly. Immunogenetics has played a major role in this knowledge explosion, and immunogenetic tools will be a significant part of the armory of the immunotherapist.

I shall examine some of the possibilities, grouping them according to the various types of medical problems where immunotherapy may apply. I shall be concerned only with specific forms of immunomanipulation, not with the shotgun approach of using drugs or xenoantisera that affect immune responses in general and that consequently have dangerous side effects.

Let us see what some of the possibilities are for immunomanipulation, starting with the area of organ transplantation. The requirement in this situation is for immunosuppression. Substantial success has already been achieved by HLA matching and the use of immunosuppressive drugs or anti-lymphocyte serum. The problem with these forms of suppression is that they are not only often inadequate but also nonspecific. If the transplant surgeon could achieve suppression aimed only at those clones, and especially those T effector cell clones, reactive with the mismatched antigens, this would represent a substantial gain.

To the best of my knowledge, the first report of a specific suppression of the cellular immune response resulted from a study of Snell et al [151], who used the passive transfer of enhancing antiserum. Mitchison [152] had already shown the importance of cellular immunity in graft rejection, and work in Bar Harbor and elsewhere had demonstrated the curious inverse effect of prior immunization on the growth of some transplantable tumors known as "immunological enhancement." Kaliss et al [153] had found enhancement to be transferable with serum from enhanced animals. The study of Snell et al brought these two lines of evidence together by showing that node and spleen lymphocytes from mice receiving antidonor serum were less suppressive of the growth of the test tumor than were the same cells from controls. Their cellular immunity had been specifically lowered.

In recent years there have been numerous studies aimed at the application of immunological enhancement to organ grafting. The results have been mixed. A major problem appears to be the number and complexity of the active agents; multiple serum factors may be involved [154–156]. In some cases of active enhancement, T suppressor lymphocytes evidently play a role [157, 158]. Classical anti-graft antibodies are sometimes effective, with anti-Ia sera being especially active in some experimental situations [159–161]. Antibodies against the major *H* antigens may work better for organs such as heart and kidney, which lack *Ia* antigens [162].

A major advance with similarities to enhancement is the discovery of antigen-specific suppressor factors released by T suppressor lymphocytes [review in 73]. The factors consist of two chains; the heavy chain carrying the specificity and perhaps being the product of an immunoglobulin locus, the light chain being an Ia molecule determined by the *I-J* subregion [163, 164]. The factors block the activation of effector clones. Thus, a suppressor factor directed against histocompatibility antigens should permit enhanced growth of a graft carrying these antigens.

The factors are produced in very small amounts, but several authors have demonstrated that this limitation can be overcome by the use of hybridomas [163, 165–167]. The hybridomas are derived by fusion of a thymona with an activated T suppressor cell and, after a period of chromosome elimination, selecting suppressor-secreting cell lines.

Suppressor factors have typically been made in mice and carry specificity against classical antigens, but there is no fundamental reason why an anti-human HLA should not be attainable. The suppressing effect is substantial. For practical use in organ transplanation, many clones would have to be produced and maintained, but their potential should be enormous.

There are other potential control mechanisms. I mention three of these to emphasize the diversity of potential approaches. The first is antigen specific; the other two, specific only in the sense that a restricted category of lymphocytes is inactivated.

Enhanced survival of skin grafts in rats was achieved by Binz and Wigzell [168] by making use of the ability of animals to form anti-antibodies or, more specifically anti-idiotypic antibodies — ie, antibodies directed against the antibody-blinding sites of other antibodies. Skin graft survival, though increased, was not permanent, but this probably was because multiple minor histoincompatibilities were present. The authors' methods were complex; in the particular system used, the active antibodies were induced as autoantibodies. This perhaps is an advantage, but it also should be possible to transfer the antibodies passively. The relevant antibodies doubtless could be obtained in quantity by the use of hybridomas or other cloning methods. This would appear to be a promising approach for the deletion or inactivation of specific lymphocyte clones.

We turn now to some less specific methods of immunosuppression.

Suppressor T cells produce nonspecific as well as antigen-specific suppressor factors. These have been less studied than the specific factors, but they could have important uses, especially as they perhaps are cell-type specific. Hybridomas that secrete these have been produced [163].

Mouse amniotic fluid contains substances with immunosuppressive activity, as shown by their effect on the mixed lymphocyte reaction. One of these substances is alpha-fetoprotein. There appears to be considerable selectivity in their action [169].

Anti-Ia antibodies and, more specifically, antibodies against Ia products of the *I-A* subregion have been shown to reduce the capacity of hosts to mount a response against isogeneic tumor transplants, presumably by killing or inactivating the relevant helper cells [170]. Again, we can expect some specificity as to the cell type attacked.

These brief summaries show the range of possibilities for antigen-specific or cell-type–specific immune suppression. Methods for producing the active agents in quantity are already in sight. While application is certainly years away, the potential appears to be enormous. The limiting factor may be the complexity, and hence the cost of some of the techniques, e.g., the maintenance of multiple antibody-producing clones. When the method is applicable, it is much simpler to put the host to work by some form of active immunization.

While this work on basic methods of immune-suppression has been going on, the surgeons engaged in organ transplantation have not been idle. Van Rood et al [171], among others, report encouraging progress. The currently used techniques, such as HLA-A and HLA-B typing and a chick for pre-existing immunity, remain important. Van Rood et al stress two new approaches. The first is serological typing for HLA-DR antigens, the counterparts of the mouse Ia. These antigens are important in graft rejection in man, as originally shown by Dausset et al [172]. The other recommended method is a single blood transfusion, with some of the buffy coat removed, prior to transplantation. This approach corresponds in methodology to active enhancement, and indeed Snell et al [173] showed that the intravenous injection of donor whole blood is a particularly effective method of enhancing the growth of tumor homografts, though the mechanism of the enhancement was not elucidated. Van Rood et al attribute the phenomenon observed in man to the induction of suppressor cells rather than to serum factors. The suppressor cell hypothesis finds support in a report by Sasportes et al [174] that two successive in vitro stimuli of human lymphocytes by HLA-incompatible cells result in the generation of suppressor cells specific for HLA-DR antigens. One of its attractive features is that, when it is employed, one HLA-DR mismatch worked almost as well as zero mismatches.

The inevitable conclusion from these studies is that over the next 50 years the percentage of success of organ transplant will increase dramatically and that the variety of organs and tissues amenable to transplantation also will expand.

In organ transplantation, immunosuppression is the desideratum; in the search for protection against viruses and tumors, most investigations are aimed at immunoaugmentation. For this reason I shall consider these last two undertakings together.

Let us first compare them in some other respects.

The targets of host defense in both cases are cells, virally infected host cells in one case and tumors in the other, and the active agents are lymphocytes and macrophages. The active lymphocytes fall into two major classes, T lymphocytes, both helper and effector, with suppressor cells playing a regulatory role, and natural killer or NK cells. Available evidence suggests that the primary targets of T cells are virus-infected cells while NK cells are principally directed at tumor cells. One study with T effector cells, for example, showed that they can adoptively transfer protection against influenza virus infections in mice [175; see Ref 73 for further evidence]. Viral antigens are clearly the targets of many T cells clones, but at least one study of NK cells found that the target was not any of the familiar viral antigens [176]. More important, there are at least hints of evidence that NK cells can protect against cancer. Thus, Kasai et al [115], by identifying Ly-5 as a surface component of NK cells and then using an anti-Ly-5 to obtain an enriched NK population, showed that NK cells can protect against the growth of transplanted isologous lymphomas, whereas other nonimmune lymphocytes do not have this property. In another study it was shown that, in athymic nude mice, which are poor in T cells but rich in NK cells, the proportion of embryos transplanted to the kidney capsule of adults which developed into teratocarcinomas was much lower than in normal controls [117]. And finally, preliminary studies in man have been cited which suggest that some degree of tumor remission has been achieved in patients whose NK cell level has been raised with interferon [177, 178].

Whereas the primary targets of T cells appear to be viruses and those of NK cells appear to be tumors, there is also evidence of overlap. Thus, several investigators have found evidence for NK cell protection against viruses [179–181], with their action perhaps being particularly important in the later stages of infection [181]. T effector cells, however, can certainly attack many transplanted isogeneic tumors, though the significance of this capacity is uncertain. They may also play a role in resistance to virus-induced leukemia [110].

To complicate the situation still further, I must also point out the existence of evidence casting doubt on claims of a role for immunological surveillance in the natural control of cancer [182].

In this complex situation, what clues can be find as to possible tools that are now, or become, available to the immunotherapist in situations requiring immunoaugmentation? Briefly stated, there is evidence for a diverse armament that is the reciprocal of that which we discussed under immunosuppression.

I turn first to factors that are not antigen specific but that probably show considerable specificity as to the cell type they activate. The best known of these factors is interferon — more properly, the interferons, since there are probably at least three of them, each produced by a distinct cell type [178]. These were originally detected because of their anti-viral effect, but they also augment NK cell activity [183, 184] and may thereby have an anti-tumor effect [177]. Another interesting effect of interferon is an increased expression of major histocompatibility antigens demonstrated on a human lymphoma cell line [185]. We have already noted a report that strains of mice resistant to a radiation-induced leukemia virus show enhanced H-2D expression on thymocytes almost immediately after infection [110]. The significance of the association of an increased MHC expression with viral resistance is thereby emphasized.

T helper cells produce both nonspecific helper factors and antigen-specific helper factors [review in 73]. It is possible that some of the nonspecific factors will prove to be the same as one or more of the interferons, identified through a different set of properties, but some of them will surely prove to be distinct entities. The factors, both specific and nonspecific, have been studied mostly in the T helper—B cell system, but an antigen-specific factor that helps the cytotoxic T cell has also been reported [186]. This form of help is the most relevant in the present context.

Just as it has been shown in Benecerraf's laboratory that anti-Ia directed against *I-A* subregion antigens blocks helper action [170], so it has been shown in the same laboratory that anti-I-J blocks suppressor action [187]. The association of *I-A* with helper action and of *I-J* with suppression was to be expected from earlier studies [review in 73].

This summary of factors available for enhancing cellular immunity, sketchy though it is, I hope gives some idea of the extensive armament that will be available in future years to the therapist seeking immunoaugmentation. It is too early to attempt many guesses as to specific applications, but a few clues are available. Certainly the interferons, and perhaps additional lymphokines, will soon move from the laboratory into practical use for viral infections, and perhaps for cancer therapy. The process will be facilitated by the development of one or another of the various cultured cell lines that can produce a single, desired biological product in quantity. There is little background for foreseeing what may be done with specific helper factors, but I can imagine that they will be used to treat specific viral infections and, if it turns out that there are any families of cancer antigens, for cancer immunotherapy. Another area for immunomanipulation may be the development of countermeasures in those situations in which the immune defenses of the host have been subverted by viruses or cancer cells. This is likely to involve methods of turning off suppressor cells that have been turned on by the parasitizing agent. I know of no very clear clues as to how this can be done, but I will venture to predict that methods can be found. The use of antigen-specific agents would be desirable, but developing these may prove difficult. Finally, our new knowledge of cellular immunity certainly will lead to improvements in active immunization of the incipient host. This may be the least spectacular area of advance, but it is perhaps the simplest and most reliable.

A final area in which recent discoveries in cellular immunity should find application is in the prevention and treatment of HLA-associated diseases. These diseases are an important group [10], and most of them defy treatment at present. An aspect of these diseases that seems to be emerging from recent studies, and that certainly will be important in plans to deal with them, is an apparent link of each disease to a specific, prior infection. Thus, besides the association with an HLA type, multiple sclerosis appears to be associated with an earlier occurrence of measles [113]; juvenile diabetes, with a coxsackie B4 virus infection [112], and ankylosing spondylitis with a Klebsiella pneumoniae infection [114]. In different areas of the world, a given autoimmune disease may show different HLA links and different infectious disease links, the relevant HLA haplotype apparently being determined by the earlier antigenic exposure [188].

One can foresee several areas in which immunogenetics and cellular immunology may be put to use in the prevention and treatment of these diseases, but at present only in general terms. One obvious area for the application of our knowledge is the use of HLA typing – particularly in families with a history of HLA-linked diseases – to detect high-risk individuals. In these cases, special care could be taken to prevent the occurrence of, or perhaps induce a favorable modification in the course of, the associated infection. With respect to therapy either of the prior infection or the subsequent autoimmune involvement, we do not really know whether we want a stimulation of helper factors or of suppressor factors, but clarification of this aspect of the problem cannot be far off. Even if many details now are obscure, I believe that this is an area where the futurist can be optimistic. The last few decades of intensive and exciting research are going to pay off.

ACKNOWLEDGMENTS

I am indebted to Drs. Donald Bailey, Hendrick Bedigian, and Michael Edidin for reading the manuscript of this paper and making valuable suggestions, and to Mrs. Jennie Jenkins and Miss Kim Parady for the typing.

REFERENCES

1. Bernstein F: Zusammenfassende Betrachtungen über die Blutstrukturen des Menschen. Z Indukt Abstamm Vererblehre 37:237, 1925.
2. Irwin MR, Cole LJ: Immunogenetic studies of species hybrids in doves and the separation of species-specific substances in the backcross. J Exp Zool 73:85, 1936.
3. Gorer PA: The detection of antigenic differences in mouse erythrocytes by the employment of immune sera. Br J Exp Pathol 17:42, 1936.
4. Irwin MR: The beginnings of immunogenetics. Immunogenetics 3:1, 1976.

5. Little CC: A possible mendelian explanation for a type of inheritance apparently non-mendelian in nature. Science 40:904, 1914.
6. Little CC: The genetics of tumor transplantation. In Snell GD (ed): "The Biology of the Laboratory Mouse." Philadelphia: Blakiston, 1941, p 279.
7. Boyse EA, Old LJ, Luell S: Antigenic properties of experimental leukemias. II. Immunological studies in vivo with C57BL/6 radiation-induced leukemias. J Natl Cancer Inst 31:987, 1963.
8. McDevitt HO, Chinitz A: Genetic control of the antibody response: Relationship between immune response and histocompatibility (H-2) type. Science 163:1207, 1969.
9. Bodmer WF (ed): The HLA system. Br Med Bull 34:213, 1978.
10. Dausset J, Svejgaard A: "HLA and Disease." Copenhagen: Munksgaard, 1977.
11. Festerstein H, Démant P: "HLA and H-2. Basic Immunogenetics, Biology and Clinical Relevance." London: Edward Arnold, 1978.
12. Gotze D (ed): "The Major Histocompatibility System in Man and Animals." New York: Springer-Verlag, 1977.
13. Katz DH, Benacerraf B: "The Role of the Products of the Histocompatibility Gene Complex in Immune Responses." New York: Academic Press, 1976.
14. Klein J: "Biology of the Mouse Histocompatibility-2 Complex." New York: Springer-Verlag, 1975.
15. McDevitt HO (ed): "Ir Genes and Ia Antigens." New York: Academic Press, 1978.
16. Snell GD, Dausset J, Nathenson S: "Histocompatibility." New York: Academic Press, 1976.
17. Svegjaard A, Hauge M, Jersild C, Platz P, Ryder LP, Staub Nielson L, Thomsen M: "The HL-A Systems: An Introductory Survey." Basel: Karger, 1979.
18. Handler P: Testimony before the Subcommittee on Science, Research and Technology, Committee on Science and Technology, House of Representatives, 6 March 1979.
19. Snell GD: Recent advances in histocompatibility immunogenetics. Adv Genet 20:291, 1979.
20. Graff RG, Brown DH: Estimates of histocompatibility differences between inbred mouse strains. Immunogenetics 7:367, 1978.
21. Bailey DW: Further studies of the brown histocompatibility complex. The Jackson Laboratory, 49th Annual Report, 1978, p 72.
22. Passmore HC, Beisel KW: Preparation of antiserum for the detection of the Ss protein and the Slp alloantigen. Immunogenetics 4:393, 1977.
23. Green MC: Genetic nomenclature for the immunoglobulin loci of the mouse. Immunogenetics 8:87, 1979.
24. Horton MA, Beverley PCL, Simpson E: Serological properties of anti-Ly-6.2 serum produced by a new immunization schedule. Immunogenetics 7:173, 1978.
25. Feeney AJ: Expression of Ly-6 on activated T and B cells: Possible identity with Ala-1. Immunogenetics 7:537, 1979.
26. Durda PJ, Gottlieb PD: Sequential precipitations of mouse thymocyte extracts with anti-Lyt-2 and anti-Lyt-3 sera. I. Lyt-2.1 and Lyt-3.1 antigenic determinants reside on separable molecules. J Immunol 121:983, 1978.
27. Taub M, Chuman L, Saier MH Jr, Sato G: Growth of Madin-Darby canine kidney epithelium cell (MDCK) line in hormone-supplemented serum-free medium. Proc Natl Acad Sci USA 76:3338, 1979.
28. Archer JR, Davies DAL: Demonstration of an alloantigen on the peritoneal exudate cells of inbred strains of mice and its association with chromosome 7 (linkage group I). J Immunogenet 1:113, 1974.
29. Snell GD: Immunogenetics: Retrospect and prospect. Immunogenetics 1:1, 1974.
30. Lemke H, Hämmerling CJ, Höhmann C, Rajewsky K: Hybrid cell lines secreting mono-

clonal antibody specific for major histocompatibility antigens of the mouse. Nature 271:249, 1978.
31. Schierman LW, McBride RA: Adjuvent activity of erythrocyte isoantigens. Science 156:658, 1967.
32. Cherry M, Chai LF: A transplantation antigen sensitive to coimmunization. The Jackson Laboratory, 49th Annual Report, 1978, p 69.
33. Mobraaten LE, Cherry M: Serology of a histocompatibility mutant. The Jackson Laboratory, 49th Annual Report, 1978, p 69.
34. Wernet D, Shafran H, Lilly F: Genetic regulation of the antibody response to $H\text{-}2D^b$ alloantigens in mice. III. Inhibition of the IgG response to noncongenic cells by preimmunization with congenic cells. J Exp Med 144:654, 1976.
35. Bailey DW: Recombinant-inbred strains: An aid to finding identity, linkage, and function of histocompatibility and other genes. Transplantation 11:325, 1971.
36. Taylor BA, Shen F-W: Location of $Lyb\text{-}2$ on mouse chromosome 4: Evidence from recombinant inbred strains. Immunogenetics 4:597, 1977.
37. Bailey DW, Kohn HI: Inherited histocompatibility changes in progeny of irradiated and unirradiated inbred mice. Genetic Res 6:330, 1965.
38. Egorov IK: A mutation of the histocompatibility-2 locus in the mouse. Genetica (Moscow) 3:136, 1967.
39. Boyse EA, Old LJ: Some aspects of normal and abnormal cell surface genetics. Annu Rev Genet 3:269, 1969.
40. Rief AE, Allen JMV: The AKR thymic antigen and its distribution in leukemias and nervous tissues. J Exp Med 120:413, 1964.
41. Raff MC, Wortis HH: Thymus dependence of Θ-bearing cells in the peripheral lymphoid tissues of mice. Immunology 18:931, 1970.
42. John M, Carswell E, Boyse EA, Alexander G: Production of Θ antibody by mice that fail to reject Θ incompatible skin grafts. Nature New Biol 238:57, 1972.
43. Hilgers J, Haverman J, Nusse R, van Blitterswijk W, Cleton F, Hageman P, van Nie R, Calafat J: Immunological, virological and genetical aspects of mammary tumor virus (TMV) cell surface antigens; the presence of these antigens as well as the Thy-1.2 antigens on murine mammary gland and tumor cells. J Natl Cancer Inst 54:1335, 1975.
44. Stern PL: Theta alloantigen on mouse and rat fibroblasts. Nature New Biol 246:76, 1973.
45. Williams AF: Many cells in rat bone marrow have cell-surface Thy-1 antigen. Eur J Immunol 6:526, 1976.
46. Thierfelder S: Haemopoietic stem cells of rats but not of mice express Th-1.1 alloantigen. Nature 269:691, 1977.
47. Acton RT, Addis J, Carl GF, McClain LD, Bridges WF: Association of Thy-1 differentiation alloantigen with synaptic complexes isolated from mouse brain. Proc Natl Acad Sci USA 75:3283, 1978.
48. Dulbecco R, Bologna M, Unger M: Role of Thy-1 antigen in the in vitro differentiation of a rat mammary cell line. Proc Natl Acad Sci USA 76:1848, 1979.
49. Barclay AN, Letarte-Muirhead M, Williams AF: Chemical characterization of the Thy-1 glycoproteins from membranes of rat thymocytes and brain. Nature 263:563, 1976.
50. Lengerová A, Selený V, Haškovec C, Hilgert I: Search for the physiological function of $H\text{-}2$ gene products. Eur J Immunol 7:62, 1977.
51. Pouysségur J, Willingham M, Pastan I: Role of cell surface carbohydrates and proteins in cell behavior studies on the biochemical reversion of an N-acetylglucosamine-deficient fibroblast mutant. Proc Natl Acad Sci USA 74:243, 1976.
52. Reif AE, Allen JMV: Specificity of isoantisera against leukemic and thymic lymphocytes. Nature 200:1332, 1963.

53. Boyse EA, Miyazawa M, Aoki T, Old LJ: Ly-A and Ly-B: Two systems of lymphocyte isoantigens in the mouse. Proc R Soc Lond B 170:175, 1968.
54. Boyse EA, Katanaki I, Stockert E, Iritani CA, Miura M: Ly-C: a third locus specifying alloantigens expressed only on thymocytes and lymphocytes. Transplantation 11:351, 1971.
55. Cherry M, Snell GD: A description of mu: a non-H-2 alloantigen in C3H/Sn mice. Transplantation 8:319, 1969.
56. David CS: The role of Ia antigen in immune response. Transplant Proc 11:677, 1979.
57. Murphy DB, Okumura K, Herzenberg LA, Herzenberg LA, McDevitt HO: Selective subpopulations. Cold Spring Harbor Symp Quant Biol 41:497, 1977.
58. Snell GD, Cherry M, McKenzie IFC, Bailey DW: Ly-4, a new locus determining a cell surface alloantigen in mice. Proc Natl Acad Sci USA 70:1108, 1973.
59. McKenzie IFC, Gardiner J, Cherry M, Snell GD: Lymphocyte antigens: Ly-4, Ly-6, and Ly-7. Transplant Proc 9:667, 1977.
60. Gani MM, Summerell JM: Ly-4-2: Doubts concerning its validity as a B-cell marker. Immunogenetics 5:569, 1977.
61. Komuro K, Itakura K, Boyse EA, John M: Ly-5; a new T-lymphocyte antigen system. Immunogenetics 1:452, 1974.
62. Woody JN, Feldman M, Beverly PCL, McKenzie IFC: Expression of alloantigens Ly-5 and Ly-6 on cytotoxic effector cells. J Immunol 118:1739, 1977.
63. McKenzie IFC, Cherry M, Snell GD: Ly-6.2: A new lymphocyte specificity of peripheral T cells. Immunogenetics 5:25, 1977.
64. Woody JN: Ly-6 is a T-cell differentiation antigen. Nature 269:61, 1977.
65. Halloran PF, Dutton D, Chance H, Cohen Z: An Ly-like specificity with extensive non-lymphoid expression. Immunogenetics 7.185, 1978.
66. Frelinger JA, Murphy DB: A new alloantigen, Ly-8, recognized by C3H anti-AKR serum. Immunogenetics 3:481, 1976.
67. Feeney AJ, Hämmerling V: Ala-1: A murine alloantigen of activated lymphocytes. II. T and B effector cells express Ala-1. J Immunol 118:1488, 1977.
68. Zeicher M, Mozes E, Lonai P: Lymphocyte alloantigens associated with X-chromosome-linked immune response genes. Proc Natl Acad Sci USA 74:721, 1977.
69. Stanton TH, Boyse EA: A new serologically defined locus, Qa-1, in the Tla-region of the mouse. Immunogenetics 3:525, 1976.
70. Stanton TH, Calkins CE, Jandinski J, Schendel DJ, Stutman O, Cantor H, Boyse EA: The Qa-1 antigenic system. Relation of Qa-1 phenotypes to lymphocyte sets, mitogen responses, and immune functions. J Exp Med 148:963, 1978.
71. Flaherty L, Zimmerman D, Sullivan KA: Qa-2 and Qa-3 antigens on lymphocyte subpopulations. I. Mitogen responsiveness. J Immunol 121:1640, 1979.
72. Hämmerling GJ, Hämmerling V, Flaherty L: Qat-4 and Qat-5, new murine T-cell antigens governed by the Tla region and identified by monoclonal antibodies. J Exp Med 15:108, 1979.
73. Snell GD: T cells, T cell recognition structures, and the major histocompatibility complex. Immunol Rev 38:3, 1978.
74. Bailey DW: Genetics of histocompatibility in mice. I. New loci and congenic lines. Immunogenetics 2:249, 1975.
75. Dalchau R, Fabre JW: Identification and unusual tissue distribution of the canine and human homologues of Thy-1 (Θ). J Exp Med 149:576, 1979.
76. Morse HC III, Chused TM, Hartley JW, Mathieson BJ, Sharrow SO, Taylor BA: Expression of xenotropic murine leukemia viruses as cell-surface gp 70 in genetic crosses between strains DBA/2 and C57BL/6. J Exp Med 149:1183, 1979.
77. Edidin M, Weiss A: Antigen cap formation in cultured fibroblasts: A reflection of mem-

brane fluidity and of cell motility. Proc Natl Acad Sci USA 69:2456, 1972.
78. Graff RJ: Histocompatibility systems, except *H-2*: Mouse. In Altman PL, Katz DD (eds): "Inbred and Genetically Defined Strains of Laboratory Animals, Part 1." Bethesda, Md: Fed Am Soc Exp Biol, 1979, p 118.
79. Koch GLE, Smith MJ: An association between actin and the major histocompatibility antigen H-2. Nature 273:274, 1978.
80. Rolink T, Eichman K, Simon M: Detection of two allotype-(Ig-1)-linked minor histocompatibility loci by the use of *H-2*-restricted cytotoxic lymphocytes in congenic mice. Immunogenetics 7:321, 1978.
81. Démant P, Ivanyi D, Nusse R, Neauport-Sautes C, Snoek M: The *H-2L* locus: alleles, products and specificities. Transplant Proc 11:647, 1979.
82. Pardee AB: Membrane transport proteins. Science 162:632, 1968.
83. Ramasamy R, Richardson NE, Feinstein A: The specificity of the Fc receptor on murine lymphocytes for immunoglobulins of the IgG and IgM classes. Immunology 30:851, 1976.
84. Rowzkowski W, Plant M, Lichtenstein LM: Selective display of histamine receptors on lymphocytes. Science 195:683, 1977.
85. Edelman GM: Surface modulation in cell recognition and cell growth. Science 192:218, 1976.
86. Gulik-Krzywicki T: Structural studies of the association between biological membrane components. Biochim Biophys Acta 415:1, 1975.
87. Loor F: Cell surface design. Nature 264:272, 1976.
88. Rothman JE, Lenard J: Membrane asymmetry. The nature of membrane asymmetry. The nature of membrane assymmetry provides clues to the puzzle of how membranes are assembled. Science 195:743, 1977.
89. Shur BD, Roth S: Cell surface glycosyltransferases. Biochim Biophys Acta 415:473, 1975.
90. Singer SJ, Nicolson GL: The fluid mosaic model of the structure of cell membranes. Science 175:720, 1972.
91. Chen HW, Kandutsch AA, Heiniger HJ: The role of cholesterol in malignancy. Prog Exp Tumor Res 22:275, 1978.
92. Boyse EA, Old LJ, Stockert E: An approach to the mapping of antigens on the cell surface. Proc Natl Acad Sci USA 60:886, 1968.
93. Flaherty L, Zimmerman D: Surface mapping of mouse thymocytes. Proc Natl Acad Sci USA 76:1990, 1979.
94. Nakayama E, Shiku H, Stockert E, Oettgen HF, Old LJ: Cytotoxic T cells: Lyt phenotype and blocking of killing activity by Lyt antisera. Proc Natl Acad Sci USA 76:1977, 1979.
95. Freund JG, Ahmed A, Dorf ME, Sell KW, Humphreys RE: Inhibition of the mixed lymphocyte culture response with anti-Lyb-4.1 serum. J Immunol 118:1143, 1972.
96. Dickler HB, Sachs DH: Evidence for identity or close association of the Fc receptor of B lymphocytes and alloantigens determined by the *Ir* region of the *H-2* complex. J Exp Med 140:779, 1976.
97. Wofsy L, McDevitt HO, Henry C: No direct association between Ia antigens and Fc receptors on the B cell membrane. J Immunol 119:61, 1977.
98. Ohno S: The original function of MHC antigens as the general plasma membrane anchorage site of organogenesis-directing proteins. Immunol Rev 33:59, 1977.
99. Bartlett PF, Edidin M: Effect of the *H-2* gene complex on rates of fibroblast intercellular adhesion. J Cell Biol 77:377, 1978.
100. Degos L, Pla M, Colombani JM: *H-2* restriction for lymphocyte homing into lymph nodes. Eur J Immunol 9:808, 1979.
101. Snell GD: The major histocompatibility complex: its evolution and involvement in cellular immunity. Harvey Lectures, Vol. 74, 1980, pp 49–80.
102. Sellens MH: Antigen expression on early mouse trophoblast. Nature 269:60, 1977.

103. Parr EL: The absence of *H-2* antigens from mouse pancreatic B-cells demonstrated by immunoferritin labeling. J Exp Med 150:1, 1979.
104. Parr EL: Diversity of expression of *H-2* antigens on mouse liver cells demonstrated by immunoferritin labeling. J Exp Med 150:45, 1979.
105. Maron R, Cohen IR: Mutation at *H-2K* locus influences susceptibility to induction of autoimmune thyroiditis. Nature 279:715, 1979.
106. Gordon RD, Simpson E: Immune response gene control of cytotoxic T-cell responses to H-Y. Transplant Proc 9:885, 1977.
107. Wettstein PJ, Frelinger JA: *H-2* effects on cell-cell interactions in the response to single non-*H-2* alloantigens. III. Evidence for a second *Ir* gene system mapping in the *H-2K* and *H-2D* regions. Immunogenetics 10:211, 1980.
108. Zinkernagel RM, Callahan GN, Althage A, Cooper S, Streilein JW, Klein J: The lymphoreticular system in triggering virus plus self-specific cytotoxic T cells: Evidence for T help. J Exp Med 147:897, 1978.
109. Meruelo D, Nimelstein SH, Jones PP, Lieberman M, McDevitt HO: Increased synthesis and expression of H-2 antigens on thymocytes as a result of radiation leukemia virus infection: A possible mechanism for *H-2* linked control of virus-induced neoplasia. J Exp Med 147:470, 1978.
110. Meruelo D: A role for elevated H-2 antigen expression in resistance to neoplasia caused by radiation-induced leukemia virus: Enhancement of effective tumor surveillance by killer lymphocytes. J Exp Med 149:898, 1979.
111. Rogers MJ, Appella E, Pierotti MA, Invernizzi G, Parmiani G: Biochemical characterization of alien H-2 antigens expressed on a methylcholanthrene induced tumor. Proc Natl Acad Sci USA 76:1415, 1979.
112. Maugh TH III: Virus isolated from juvenile diabetic. Science 204:1187, 1979.
113. Neighbour PA, Bloom BR: Absence of virus-induced lymphocyte suppression and interferon production in multiple sclerosis. Proc Natl Acad Sci USA 76:476, 1979.
114. Saeger K, Bashir HV, Geczy AF, Edmonds J, de Vere-Tyndall A: Evidence for a specific B27-associated cell surface marker on lymphocytes of patients with ankylosing spondylitis. Nature 277:68, 1979.
115. Kasai M, Leclerc JC, McVay-Boudreau L, Shen FW, Cantor H: Direct evidence that natural killer cells in nonimmune spleen populations prevent tumor growth in vivo. J Exp Med 149:1260, 1979.
116. Roder J, Duwe A: The beige mutation in the mouse selectively impairs natural killer cell function. Nature 278:451, 1979.
117. Solter D, Damjanov I: Teratocarcinomas rarely develop from embryos transplanted into athymic mice. Nature 278:554, 1979.
118. Petranyi GG, Kiessling R, Klein G: Genetic control of "natural" killer lymphocytes in the mouse. Immunogenetics 2:53, 1975.
119. Phillips SM, Hirsch MS, Andre-Schwartz J, Solnick C, Black P, Schwartz RS, Merrill JP, Carpenter CB: Cellular immunity in mice. V. Further studies on leukemia virus activation in allogeneic reactions of mice: Stimulatory parameters. Cell Immunol 15:169, 1975.
120. Anderson PR, Barbacid M, Tronick SR, Clark HF, Aaronson SA: Evolutionary relatedness of viper and primate retroviruses. Science 204:318, 1979.
121. Cohen JC, Varmus HE: Endogenous mammary tumour virus DNA varies among wild mice and segregates during inbreeding. Nature 278:418, 1979.
122. Klein J: The major histocompatibility complex of the mouse. Science 203:516, 1979.
123. Wakeland EK, Klein J: The histocompatibility-2 system in wild mice. VII. Serological analysis of 29 wild-derived *H-2* haplotypes using antisera to inbred I-region antigens. Immunogenetics 8:27, 1979.

124. Shonnard JW, Cramer DV, Polosky PE, Kunz HW, Gill TJ Jr: Polymorphism of the major histocompatibility locus in the wild Norway rat. Immunogenetics 3:193, 1976.
125. Cramer DV, Davis BK, Shonnard JW, Stark O, Gill TJ Jr: Phenotypes of the major histocompatibility complex in wild rats of different geographic regions. J Immunol 120:179, 1978.
126. Phillips JT, Streilein JW, Duncan WR: The biological characterization of Syrian hamster cell-surface antigens. I. Analysis of allogeneic differences between recently wild and highly inbred hamsters. Immunogenetics 7:445, 1978.
127. Aaronson SA, Stephenson JR: Endogenous type-C RNA viruses of mammalian cells. Biochim Biophys Acta 458:323, 1976.
128. Tennant JR: Susceptibility and resistance to viral leukemogenesis in the mouse. II. Response to the virus relative to histocompatibility factors carried by the prospective host. J Natl Cancer Inst 34:633, 1965.
129. Streilein JW: Summation of the symposium. Fed Proc 37:2108, 1978.
130. East J: Molecular relationships of RNA tumor viruses and human cancer, in Pagel WJ (ed): "Research Report, University of Texas System Cancer.Center and MD Anderson Hospital and Tumor Institute," Houston: University of Texas Press, 1978, p 386.
131. Švec J, Michalides R: Biochemical properties of endogenous rat C-type viruses. Neoplasma 24:601, 1977.
132. Stephenson JR, Devare SG, Reynolds FH Jr: Translational products of type-C RNA tumor viruses. Adv Cancer Res 27:1, 1978.
133. Jerne NK: The somatic generation of immune recognition. Eur J Immunol 1:1, 1971.
134. Von Boehmer H, Hass W, Jerne NK: Major histocompatibility complex-linked immune-responsiveness is acquired by lymphocytes of low-responder mice differentiating in thymus of high-responder mice. Proc Natl Acad Sci USA 75:2439, 1978.
135. Zinkernagel RM, Althage A, Cooper S, Kreeb G, Klein PA, Sefton B, Flaherty L, Stimpfling J, Shreffler D, Klein J: *Ir*-genes in *H-2* regulated generation of anti-viral cytotoxic T cells: Mapping to *K* or *D* and dominance of unresponsiveness. J Exp Med 148:592, 1978.
136. Lewontin RC, Ginzberg LR, Tuljapurkar SD: Heterosis as an explanation for large amounts of genetic polymorphism. Genetics 88:149, 1978.
137. Hurme M, Hetherington CM, Chandler PR, Simpson E: Cytotoxic T-cell responses to H-Y: mapping of the *Ir* genes. J Exp Med 147:758, 1978.
138. Shultz FT, Briles WE: The adaptive value of blood group genes in chickens. Genetics 38:34, 1953.
139. Gilmore DG: Current status of blood groups in chickens. Ann NY Acad Sci 97:166, 1962.
140. Morton JR, Gilmore DG, McDermid EM, Ogden AL: Association of blood-group and protein polymorphisms with embryonic mortality in the chicken. Genetics 51:97, 1965.
141. Palm J, Ferguson FW: Immunogenetic studies of cellular antigens. Biennial Research Report. The Wistar Inst Anat and Biol, 1969, p 64.
142. Converse PJ, Williams DRR: Increased HLA-B heterozygosity with age. Tissue Antigens 12:275, 1978.
143. Tennant JR, Snell GD: The *H-2* locus and viral leukemogenesis as studied in congenic strains of mice. J Natl Cancer Inst 41:597, 1968.
144. Snell GD: The *H-2* locus of the mouse: observations and speculations concerning its comparative genetics and its polymorphisms. Folia Biol (Prague) 14:335, 1968.
145. Bloom BR: Games parasites play: how parasites evade immune surveillance. Nature 279:21, 1979.
146. Kumar V, Bennett M, Eckner RJ: Mechanisms of genetic resistance to Friend virus leukemia in mice. I. Role of ^{89}Sr-sensitive effector cells responsible for rejection of bone marrow allografts. J Exp Med 139:1093, 1974.

147. Bansal SC, Bansal BR, Boland JP: Blocking and unblocking serum factors in neoplasia. Current Top Microbiol Immunol 75:45, 1976.
148. James K: The influence of tumour cell products on macrophage function in vitro and in vivo: A review. In James R, McBride WH, Stuart AE (eds): "The Macrophage," Edinburgh: James, McBride and Stuart, 1977, p 225.
149. Rhodes J, Bishop M, Benfield J: Tumor surveillance: How tumors may resist macrophage-mediated host defense. Science 203:179, 1979.
150. McBride WH, Bard JBL: Hyaluronidase-sensitive halos around adherent cells: Their role in blocking lymphocyte-mediated cytolysis. J Exp Med 149:507, 1979.
151. Snell GD, Winn HJ, Stimpfling JH, Parker SJ: Depression by antibody of the immune response to homografts and its role in immunological enhancement. J Exp Med 112:293, 1960.
152. Mitchison NA: Studies on the immunological response to foreign tumor transplants in the mouse. I. The role of lymph node cells in conferring immunity by adoptive transfer. J Exp Med 102:157, 1955.
153. Kaliss N, Molomut N, Harriss JL, Gault SD: Effect of previously injected immune serum and tissue on the survival of tumor grafts in mice. J Natl Cancer Inst 13:847, 1953.
154. Cohen IR, Wekerle H: Autoimmunity, self-recognition and blocking factors. In Talal N (ed): "Autoimmunity." New York: Academic Press, 1977, p 231.
155. Nepom JT, Hellström I, Hellström KE: Antigen-specific purification of blocking factors from sera of mice with chemically induced tumors. Proc Natl Acad Sci USA 74:4605, 1977.
156. Tilney NL, Bancewicz J, Rowinski W, Notis-McConarty J, Finnegan A, Booth D: Enhancement of cardiac allografts in rats. Comparison of host responses to different protocols. Transplantation 25:1, 1978.
157. Batchelor JR, Welsh KI, Burgos H: Immunological enhancement. Transplant Proc 9:931, 1977.
158. Hendry WS, Tilney NL, Baldwin WM III, Graves MJ, Milford E, Strom TB, Carpenter CB: Transfer of specific unresponsiveness to organ allografts by thymocytes. Specific unresponsiveness by thymocyte transfer. J Exp Med 149:1042, 1979.
159. Duc HT, Kinsley RG, Voisin GA: Ia versus K/D antigens in immunological enhancement of tumor allografts. Transplantation 25:182, 1978.
160. McKenzie IFC, Henning MM: The differential destructive and enhancing effects of anti-H-2K, H-2D, and anti-Ia antisera on murine skin allografts. J Exp Med 147:611, 1978.
161. Prud'homme GJ, Sohn U, Delovitch TL: The role of H-2 and Ia antigens in graft-versus-host reactions (GVHR). Presence of host alloantigens on donor cells following GVHR and suppression of GVHR with an anti-Ia antiserum against host Ia antigens. J Exp Med 149:137, 1979.
162. Davis WC: Enhancement of heart allograft survival across *H-2* complex. Transplant Proc 9:937, 1977.
163. Taniguchi M, Miller JFAP: Specific suppressive factors produced by hybridomas derived from the fusion of enriched suppressor T cells and a T lymphoma cell line. J Exp Med 148:373, 1978.
164. Taussig MJ, Holliman A: Structure of an antigen-specific suppressor factor produced by a hybrid T-cell line. Nature 277:308, 1979.
165. Kontianen S, Simpson S, Bohrer E, Beverley PCL, Herzenberg LA, Fitzpatrick WC, Vogt P, Torano A, McKenzie IFC, Feldman M: T-cell lines producing antigen-specific suppressor factor. Nature 274:477, 1978.
166. Taniguchi M, Saito T, Tada T: Antigen specific suppressive factor produced by a transplantable I-J bearing T-cell hybridoma. Nature 278:255, 1979.

167. Taussig MJ, Corvalán JRF, Binns RM, Holliman A: Production of an H-2-related suppressor factor by a hybrid T-cell line. Nature 277:305, 1979.
168. Binz H, Wigzell H: Specific transplantation tolerance induced by autoimmunization against the individuals own, naturally occurring idiotypic, antigen-binding receptors. J Exp Med 144:1438, 1976.
169. Krco CJ, Johnson E, David CS, Tomasi TB Jr: Differences in the susceptibility of MHC- and non-MHC mixed lymphocyte reactions to suppression by murine amniotic fluid and its components. J Immunogenet 6:439, 1979.
170. Perry LL, Dorf ME, Benacerraf B, Greene MI: Regulation of immune response to tumor antigens: Interference with syngeneic tumor immunity by anti-IA alloantisera. Proc Natl Acad Sci USA 76:920, 1979.
171. Van Rood JJ, Persijn GG, van Leeuwen A, Goulmy E, Gabb BW: A new strategy to improve kidney graft survival: The induction of CML non-responsiveness. Transplant Proc 11:736, 1979.
172. Dausset J, Contu L, Legrand L, Rapaport FT: The role of HLA-DR antigens in transplantation — survival of skin allografts in HLA-haploidentical donor-recipient combinations. Transplant Proc 10:995, 1978.
173. Snell GD, Smith P, Fink MA: An enhancement of the growth of tumor homiotransplants in mice produced by the intravenous injection of donor whole blood (abstract). Proc Am Assn Cancer Res 2:47, 1955.
174. Sasportes M, Fradelizi D, Dausset J: HLA-DR specific human suppressor lymphocytes generated by repeated in vitro sensitization against allogeneic cells. Nature 276:502, 1978.
175. Yap KL, Ada GL, McKenzie IFC: Transfer of specific cytotoxic T lymphocytes protects mice inoculated with influenza virus. Nature 273:238, 1978.
176. Roder JC, Rosén A, Fenyö EM, Troy FA: Target-effector interaction in the natural killer cell system: Isolation of target cell structures. Proc Natl Acad Sci USA 76:1405, 1979.
177. Welsh RM: Mouse natural killer cells: Induction, specificity and function. J Immunol 121:1631, 1978.
178. Marx JL: Interferon (I): On the threshold of clinical application. Science 204:1183, 1979.
179. Kumar V, Bennett M: *H-2* compatibility requirements for T suppressor cell function induced by Friend leukemia virus. Nature 265:345, 1977.
180. Kende M, Hill R, Dinowitz M, Stephenson JR, Kelloff GJ: Naturally occurring lymphocyte-mediated immunity to endogenous type-C virus in the mouse. Blocking of the lymphocyte reactivity with antisera to the virus. J Exp Med 147:358, 1979.
181. Minato N, Bloom BR, Jones C, Holland J, Reid LM: Mechanisms of rejection of virus persistently infected tumor cells by athymic nude mice. J Exp Med 149:1117, 1979.
182. Prehn RT: Do tumors grow because of the immune response of the host? Transplant Rev 28:34, 1976.
183. Djeu JY, Heinbaugh JA, Holden HT, Herberman RB: Role of macrophages in the augmentation of mouse natural killer cell activity by Poly I: C and interferon. J Immunol 122:182, 1979.
184. Herberman RR, Ortaldo JR, Bonnard GD: Augmentation by interferon of human natural and antibody-dependent cell-mediated cytotoxicity. Nature 277:221, 1979.
185. Heron I, Hokland M, Berg K: Enhanced expression of β_2-microglobulin and HLA antigens on human lymphoid cells by interferon. Proc Natl Acad Sci USA 75:6215, 1978.
186. Kilburn DG, Talbot FO, Teh H-S, Levy JG: A specific helper factor which enhances the cytotoxic response to a syngeneic tumour. Nature 277:474, 1979.
187. Pierres M, Germain RN, Dorf ME, Benacerraf B: The vivo effects of anti-Ia alloantisera. I. Elimination of specific suppression by in vivo administration of antisera specific for

I-J controlled determinants. J Exp Med 147:656, 1978.
188. Kurdi A, Abdallat A, Ayesh I, Maayta U, McDonald WI, Compton DAS, Batchelor JR: Different B lymphocyte alloantigens associated with multiple sclerosis in Arabs and North Europeans. Lancet 1:1123, 1977.

ADDENDUM

Since this paper was written, two reports have appeared of sufficient relevance to warrant an addendum.

Curtis and Rooney [1] have studied the extent of contact inhibition occurring when outgrowths of H-2-matched and H-2-mismatched mouse kidney epethelial cells encounter one another in tissue culture. The most pronounced inhibition was seen in the matched combinations. The active loci mapped in the $K,D,L,$ regions. This confirms the importance of these loci in cell interactions.

Streilein and Duncan [2] have immunized two partly inbred strains of Syrian hamsters, recently derived from wild animals, with cells from the existing inbred strains. High titer cytotoxic alloantibodies were produced. The reciprocal combinations gave similar results. The strength of the antibodies suggests that they were formed against the hamster homologs of the mouse K,D,L antigens. If so, at least two alleles must exist at the determining loci. Further tests are needed, but earlier results suggesting a total lack of allelic diversity may have to be reinterpreted. However, that hamsters show a low order of MHC polymorphism still seems likely.

REFERENCES

1. Curtis ASG, Rooney P: H-2 restriction of contact inhibition of epithelial cells. Nature 281:223, 1979.
2. Streilein JW, Duncan WR: Alloimmune reactions among recently wild Syrian hamsters and classical inbred strains include alloantibody production. Immunogenetics 9:563, 1979.

Immunogenetics: Chairman's Summary

Dorothea Bennett

The three richly diverse talks of Drs. Bodmer, Klein, and Snell exemplify the scope of modern immunogenetics which, by its very name, is clearly a hybrid discipline. Actually, immunogenetics is comprised of two almost separate fields, depending on the bias of the participant. On one hand, it consists of the use of immunological methods to define the phenotype associated with specific genes, whether or not those genes and their products have anything at all to do with the immune system; on the other hand it is the study of genes involved specifically in immune functions.

As this symposium amply demonstrates, much of immunogenetics as it exists today not only relates to both these areas, but is also heavily concerned with tumor biology. These concentrations of interest are, in a sense, an interesting historical accident stemming from the origins of the field of immunogenetics. "Immuno" is really just an operational term, indicating in the context of the field that cell surface components can be detected by immunological means. It does not mean that cell components thus detected either behave as antigens in vivo or are involved in immune reactions in vivo. Interestingly enough, however, much of the work that has been done on defining genetically determined "alloantigens" has been done on lymphocytes. No doubt, the original reason for this was that lymphocytes are readily available as free-cell populations, and it is under just such conditions that cell membrane antigens are most readily studied by immunological means. But then immunogenetics became truly fixated on lymphocytes: first, when it was found that these cells are highly interactive and responsive to one another and to genetically foreign cells, and second, when the fortuitous circularity of this approach became strikingly evident with discoveries that the interesting interactions shown by lymphocytes are actually controlled by some of their most readily detectable antigens! *But* it is important to remember that recognition and response mechanisms in the cell membrane are almost certainly equally important for many other situations where cells interact with one another. One such situation, an important province of immunogenetics, is *tumor immunology*, where immunological methods can be used to define specific antigens associated with specific tumors produced by chemicals or viruses, and to follow the expression, or lack of expression, of normal cellular genes. Actually, this is another part of the circular, nepotistic, and incestuous interrelationships amongst the various important areas of immunogenetics as currently practiced — since the construction of congenic strains and the working out of the "Laws" of transplantation which were directly responsible for the definition of histocompatibility antigens all stemmed from a primary interest in tumor transplantation. So far though, the application of immunogenetics to tumor biology has not thoroughly fulfilled the promise that so many of us think it contains. Although it seems so painfully clear to many of us that tumor cells differ from the normal cells of their hosts in ways that must involve differences in their cell surface, since their abnormal activities such as metastasis are all most readily explained by abnormal cell-cell interactions, there have been virtually no clear demonstrations of important alterations in the antigenicity of spontaneously occurring tumors. The situation in autochthonous tumors may not be so simple as chemically and virally induced tumors led us to believe. It may well be that spontaneous tumors result from subtle alterations in complexes of existing antigens (what George Snell referred to as "islands") that result in profound functional differences but not in differences in antigenicity that are detectable with current methods.

By way of summary, it can be said immunogenetics represents a kind of microcosm containing virtually everything the mammalian geneticist would like to approach, that is: to identify genes with a functionally important role in the activities of a cell, not just in terms of the existence of that cell as an isolated entity but, much more specifically and interestingly, in how that cell interacts with others to create and maintain a functioning organism; to understand how those genes are organized and how they have evolved and become controlled, not only in the nuclear genome but in populations; and finally, to define not only the kind of molecule those genes produce but how that molecular structure relates to function.

SESSION V. THE ETIOLOGY OF CANCER

RICHMOND T. PREHN, Chairman

Introduction to Session V

INTRODUCTION

The three papers presented in this session are concerned with the role of genes and viruses in the causation of cancer. Some of the research described started even before the foundation of The Jackson Laboratory, while other contributions come from the latest research efforts at the National Cancer Institutes and at Massachusetts Institute of Technology.

Particularly pertinent for the 50th Anniversary Symposium of The Jackson Laboratory is Walter E. Heston's fascinating account of Dr. C.C. Little and his early co-workers, of the development of inbred strains of mice, of the impact of these strains in demonstrating the importance of genetic factors in determining both cancer incidence and the types of cancer produced, and the discovery of the mammary tumor virus. He goes on to emphasize the importance of mouse inbred strains for advances in immunogenetics, cancer, virology, carcinogenesis testing, and chemotherapy.

There is a long but clear pathway from mousedom's early contributions to cancer research to Wallace Rowe's recent studies on viruses as chromosomal genes. For a long time vertically transmitted oncogenic viruses were very puzzling, but Rowe's paper shows clearly that all or major portions of C-type RNA viral genomes do become incorporated (presumably as DNA), into the chromosomes of their murine hosts at specific sites, and are transmitted from generation to generation in mendelian fashion. They also replicate independently in the host, and the inherited unit can become altered by somatic recombination followed sometimes by re-insertion into the mouse genome. As Rowe says, this appears to be a biologic system functioning "at the interface of virology and mammalian genetics."

David Baltimore asks different questions. How do viruses transform cells? What part of the viral genome effects transformation? What kind of cells are targets for a particular transforming virus? Baltimore describes the origin of the lymphoma-inducing Abelson virus in a mouse infected with the thymic-lymphoma-inducing Maloney virus. The cellular target had changed! Also, the Abelson virus is defective, requiring helper virus for replication. Structural studies of the Abelson virus indicate that both its ends are homologous with ends of the Maloney genome, but the mid-portion seems to be normal mouse cellular DNA coding for a transmembrane protein known to be expressed in T-lymphocytes. This Abelson virus can transform cells at "specific stages along the pathway to B-lymphocyte differentiation." The very extensive research leading to these fascinating findings may lead the way toward understanding of the complicated program controlling lymphocyte differentiation.

Development and Utilization of Inbred Strains of Mice for Cancer Research

Walter E. Heston

INTRODUCTION

The history of the inbred strains of mice and their use in cancer research is, in large measure, an account of the history of the Jackson Laboratory and especially of its founder and first director, Dr. Clarence Cook Little. Many of our younger researchers today did not have the good fortune to know Dr. Little personally, but I knew him well. He was the greatest man I ever knew. He had a fertile mind that inspired every investigator with whom he came in contact.

Two of Dr. Little's principal aims in life were to develop proper experimental animals for cancer investigators and to provide the funds to support their research. He accepted the Presidency of the University of Michigan only after being assured ample funds to support his own cancer research program. He organized the Women's Field Army, whose members rang doorbells to collect dollars from the American public for the support of cancer research, an endeavor that later developed into the grant program of the American Cancer Society. He helped write the National Cancer Institute Act and defended it before congressional hearings to establish the National Cancer Institute with federal funds to support cancer research, not only at the Institute but in other laboratories throughout the country. He viewed his association with the tobacco industry as an opening wedge into the almost unlimited funds of private industry that had hardly been tapped for medical research.

The publication of Mendel's laws of inheritance was rediscovered in 1900. In 1903, Johannsen [1] postulated that inbreeding would result in a pure line. In 1907, an extensive program on the effects of inbreeding in the guinea pig was started by the US Department of Agriculture in its laboratory in Bethesda,

Maryland. Later this program was transferred to Beltsville and taken over by Dr. Sewall Wright who for many years collected data resulting in his monumental publications [2–4] on the principles of inbreeding and the results of crosses between two inbred strains. These publications have been of great value in understanding the use of inbred strains and the interpretation of data obtained from them.

DEVELOPMENT OF INBRED STRAINS OF MICE

It was C.C. Little who had the foresight to realize the importance of using inbred strains of experimental animals in medical research. He saw that, just as the chemist needed pure chemicals, the biologist needed genetically pure strains of animals if he was going to succeed in analyzing the causes of the various complex medical traits, especially cancer. Characteristic of his always being ahead of the crowd in his thinking was the fact that he began to foresee this while still an undergraduate student. In 1909, he started to inbreed strain DBA from some mice he was carrying out coat-color studies on at the newly-founded Bussey Institution of Harvard University under the tutelage of the father of mammalian genetics, Dr. William E. Castle. He initially selected for the three recessive coat-color genes, dilute, brown, and nonagouti. Hence, the designation dba, later changed to DBA. Selection for mammary tumors established the strain as a high mammary tumor strain. From this beginning until the day he retired from the directorship of the Jackson Laboratory in 1956, Dr. Little was the guiding force in the development of the inbred strains and the education of biomedical scientists in the importance of using them in their research. Many others have been involved in the actual inbreeding of the many strains, but almost all of these persons initially came under the influence and inspiration of Dr. Little.

During the first decade, Dr. Little was involved in inbreeding of strain DBA and delving into the inheritance of cancer [5]. In contrast with his work was that of Dr. Maud Slye [6, 7], the other leading investigator of the inheritance of cancer during that time. She bred thousands of mice on which she kept complete records and from which pathologic descriptions of practically all kinds of tumors ever observed in mice were made by her associate pathologist, Dr. H.G. Wells. She initially concluded all cancer was inherited as a single recessive trait, a conclusion she later modified several times. But none of her conclusions were valid, as Dr. Little [8] was always ready to point out. Today, many of the younger investigators in cancer have not even heard of Maud Slye and her work and none of our hundreds of strains of inbred mice can be traced back to her stocks. Why? Because she did not inbreed her mice.

In 1919, Leonell C. Strong, at that time a graduate student of Dr. Thomas Hunt Morgan at Columbia University, became associated with Dr. Little at The Carnegie Institution of Washington at Cold Spring Harbor on Long Island. While working with Little there, Strong mated strain DBA with a Bagg Albino stock and from this mating, inbred several strains, the best known of which are C3H and CBA. He also mated some Bagg Albinos with an albino stock that Little had and, from this mating, he developed inbred strain A. Dr. E.C. MacDowell (a graduate student with Little at Harvard and later associated with him at Cold Spring Harbor) started to inbreed the Bagg Albinos, thus developing the strain later known as BALB/c. Continuation of inbreeding this strain was carried out later by Dr. George Snell with Dr. Little at the Jackson Laboratory.

In 1921, Dr. Little mated two females, numbers 57 and 58, with a male, number 52 — all litter mates of one of Lathrop's stocks. The offspring of female 57, which was black, segregated for black and brown. By inbreeding and selecting for black, Little originated strain C57BL, and for brown, strain C57BR. MacDowell inbred the offspring of female 58, thus developing the high leukemia strain C58, which is also black. In 1933, Dr. J.M. Murray discovered in the C57BR strain a coat-color mutation he designated leaden and from the leaden segregants he originated strain C57L.

In 1922, Dr. Little moved with his mouse stocks from Cold Spring Harbor to Orono, Maine, to become President of the University of Maine. There he gathered around him three graduate students, Joseph M. Murray, William S. Murray, and Arthur M. Cloudman. When Dr. Little was appointed President of the University of Michigan in 1925, these three students transferred to Ann Arbor with him, bringing along the mouse stocks. There John J. Bittner, Elizabeth Fekete, and Charles V. Green also came under his tutelage as graduate students. There also, Dr. Strong rejoined him with his stocks. This whole group came with Dr. Little to Bar Harbor in 1929 when he established the Jackson Laboratory there. They brought with them all their strains to form the foundation of the Jackson Laboratory inbred strains. All of these young investigators were involved in some way with the inbreeding of these early strains.

Since that time, many others have joined the staff of the Jackson Laboratory and all have become involved in the inbreeding and use of these and many other strains, and in the development of new types of strains such as coisogenic strains and congenic lines. During these 50 years the Jackson Laboratory has held the lead in the development of inbred strains of mice and in supplying them to biomedical researchers throughout the world.

The Jackson Laboratory also has been very influential in establishing colonies of inbred mice in many other laboratories. Dr. Andervont became associated with Dr. Little in the early 1930s through summer visits to the Jackson Laboratory. He took some of the Jackson Laboratory strains to his laboratory in Boston, and later transferred them to Bethesda when the National Cancer Institute was opened there. Dr. Lloyd Law and I joined Dr. Little and his staff in 1938, and when I transferred to the National Cancer Institute in 1940, I brought the Jackson Laboratory strains with which I had been working. When Dr. Law came to the National Cancer Institute in 1947 he intended to bring his Jackson Laboratory strains but unfortunately, they were caught in the fire at the laboratory, packed in their cases ready to be shipped the following morning. Dr. George Jay followed us to Bar Harbor and then to Bethesda, and he also brought Jackson Laboratory strains to the National Institutes of Health.

Soon after the establishment of the Jackson Laboratory in Bar Harbor Dr. R. Korteweg, Director of the National Cancer Institute in Amsterdam, visited Dr. Little and his laboratory. Dr. Korteweg took back with him two of the Jackson Laboratory strains, DBA and C57BL, to start a colony of inbred strains in his laboratory in the Netherlands. His successor, Dr. Otto Mühlbock, also became associated with Dr. Little and retained that association with the Jackson Laboratory subsequent to Dr. Little's retirement. Dr. Mühlbock was very instrumental in the establishment of an adequate source of inbred mice for medical research in Europe. He also developed the very interesting strain GR.

One could continue to emphasize the Jackson Laboratory's role in development and utilization of inbred strains of mice and their distribution throughout the world, but Dr. Little's influence extended beyond mice. He brought Dr. Paul Sawin to the Jackson Laboratory for the primary purpose of developing inbred strains of rabbits. Dr. Maynie R. Curtis became associated with Dr. Little in the New York area in 1919 and, in 1920, began inbreeding strains of rats at the Crocker Laboratories of Columbia University; for many years she and Dr. Wilhelmina F. Dunning pioneered the development of inbred strains of rats. The last of Dr. Wright's inbred strains of guinea pigs still at Beltsville were about to be discarded because of budgetary and space restrictions when Dr. Andervont and I, indoctrinated in the value of the inbred strains at the Jackson Laboratory, and Dr. Shimkin, indirectly so indoctrinated, rushed in to salvage them. We succeeded in getting two of the strains, numbers 2 and 13, established at the National Cancer Institute and, today, they are being widely used in biomedical research.

I shall not go into a discussion of the breeding systems used in the production of the various kinds of strains. Most strains have been produced by brother-sister matings, but Dr. Little's successor, Dr. Earl Green, is a master at presenting this subject and has very eloquently done so on several occasions [9]. I will, therefore, devote the rest of my time to how these inbred strains of mice have been utilized. Since a very large part of modern basic biomedical research is now based on the inbred mouse, I cannot begin to cover the subject. I will only point out how some of the more important, and possibly more interesting, advancements have been brought about through the use of the inbred strains.

UTILIZATION IN IMMUNOGENETICS

In the early years the inbred strains were used largely for studies on the genetics of tumor transplantation, a subject that could not be studied without inbred strains. In graduate school Dr. Little became interested in this subject through the influence of Dr. E.E. Tyzzer, with whom he carried out experiments on the transplantation of tumors in hybrids between a stock of Japanese waltzing mice and the strain DBA [10]. Their data suggested heredity was involved but did not give clear-cut evidence of mendelizing genes. As a graduate student in 1914, Little [11] published a brief note hypothesizing that certain characters in an organism might be dependent upon the presence of two or more genes. When applied to tumor transplantation this would mean growth of a transplanted tumor would depend upon the simultaneous presence, in the tumor and the host, of multiple, dominant mendelizing genes. The number of these genes could theoretically be determined from the percentage of susceptible animals in the F_2 generation from a cross between the strain of origin of the tumor and a resistant strain, and in the first backcross to the resistant strain. The percent susceptible animals expected in the F_2 would be 75% for one gene, 56.2% for two genes, etc, and in the backcross to the resistant parent would be 50% for one gene, 25% for two, etc.

This hypothesis was successfully applied to many studies on genetics of tumor transplantation in the 1920's by Little and by Strong. Little and Beatrice Johnson, who was later to become Mrs. Little, also carried out such studies on the transplantation of normal tissue, spleen. This group was later joined by Bittner, who in the early 1930s published many papers in this area, and by Cloudman who also carried out such studies. Dr. Andervont published some papers in this area in the early 1930s. The general conclusions from these studies were that malignant tumors arising in an inbred strain could be successfully transplanted into practically all other animals of that strain but not into animals of another strain; that growth of the tumor was determined by multiple dominant genes; and that the number of these genes could be determined by the percentage of susceptible animals in the F_2 and backcross hybrids. In certain tumors, through successive generations of transplantation, this number of genes would change to a lesser number. The number of genes involved was specific for a given tumor. Two tumors of the same histologic type arising in the same animal could be controlled by a different number of genes (for a review, see Little [12]).

In the 1940s, Dr. George D. Snell, also a student of Castle and by now on the staff of the Jackson Laboratory, began a series of very significant studies on the genetics of tumor transplantation, and the development of new methods for studying histocompatibility genes [13]. Immediately following World War II, Dr. Peter Gorer, a pathologist at Guy's Hospital in London, who had been studying tumor transplantation from the viewpoint of an immunologist [14], came to the Jackson Laboratory for a year of study. While there, he established two very significant relationships. He became associated with Miss Elizabeth Keucher, one of the most important members of the Jackson Laboratory staff, and took her back to London as Mrs. Gorer. Also, he became associated with Dr. Snell and they soon discovered that the antigens Gorer had been concerned with were the same as the genes Snell had been dealing with. More specifically stated, the alloantigens were the end products of the histocompatibility genes. This scientific marriage between these two investigators working with Jackson Laboratory inbred mice gave birth to the modern science of immunogenetics (for reviews, see Snell [15, 16]). None of the studies reported at this conference in this area would have or could have been done, had it not been for the inbred strains of mice. There probably would not even be any modern science of immunogenetics had we not had inbred strains of mice.

UTILIZATION IN GENETICS OF CANCER

The first evidence that heredity is involved in the occurrence of cancer was published by Bashford and Murray [17, 18] in London at the end of the first decade of this century, and this study was carried out on non-inbred mice. However, no significant analysis of the genetics of cancer was possible before the development of the inbred strains. Actually, the inbreeding of the strains themselves and the recording of the incidences of tumors in each strain were the most significant experiments on the genetics of cancer that have yet been carried out. The fact that inbreeding with selection established strains with certain incidences of tumors that were constant, generation after generation, was proof that genetic factors were involved. The observation that strains could not be divided into cancer strains and non-cancer strains, but that each strain had its own specific types of cancer with its specific incidence for each type, showed cancer cannot be inherited as a single disease but that each type of cancer is a separate entity with its own genetics. When the incidences of specific tumors were recorded for the various strains, it was found these incidences did not have to be either 100% or 0% but that there were intermediate incidences. With respect to lung tumors strain A had an incidence of 90%, SWR 40%, BALB/c 20%, and C57L less than 1%. This indicated that the different types of cancer were not controlled by single genes.

We had to go back to Wright's studies [4] of crosses between two inbred strains of guinea pigs to get an explanation of the type of inheritance involved. Just as his polydactyly in the guinea pig is inherited as a threshold character, so are the different types of cancer. That is, they are controlled by multiple genetic and nongenetic factors, the tumor appearing when the combined effects of the two sets of factors surpass a certain physiologic threshold which, in this case, is the malignant transformation. Thus, the genotype of the strain establishes the incidence of tumor; whether or not an individual within the strain is one that gets the tumor, or one that does not, is determined by the variable nongenetic factors. Age is one of the important nongenetic factors.

There can be a gene that will cause all animals to develop a certain kind of cancer, as Dr. Elizabeth Russell [19] showed in her studies of the W-series. One of the genes in this series causes all females to develop ovarian tumors. This same type of tumor, however, occurs in certain strains without this gene and in different incidences in different strains so it, too, would fall in line with the general type of inheritance of other tumors except it does have one gene that always brings the animal above the threshold.

With the use of the inbred strains, the carcinogens were placed in proper perspective in relation to the genetics. They are some of these nongenetic factors that have positive effects. Lynch took three inbred strains with high, medium, and low incidences of spontaneous lung tumors, respectively, and to each gave the same dose of a well-known carcinogen. She obtained a greater tumor response in each strain, but the strains retained their relative positions. Andervont has published similar results.

Using inbred strains and a carcinogen that induced multiple tumors, it was possible to get a quantitative measure of response of animals of a high- and a low-lung tumor strain and of their hybrids, thus confirming multiple-factor inheritance for this type of tumor. This, in turn, was confirmed for spontaneous lung tumors using the same inbred strains and their hybrids and measuring susceptibility not only by tumor incidence but also by tumor age. Burdette and Strong demonstrated multiple-factor inheritance for induced subcutaneous sarcomas using high- and low-susceptible inbred strains and their hybrids with one of the polycyclic hydrocarbon carcinogens.

All of the work showing the effects of specific-known genes on specific types of tumors — and there are quite a number of known genes that do increase or decrease the occurrence of tumors in the mouse — could be carried out only with the use of inbred strains (for review of genetics of cancer, see Heston [20]).

In retrospect, it is inconceivable that anyone could carry out work on the genetics of experimental tumors without genetically controlled, inbred strains of experimental animals. It is no wonder that none of Slye's conclusions were valid.

UTILIZATION IN CANCER VIROLOGY

The viral approach to the cancer problem had its beginning at about the same time as the genetic approach. In 1911, Dr. Peyton Rous described the chicken tumor virus that now bears his name [21]. For the next 30 years, however, relatively little progress was made in this field. This was due, in part, to the strong opposition Dr. Rous met from his fellow staff member of the Rockefeller Institute, Dr. James B. Murphy, a strong supporter of Dr. Little and for many years a Trustee of the Jackson Laboratory. Dr. Murphy [22] published this opposition (1931) arguing since the fowl tumor group was composed of many different types, each with a specific agent, one must assume an infinite variety of such agents. Therefore, he proposed these agents could not be considered extraneous infectious agents as were viruses. Rather, they must be of *endogenous origin*. He also compared their transformation of cells with the transforming agent of pneumococcus and, thus, proposed they not be considered as viruses. He suggested they be called "transmissible mutagens." These ideas of Dr. Murphy, recorded at about the time of establishment of the Jackson Laboratory 50 years ago, sound very modern today. Unfortunately, he and Rous could never join forces. However, I do not know just what could have been done if they had brought their ideas together, for the non-inbred chickens Rous was working with were hardly suitable material for studying endogenous oncogenic viruses. Thus, the viral approach to cancer went into decline.

The group of geneticists led by Little, and their inbred strains of Jackson Laboratory mice, were largely responsible for reviving the viral approach. After developing their inbred mouse strains, some of which had high incidences of mammary tumors while others had low incidences, their next logical step was to cross these high- and low- tumor strains. Fortunately, they made reciprocal crosses. The discovery was that the F_1 females had an incidence of mammary tumors like that of the strain of the mother. Since the reciprocal F_1 females were identical genetically, this causative agent had to be an extrachromosomal factor that was transmitted maternally. This important discovery was published by the staff of the Jackson Laboratory in 1933 [23]. A year later Korteweg [24] published similar results from his reciprocal crosses between strains DBA and C57BL he had gotten from Little. The steps are well known through which first the geneticists, and later the virologists who joined them, demonstrated the agent was transmitted through the milk and finally identified it as the mouse mammary tumor virus (MMTV), an RNA virus seen in its mature form as the B particle. This was the first virus to be recognized universally in this country as a cancer-causing virus. Prior to this the situation had developed in which, almost by definition, if it were cancer it could not be caused by a virus and if there was a causative virus it could not be cancer. But the mammary cancer of the mouse had been recognized as cancer for decades, so this virus had to be considered as cancer-causing.

It was recognized from the outset, however, that this virus was not a classical virus. For many years Bittner persisted in calling it the "milk agent" and never considered it the sole cause of the cancer, but insisted there were three factors of equal importance: the genotype of the host, the hormonal influence, and the milk agent. As early as 1945, Heston, Deringer, and Andervont published data showing close association between this agent (or virus) and the genes of the host and the discussion suggested the virus may have originated in some way from the genes.

The inbred strains could be manipulated. The virus in strain C3H, then known to be transmitted through the milk, could easily be removed by foster-nursing the young on strain C57BL females without the virus. This fostered line was designated as C3HfB. But the females of this line also had mammary tumors, although they occurred in a lower incidence and at an older age. These tumors also had B particles that were not transmitted through the milk but by the sperm and egg.

By introducing the cancer-enhancing gene A^{vy}, producing line C3H-A^{vy}fB, the incidence of mammary tumors was raised to 90%; still, the virus of these tumors was not transmitted through the milk but by both, sperm and egg (For more recent work and references, see Heston, Smith, Parks [25] and Heston, Parks [26].

This line of investigation was climaxed, in 1968, by Bentvelzen's proposal that the MMTV was genetically transmitted [27, 28]. Again, this advance in thinking was made because of a special inbred mouse strain GR developed by Mühlbock [29]. This strain had a very high incidence of mammary tumors that arose at a relatively early age and were caused by a potent MMTV transmitted in the milk and also, definitely, in the egg and in the sperm.

These, and even greater advancements with the mouse leukemia viruses that Dr. Rowe [30] will be discussing in this conference, opened up a new perspective of the oncogenic viruses as being endogenous and genetically transmitted. This whole field of research could not have developed had it not been for the inbred strains of mice. I do not believe the endogenous viruses could have been demonstrated, as such, without the inbred strains.

UTILIZATION IN TESTING FOR CARCINOGENS

Testing possible carcinogens in our environment today is a very important program. It is also one in which our inbred strains of mice are very widely and very beneficially used. In fact, it is hard to conceive carrying out this program without the inbred strains of mice, which are mammals, as is the human being, and provide very sensitive tests for induction of so many types of neoplasms.

Yet, in probably no other area are there so many misunderstood points regarding the use of inbred strains of mice. Possibly, no one in this conference would think the question would be raised anymore as to the appropriateness of using an inbred strain for testing the carcinogenicity of a substance and applying the information obtained to the genetically heterogenous population of human beings. I warn you, however, that if any of you are asked to appear at an Environmental Protection Agency hearing as an expert witness, you may well get such a question put to you by those trying to defend their products as being safe. The answer to the question, of course, is that we are not trying to duplicate the population of human beings. That would be impossible, for mice are mice and men are men. What we are trying to do is establish, as a fact, whether or not the substance tested will cause the malignant transformation in cells of a mammalian organism. This can be done much more effectively and with fewer animals if we use properly selected inbred strains. Then, we apply these facts to the genetically heterogenous population of human beings. If a substance is in this way shown to be carcinogenic we can reasonably expect, in a population of human beings exposed to this substance, certain individuals will get cancer in some organ they would not have, had they not been exposed to the substance. I remember, as a young geneticist in the early days of the National Cancer Institute, having to similarly defend the use of a highly inbred strain to obtain data to be applied to human beings. It was reassuring to me to see Dr. Sewall Wright, who happened to be attending the seminar, turn around and listen intently to my explanation and then give his characteristic quick little nod of approval.

I do not go along with the thinking held by some that, after the fact the substance is carcinogenic has been established in an inbred strain of mice, it is necessary this also be established in a second species before attempting to apply the information to the population of human beings. This seems to me to be wasted effort unless, of course, the second species is man himself.

There is also some confusion as to the most appropriate strain to use. Should it be a strain very susceptible to the type of tumor with which one is concerned, very resistant, or intermediate? This misunderstanding can be cleared up by going back to Wright's early publication [4] and getting a clearer understanding of threshold characters. From Wright's bell-shaped curves, showing the distribution of animals about the threshold one can see a carcinogen is a substance that shifts this population curve a certain distance to the right, bringing more of the animals above the threshold and, thus, increasing the incidence of tumor in the population. If a strain has a very susceptible genotype so that the curve is already so far to the right only a few animals in the left tail end of the curve will not exceed the threshold and get cancer without the carcinogen, then a further shifting of the distribution curve to the right by this distance with a carcinogen is going to make very little difference in the final incidence of tumor. I could not demonstrate carcinogenicity of carbon tetrachloride in the liver of C3H mice because C3H mice are genetically very susceptible to hepatomas. On the other hand, if the strain is genetically very resistant so that only a few animals in the right tail end of the distribution curve normally exceed the threshold and get cancer, this same shift of the population curve to the right by the carcinogen is also going to make very little difference in the final incidence of the cancer. Had dibenzanthracene been tested only for the induction of lung tumors in strain C57L, it probably would have been concluded to be noncarcinogenic. With a large dose of this carcinogen injected intravenously in C57L mice, I was able to induce only a very few lung tumors and they were so small as to hardly be detectable. Thus, by insisting on using very resistant strains many carcinogens will be missed. In a strain with an intermediate degree of susceptibility where the threshold is in the vicinity of the peak of the distribution curve, the shift to the right of the same magnitude by the carcinogen will bring a much greater portion of the population above the threshold, making a very significant difference in incidence of tumor. Thus, it is the strain with an intermediate degree of susceptibility to the tumor that one suspects will be induced that makes the ideal test strain. This is not always understood by those in the testing program.

Nevertheless, with the multitude of inbred strains, proper test strains can be selected. The inbred strains of mice have contributed greatly to this program and, despite a search for more efficient tests through induction of mutations in lower organisms, their use will probably continue in this program for years to come.

UTILIZATION IN CHEMOTHERAPY

Some mention also must be made of the immense contribution the inbred strains have made to the advancement of the chemotherapy program. Since it was hardly feasible to use spontaneous tumors in the testing program, the next best thing — a transplanted tumor in the inbred strain of origin or in an F_1 hybrid between two inbred strains — was used. The many thousands of inbred mice, and of these F_1 hybrids, used have made the immense testing program possible.

But inbred mice have also been used in basic research that has guided the program. For example, the basic study by Law [31] on the chemotherapy of leukemia in mice. In this study, which could be done only with inbred strains, he demonstrated the advantage of using combinations of antileukemic agents. It took years for these basic findings to find application in human patients, but now children who get leukemia owe the fact that today they can expect to recover to this basic study and, in part, to the inbred mice that made the study possible.

On this 50th anniversary, the Jackson Laboratory can be proud of the contribution it has made to biomedical research through the inbred mouse. I have always thought Dr. Little should have gotten the Nobel Prize for having conceived of the inbred strains, for his leadership in their development, and for educating the research community in the value of using inbred strains. Possibly, one of the reasons he did not was that it would have had to be shared with so many of his colleagues who also contributed so much to this endeavor. I am sure, however, that seeing these strains develop and observing the great contributions they were making to biomedical research was of far greater value to Dr. Little than any prize would have been.

REFERENCES

1. Johannsen W: Uber Erblichkeit in Populationen und in Reinen Linien. Jena, 1903.
2. Wright S: Systems of mating. Genetics 6:111–178. 1921.
3. Wright S: The effects of inbreeding and crossbreeding on guinea pigs. III. Crosses between highly inbred families. US Dept Agri Bull No 1121, 1922.
4. Wright S: The results of crosses between inbred strains of guinea pigs, differing in number of digits. Genet 19:537–551, 1934.
5. Little CC: The relation of heredity to cancer in man and animals. Sci Monthly 3:196–202, 1916.
6. Slye M: The incidence and inheritability of spontaneous cancer in mice. Third report. J Med Res 32:159–200, 1915.
7. Slye M: The inheritance behavior of cancer as a simple Mendelian recessive. Twenty-first report. J Cancer Res 10:15–50, 1926.

8. Little CC: Evidence that cancer is not a simple Mendelian recessive. J Cancer Res 12:30–46, 1928.
9. Green EL: Breeding systems, in Green EL (ed): Biology of the Laboratory Mouse, Ed 2. New York: McGraw Hill, 1966, p 11.
10. Little CC, Tyzzer EE: Further studies on inheritance of susceptibility to a transplantable tumor of Japanese waltzing mice. J Med Res 33:393–425, 1916.
11. Little CC: A possible Mendelian explanation for a type of inheritance apparently non-Mendelian in nature. Science 40:904–906, 1914.
12. Little CC: The genetics of tumor transplantation, in Snell GD (ed): Biology of the Laboratory Mouse. Philadelphia, Blakiston, 1941, p 279.
13. Snell GD: Methods for the study of histocompatibility genes. J Genet 49:87–108, 1948.
14. Gorer PA: The antigenic basis of tumour transplantation. J Pathol Bacteriol 47:231–252, 1938.
15. Snell GD: The immunology of tissue transplantation, in Conceptual Advances in Immunology and Oncology. 16th Annu Symp Fundamental Cancer Res (Univ Texas, Houston) New York, Hoeber-Harper, 1963, p 323.
16. Snell GD: Genetics of tissue transplantation, in Green EL (ed): Biology of the Laboratory Mouse, Ed 2. New York, McGraw-Hill, 1966, p 457.
17. Bashford EF Murray JA: The incidence of cancer of the mamma in female mice of known age. Proc Roy Soc B 81:310–313, 1909.
18. Murray JA: Cancerous ancestry and incidence of cancer in mice. Scient Rep Invest Imp Cancer Res Fund 4:114–131, 1911.
19. Russell ES: Review of the pleiotropic effects of W-series genes on growth and differentiation, in Rudnick D (ed): Aspects of Synthesis and Order in Growth. Princeton, NJ, Princeton Univ Press, 1955, p 113.
20. Heston WE: Genetics: Animal tumors, in Becker FF (ed): Cancer a Comprehensive Treatise. New York, Plenum Press, 1975, vol 1, p 33.
21. Rous P: A sarcoma of the fowl transmissible by an agent separable from the tumor cells. J Exper Med 13:397–411, 1911.
22. Murphy JB: Discussion of some properties of the causative agent of a chicken tumor. Trans Assoc Am Physicians 46:182–187, 1931.
23. Jackson Laboratory Staff: The existence of nonchromosomal influences in the incidence of mammary tumors in mice. Science 78:465–466, 1933.
24. Korteweg R: Proefondervindelijke onderoekingen aangaande erfelijkheid van kanker. Nederlandsch Tijdschrift voor Geneeskunde 78:240–245, 1934.
25. Heston WE, Smith B, Parks WP: Mouse mammary tumor virus in hybrids from strains C57BL and GR: Breeding test of backcross segregants. J Exper Med 144:1022–1030, 1976.
26. Heston WE, Parks WP: Mammary tumors and mammary tumor virus expression in hybrid mice of strains C57BL and GR. J Exper Med 146:1206–1220, 1977.
27. Bentvelzen PAJ: Genetical control of the vertical transmission of the Mühlbock mammary tumor virus in the GR mouse strain. Amsterdam, Hollandia Publ Co, 1968.
28. Bentvelzen P: Hereditary infections with mammary tumor viruses in mice, in Emmelot P, Bentvelzen P (eds): RNA Viruses and Host Genome in Oncogenesis. Amsterdam, North Holland Publishing Co, 1972, p 309.
29. Mühlbock O: Note of a new inbred mouse strain GR/A. Eur J Cancer 1:123–124, 1965.
30. Rowe CE, Hartley JW, Kozak CA: Murine leukemia viruses as chromosomal genes of the mouse. Jackson Laboratory 50th Anniversary, 1980 Symposium.
31. Law LW: Effects of combinations of antileukemic agents on an acute lymphocytic leukemia of mice. Cancer Res 12:871–898, 1952.

Murine Leukemia Viruses as Chromosomal Genes of the Mouse

Wallace P. Rowe, Janet W. Hartley, and Christine A. Kozak

One of the many exciting and unanticipated discoveries in genetics made possible by the availability of inbred mouse strains is the discovery that C-type Retrovirus genomes are carried as normal, chromosomal, cellular genes in the mouse and, by inference, in many other species of vertebrate as well.

Studies leading to this conclusion began with the paradox raised by strain AKR. On the one hand, the incidence of leukemia in crosses between AKR and low-leukemia strains of mice was clearly under mendelian genetic control; in contrast, the classical experiments of Gross showed the AKR disease had the major hallmark of a virus disease in that it was transmissible by cell-free filtrates. Subsequent to Gross's finding, a massive amount of evidence was acquired during the 1950s and 1960s that showed that hematopoietic neoplasms in mice were almost invariably associated with what is now termed C-type Retrovirus infection.

As with most problems in understanding the natural history of virus infections, success depended primarily on finding a meaningful, sensitive, and rapid assay for the virus. In the case of the murine C-type viruses, our laboratory (initially under Dr. Robert Huebner, and subsequently led by Drs. Hartley and Rowe) was able to investigate the AKR system intensively by virtue of having developed simple, quantitative, and sensitive tissue culture assays for the major type of AKR C-type virus. This virus type is now called "ecotropic virus," expressing the concept that the virus is able to enter into and replicate in mouse cells. On examining tissues of AKR mice for the presence of this virus we made several surprising observations. First, virus is present in all AKR mice; it is first detectable about the time of birth and is present in high titer throughout the life of all mice of this strain. Second, AKR mouse embryo cells grown in tissue culture are initially noninfectious, but all have the capacity to begin producing the ecotropic C-type virus. They do so spontaneously at a very low rate—perhaps one cell per million per generation—but the induction rate can be increased four to five orders of magnitude by treating the cells with the thymidine analogs BUdR or IUdR. As in the mouse, once ecotropic virus is induced from a single cell, it spreads throughout the tissue culture and a high titer carrier state is established. These tissue culture induction studies made it clear the complete genetic material of the AKR ecotropic virus is present in unexpressed form in all cells of the AKR mouse. The same kind of virus can be induced by IUdR treatment of the cells of many other mouse strains, including many low-leukemia strains. In general, high-virus/high-leukemia strains show a high degree of inducibility similar to AKR, while only trace amounts of virus can be induced from mouse strains showing low frequency expression of virus in vivo. A few strains of mice do not produce ecotropic virus, either in vivo or following IUdR treatment of the cells in vitro.

We further analyzed these patterns of spontaneous and IUdR-induced expression of virus by means of classical mendelian genetic analysis. AKR mice were mated with various low-leukemia, low- or negative-virus strains and appropriate backcrosses carried out. Provided the crosses were done with inbred strains that do not carry the virus-inhibitory allele of the *Fv-1* gene, F$_1$ animals were always positive for virus, regardless of the direction of the cross. Analysis of the backcrosses confirmed that the capacity to produce virus was inherited as a dominant genetic trait, and showed the AKR mouse carries two unlinked loci, either of which results in the high-virus phenotype in descendents inheriting that locus. Further, we were able to establish the chromosomal map location of these two loci. One locus, termed *Akv-1*, was mapped by standard mendelian analysis by the observation that it shows distant linkage to *c;* this allowed the subsequent demonstration of relatively close linkage to *Gpi-1*, with gene order *Akv-1/Gpi-1/c*. The second locus, *Akv-2*, was recently mapped by Dr. Kozak using somatic cell hybrid techniques; her studies indicated it was on chromosome 16.

By means of nucleic acid hybridization studies, we were able to demonstrate that the chromosomal region containing *Akv-1* actually contains the genetic sequences of the ecotropic C-type virus. Our associate, Dr. Sisir Chattopadhyay, showed that chromosomal DNA of mouse strains having the capacity to yield ecotropic MuLV contains sequences that react with 100% of the cDNA probe prepared from the RNA of the AKR-type ecotropic virus. In contrast, the DNA of the mouse strains such as the NIH Swiss, which do not have the capacity to produce this class of virus, reacts with a portion of the AKR virus probe but lacks about 20% of the AKR viral genome sequences. When the *Akv-1* virus-inducing locus was bred into the NIH Swiss mouse, we found that the presence or absence of this 20% portion of the viral genome segregated with the same chromosomal markers as did the virus-inducing locus. Thus, the genome sequences of the virus are carried in that portion of mouse chromosome 7.

Ecotropic viruses are not the only murine leukemia viruses present in the mouse genome. There are two other classes of murine C-type viruses: the xenotropic and amphotropic viruses. These have the same virion internal structural proteins as the ecotropic virus but different envelope glycoprotein molecules that confer distinct host range, serologic, and interference specificities. Xenotropic viruses, which do not have the capacity to induce exogenous infection of mouse cells, are carried by all mice but there are marked differences between mouse strains with regard to the ease with which the xenotropic C-type viruses are induced, either spontaneously or by IUdR. From nucleic acid hybridization assays it is clear the entire genome of xenotropic viruses is carried, in multiple copies, in the chromosomal DNA of all inbred mouse strains and in wild house mice, as well. Also, the genetic material of the amphotropic viruses is present, again in multiple copies, in the genome of all inbred mouse strains but, interestingly, these viruses cannot be induced from them. Overall, the biochemical studies indicate there are on the order of 50 partial or complete C-type genomes per haploid genome in both laboratory and wild-type Mus musculus.

As yet, little is known of the arrangement and organization of these genomes; that is, whether they are concentrated at certain regions of particular chromosomes; or are present as tandem copies, or contain intervening sequences; also, it is not known if the endogenous sequences are the sites at which the genomes of exogenously infecting C-type viruses integrate. It seems likely that these viral genomes are present on multiple chromosomes. Between different inbred strains, the virus-inducing loci are clearly present at diverse sites in the genome. In our studies of the chromosomal map location of the endogenous viruses of various inbred mouse strains, we have detected ecotropic viral genomes on five different chromosomes (Nos. 6, 7, 8, 11, and 16) and xenotropic virus on another (Chromosome 1).

A most intriguing and important outcome of the existence of multiple endogenous viral genomes is the emergence of recombinants between them. Endogenous viral genomes in the form of chromosomal genes at nonallelic sites have virtually no chance of recombining; in contrast, when they form viruses these genomes can undergo genetic interactions in infected somatic cells. Three classes of somatic recombinant have now been recognized; in all cases they involve an endogenous ecotropic virus. One type of recombinant, the so-called mink cell focus forming (MCF) viruses, is a recombinant in which a portion of the envelope gene of a xenotropic virus has recombined into the envelope gene region of the AKR ecotropic virus. This recombination, which often involves multiple crossovers, results in the formation of large numbers of unique new viruses, each producing its own distinctive glycoprotein molecule; these molecules are expressed as both viral envelope components and constituents of the cell membrane of infected cells.

These MCF viruses appear to play a major role in the spontaneous leukemogenesis of the AKR mice. They regularly appear in the AKR thymus shortly before onset of lymphoma, and inoculation of the AKR-MCF viruses into young AKR mice markedly accelerates the onset of thymic lymphoma.

Another class of somatic cell recombinant is the so-called B-tropic viruses; these are ecotropic viruses that have acquired an altered host range with respect to ability to infect mouse cells carrying a normally restrictive allele at the *Fv-1* locus. The B-tropic derivatives of the normal ecotropic virus arise by recombination in the gene for the major internal virion protein p30. There is suggestive evidence that, like MCF viruses, these arise by recombination of the ecotropic virus with some virus of the xenotropic family but they involve different viral genes, a different selection pressure, and they arise in a different cell type.

The third class of recombinant is of major importance in the study of cancer. These are the highly pathogenic C-type virus variants such as sarcoma viruses and the Friend and Abelson type leukemia viruses. These are defective, helper-dependent viruses which have acquired new genetic information by recombination with some type of differentiation-specific cellular DNA sequence, which may well represent genes of another endogenous virus.

The endogenous C-type viral genomes constitute a unique gene system in being a dynamic system. Because of the infectious nature of the viruses that are the gene product of these chromosomal genes, on occasion, there is reinsertion of the viral genome into new sites in the mouse chromosome. In the course of the mouse breeding experiments in which we bred the *Akv-1* ecotropic virus locus into the NIH Swiss genetic background, we have now observed repeatedly the formation of new genes for virus induction, unlinked to *Gpi-1* and separable by breeding from the *Akv-1* locus. These genomes are phenotypically identical to *Akv-1* and represent chromosomal integration of a copy of the viral genome following infection of germ cells.

Another type of reinsertion we have encountered is a recombinant virus – a B-tropic virus that has become a chromosomal gene in the *H-2* congenic strain B10·BR/SgLi. This insertion is of particular interest in that B-tropic virus has never been found previously as an endogenous genome – only as a virus emerging sporadically in somatic cells. Whether the more highly leukemogenic and somewhat cytopathic MCF viruses can ever become integrated into germ cells and passed on as chromosomal loci is not known; it may well be that their effects on cell membranes are incompatible with fetal viability.

A further striking feature of the endogenous C-type viral genomes is that they can act in certain instances as classical genes; ie, their expression as intracellular p30 antigen, cell surface glycoprotein antigens, or virus particles is programmed. The expression is seen at precise stages of early fetal development and in certain differentiated cell types in later life.

In summary, we have outlined here a biologic system at the interface of virology and mammalian genetics. With inducible endogenous virus genomes, we have a gene system where the gene products are neatly packaged along with transcripts of the gene itself, where we can readily discriminate members of a class of moderately abundant cellular genes, witness formation of new chromosomal loci, and follow a unique class of genetic recombinations and reinsertions in somatic cells.

REFERENCES

1. Chattopadhyay SK, Rowe WP, Teich NM, Lowy DR: Definitive evidence that the murine C-type virus inducing locus Akv-1 is viral genetic material. Proc Natl Acad Sci USA 72:906, 1975.
2. Rowe WP: Genetic factors in the natural history of murine leukemia virus infection. Cancer Res 33:3061, 1973.
3. Rowe WP: Leukemia viral genomes in the chromosomal DNA of the mouse. Harvey Lectures, Series 71:173, 1978.
4. Rowe WP, Hartley JW: Chromosomal location of C-type viral genomes in the mouse, in Morse HC (ed): Origins of Inbred Mice. New York, Academic Press, 1978, p 289.
5. Rowe WP, Cloyd MW, Hartley JW: The status of the association of MCF viruses with leukemogenesis. Cold Spring Harbor Symp on Quant Biol 44:1265–1268, 1979.
6. Steeves R, Lilly F: Interactions between host and viral genomes in mouse leukemia. Ann Rev Genet 11:277, 1977.
7. Old LJ, Stockert E: Immunogenetics of cell surface antigens of mouse leukemia. Ann Rev Genet 11:127, 1977.

Abelson Murine Leukemia Virus-Induced Transformation of Immature Lymphoid Cells

David Baltimore

INTRODUCTION

Abelson murine leukemia virus (A-MuLV) arose during an experiment in which Abelson and Rabstein [1, 2] infected steroid-treated BALB/cCr mice with Moloney murine leukemia virus (M-MuLV) (Fig. 1). One of the mice in the study developed a lymphosarcoma 37 days after virus infection. This tumor involved the cervical and inguinal lymph nodes, the lower vertebral column, the marrow of the calvaria, and the meninges. No evidence of a thymic tumor was present. Histologically, the tumor cells were indistinguishable from other virus-induced lymphoma cells. The lymphoblastic cells were large in size with a high nucleus-to-cytoplasm ratio. The nucleus was characterized by diffuse chromatin and a prominent nucleolus, and the cytoplasm had abundant free ribosomes [3, 4].

When extracts of the original tumor were injected into normal adult or neonatal mice, a similar disease syndrome was observed. In particular, the thymus – the organ required for disease induction by most murine leukemia viruses – was spared in all of the mice with Abelson lymphosarcoma [1–4]. The unique pathology of Abelson lymphosarcoma and the short three to four-week-latent-period characteristic of Abelson disease distinguished this new syndrome from M-MuLV-induced tumors.

Scher and Siegler in 1975 [5] discovered the very important characteristic of A-MuLV that it could transform the growth properties of certain continuous lines of murine fibroblastic cells. Thus, they demonstrated that the virus has a direct transforming ability and, in addition, they provided a focal assay for the enumeration of virus particles. Furthermore, they were able to isolate nonproducer, transformed fibroblasts that could be rescued with defined helper viruses. The isolation

This article is a modified version of Baltimore et al, *Immunological Reviews* (1979).

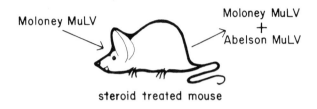

	Moloney MuLV	Abelson MuLV
Latency	4-6 months	3-5 weeks
Lymphoid target cell	"T"	"B" stem cell
Fibroblast transforming	−	+
Genome size	9 kb	5.5 kb
Requires helper	−	+

Fig. 1. Derivation of Abelson murine leukemia virus.

of nonproducers, as well as many later observations of a similar sort [6], have demonstrated that A-MuLV is a defective virus and can only propagate in the presence of a helper virus. In fact, as will be evident later, A-MuLV is unable to code for any replicative functions and all such functions must be provided in *trans* by a helper virus.

Structure of A-MuLV Genome

To characterize the A-MuLV RNA genome it was necessary to separate it from the genome of its helper virus. Such an experiment became possible only recently when stocks of A-MuLV were developed that had approximately equal ratios of helper virus and transforming agent [7]. When the RNA is extracted from such a viral stock, two different sizes of molecules are recovered: one has the size and translation specificity of the helper virus RNA while the other, smaller RNA is unique to A-MuLV. We have proved that the smaller RNA is the viral genome by showing that its translation product is the protein characteristic of A-MuLV [7].

Because A-MuLV arose in a mouse that was injected with M-MuLV, it was reasonable to expect that at least a portion of A-MuLV would be derived from M-LV. Liquid hybridization experiments, using an M-MuLV cDNA probe, had suggested that about 25% of the M-MuLV genome was present in the A-MuLV genome [8]. To examine this question more precisely we utilized heteroduplex methods and found, as illustrated in Figure 2, that 1,320 bases from the 5'-end of M-MuLV and 730 bases from the 3'-end of M-MuLV were located, respectively, at the 5'- and 3'-ends of the A-MuLV genome [7]. About 3.6 kilobases (kb) of the

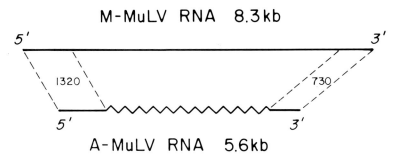

Fig. 2. Sequence contribution of the Moloney murine leukemia virus genome to that of Abelson murine leukemia virus. Using heteroduplex methods, Shields [7] found that 1,320 bases from the 5'-end and 730 bases from the 3'-end of the M-MuLV genome were contributed to the ends of the A-MuLV genome. The central 3.6 kb of the A-MuLV genome has no homology with the M-MuLV genome.

central portion of the A-MuLV genome, however, had no detectable homology to the M-MuLV genome and must have had a separate origin. Another consequence of these experiments was to demonstrate directly that the cloned M-MuLV, called in our laboratory clone 1 virus [9], was very closely related to the parent of A-MuLV. Clone 1 virus was derived from a cell line infected by the original stocks of M-MuLV but is certainly not the only virus contained in those stocks. Thus, it was possible before these results, that A-MuLV could have had a parentage quite different from clone 1 M-MuLV.

To examine the origin of the 3.6 kb central region of the A-MuLV genome we have used the methodology pioneered by Stehelin et al [10]. We prepared a ^{3}H-cDNA probe specific for this region by making cDNA with a mixed M-MuLV stock and then removing the M-MuLV specific sequences by subtractive hybridization [7]. With this probe for the A-MuLV unique region, we asked first whether these sequences are also found in other murine transforming viruses. We found them absent in both the Moloney murine sarcoma and the Harvey sarcoma viruses, suggesting that the sequences contained in A-MuLV are unique to that virus.

Because A-MuLV appeared during passage of M-MuLV through a mouse, it seemed possible that the unique sequences of the A-MuLV genome were derived from the genome of the mouse. To examine this possibility we hybridized DNA from infected mouse cells. We found A-MuLV-related sequences in both uninfected and infected cells [7]. In uninfected cells the sequences hybridized with the same $Cot_{1/2}$ that characterizes unique mouse-cell DNA, suggesting that the copy number of the relevant gene in the mouse genome is very small. The DNA from infected cells had more copies of the A-MuLV unique region as evidenced by the lower $Cot_{1/2}$ with which infected-cell DNA drove the probe into hybrid.

Thus, it appears that A-MuLV arose by some form of recombination between the M-MuLV genome and normal mouse-cell DNA. The low copy number of the relevant mouse gene suggests that it is probably not a gene contained in a virus-related sequence because all such genes occur in multiple copies in animal genomes [11]. A number of other transforming retroviruses appear to have arisen in a manner similar to A-MuLV including the murine sarcoma virus, the Rous sarcoma virus, the feline sarcoma virus, and a number of avian leukosis viruses [12–16]. In fact, it almost appears a general rule that viruses able to directly transform cells in culture have a hybrid structure in which a central portion of the genome is derived from normal cellular information.

Structure of the A-MuLV Protein

Only a single, A-MuLV-related polypeptide has been reported [17, 18]. As might be expected from the structure of the genome, this protein is a hybrid consisting of amino acid sequences shared partly with M-MuLV and partly with a normal host cell protein (Fig. 3). The hybrid structure of this protein has been demonstrated using antisera. The M-MuLV-related portion of the polypeptide reacts with antisera specific for the M-MuLV internal structural proteins. There are four such proteins that are made as a single polyprotein. Antisera that react with the proteins that map in the 5′-proximal region of this polyprotein are the ones that precipitate the A-MuLV protein. Antisera to the other end of the polyprotein are nonreactive suggesting that the recombination event that gave rise to the A-MuLV genome split this polyprotein.

The demonstration of determinants of a normal cell protein in the A-MuLV protein required a more elaborate experimental system. It was first necessary to derive antibodies that could react with the A-MuLV-unique region of the protein. This was accomplished when it was discovered that C57L mice are able to reject syngeneic bone marrow cells transformed by A-MuLV [19]. C57L is the only

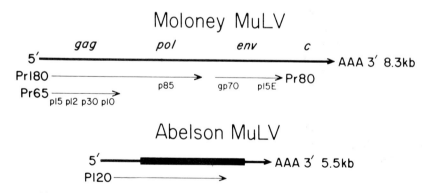

Fig. 3. Relationship of coding sequences in the M-MuLV and A-MuLV genomes.

mouse strain we have found that reproducibly rejects a syngeneic challenge but Risser et al [20] found that C57BL/6 would, on occasion, reject certain tumor cell lines. Almost all other inbred strains of mice are killed by syngeneic A-MuLV-transformed cells. The survival of C57L mice may be mediated by syngeneic T-lymphocyte killer cells that appear in the spleens of C57L mice rejecting A-MuLV-transformed cells (VL Sato, personal communication).

Antisera from C57L mice that have rejected multiple doses of A-MuLV-transformed cells are able to immunoprecipitate the A-MuLV-related protein even in the presence of an excess of M-MuLV virion proteins (Fig. 4, lane D). Such antisera also precipitate a normal lymphoid cell protein (Fig. 4, lanes B and C). This precipitation is blocked by extracts that contain the A-MuLV protein [21]. Thus, there is shared antigenicity between a normal cell protein and the A-MuLV protein.

The normal cell protein has a molecular weight (mol wt) of 150,000 (NCP150). The hybrid protein made by A-MuLV can have a variety of sizes from 90,000 mol wt, depending on the specific strain of A-MuLV [22]. The strain whose protein has been most widely studied makes a protein of 120,000 mol wt (P120). Because no biological differences have yet been demonstrated between these various strains of the virus, it is not necessary to distinguish between them for this discussion.

The M-MuLV-related region of the A-MuLV protein constitutes about 30,000 daltons of the protein. Thus, the A-MuLV-specific region varies from 60,000 to 130,000 mol wt but all of these are smaller than NCP150. The exact relationship of the various proteins to each other is still under investigation.

The A-MuLV protein as well as NCP150 are phosphoproteins; thus, with the discovery that the Rous sarcoma virus-transforming phosphoprotein is associated with a protein kinase [23, 24], it was natural to investigate whether the A-MuLV protein might also be associated with a protein kinase. This possibility has been supported by the recent demonstration that immunoprecipitates of the A-MuLV protein are able to transfer phosphate from γ-^{32}P-ATP to the A-MuLV protein itself [25]. By heating extracts that contain the A-MuLV protein and then mixing them with unheated extracts, it has been possible to show that the A-MuLV protein has the properties of an auto-kinase. Because, however, these experiments have all been performed by incubating immunoprecipitates of the protein, a direct demonstration of kinase activity in the protein must await the purification of the protein to homogeneity. Also, because the level of NCP150 is only 1% that of the A-MuLV-related protein, thus far it has been impossible to demonstrate whether NCP150 is or is not a kinase.

Localization of the A-MuLV Protein

To examine where in the cell the A-MuLV protein is localized, we used immunofluorescence methods [19]. Hyperimmune serum from C57L mice that had rejected multiple doses of syngeneic A-MuLV-transformed cells (anti-AbT sera)

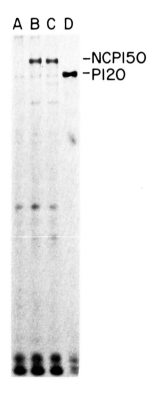

Fig. 4. Detection of a normal cellular protein cross-reactive to the A-MuLV protein. Single cell suspensions of C57BL6/J thymus (2×10^7 cells/ml) and 2M3 cells (5×10^6 cells/ml) – an A-MuLV-transformed BALB/c lymphoid cell line – were labeled with 200 μCi of ^{35}S-methionine in methionine-free media supplemented with 10% dialyzed fetal calf serum for one hour. Cells were pelleted, extracted, clarified, and immunoprecipitated as previously described [17]. Lane A: 2×10^7 thymus cells with 5 μl of anti-AbT serum. Lane B: 2×10^7 thymus cells with 5 μl of anti-AbT serum. Lane C: 2×10^7 thymus cells with 5 μl of anti-AbT serum and 200 μg of unlabeled M-MuLV virion proteins. Lane D: 5×10^5 2M3 cells with 5 μl of anti-AbT serum and 200 μg unlabeled M-MuLV proteins. All samples were collected with S aureus, denatured and analyzed on an SDS-10% polyacrylamide gel developed by fluorography. The exposure time was ten days.

were used in these studies. It was found that when live, A-MuLV-transformed cells were exposed to anti-AbT serum and then to fluoresceinated rabbit anti-mouse serum, cell surface fluorescence was evident. Such fluorescence was evident neither on cells transformed by other means nor with normal mouse serum. A-MuLV-transformed fibroblasts also demonstrated fluorescence, implying strongly that the fluorescence is due to the A-MuLV protein. Thus, it would appear that at least a portion of the A-MuLV protein is localized on the exterior surface of the cell. It is not, however, a completely exterior protein because antisera to the M-MuLV-related portion of the protein do not produce fluorescence on A-MuLV-transformed cells. Thus, the protein would appear to be a transmembrane protein. In fact, the C57L sera used to detect fluorescence may selectively react with the determinants on the exterior of the cell because the sera were raised in animals that were rejecting live tumor cells.

Cell fractionation experiments are in accord with the idea that the A-MuLV protein has a surface localization. A large fraction of the A-MuLV protein is recovered in membranous fractions from the cell and has the low density, during isopycnic centrifugation, characteristic of plasma membranes (Witte et al, unpublished results).

In spite of its surface localization, the A-MuLV protein is not a glycoprotein. Its size is not affected by endoglycosidase H and it is not labeled by radioactive sugars [19]. The protein is also not labeled by lactoperoxidase-catalyzed iodination of intact cells. Thus, it does not appear to have an exposed tyrosine residue. It is interesting to note that neither the Rous sarcoma virus nor the feline sarcoma virus proteins are glycoproteins, although both have a close association with a plasma membrane and the feline sarcoma virus protein appears on the surface of the cell (data from various laboratories presented at Cold Spring Harbor Symposium, 1979). Thus, although the various transforming proteins appear to have independent genetic origins, they may have important functional homologies. That the proteins localize to the plasma membrane and have the ability to kinase other proteins is strongly suggestive of surface receptor proteins that can recognize hormones. It is possible that these proteins act to stimulate cell growth in ways analogous to those used by normal surface-localized, growth-controlling elements. Possibly, these proteins transmit to the cells a growth signal in the absence of an appropriate ligand; thus, the proteins would stimulate cell growth in the absence of an external growth signal. In the case of the A-MuLV protein this growth signal seems to be, at least partially, cell lineage-specific. In animals and in culture, many of the A-MuLV-transformed cells are related to the B-lymphocyte lineage and it could be that the growth signal is only recognized as such, in the context of the differentiated program of the B-lymphocyte. This cannot, however, be the entire story because A-MuLV can transform continuous lines of cells (although not primary fibroblasts), and transformed macrophage-like cells have been observed in A-MuLV-infected animals.

Lymphoid Cell Markers Expressed by A-MuLV-Transformed Cells

Study of the phenotype of transformed hematopoietic cells has proven useful in confirming the cell lineage of the transformants in the case of both thymic lymphomas and Friend virus erythroleukemia. In addition, differentiation studies with Friend cells have supplied valuable insights into some of the steps of erythrocyte differentiation. In the case of A-MuLV-transformed lymphoid cells, such studies may be of even greater importance because of a lack of pathologic and histologic clues pinpointing the nature of the cells. A critical assumption made in all investigations of this sort is that, while the cells under study are malignant, the differentiated properties they express bear a reasonably close relationship to the phenotype of the normal cell that was originally transformed.

An important clue to the lymphocyte lineage of A-MuLV-transformed cells was provided by Potter and co-workers [26]. They showed that the Abelson disease syndrome was modified when prestane-primed adult BALB/c mice were infected with A-MuLV. Five to ten percent of these mice developed typical Abelson lymphosarcomas. Pristane treatment alone induced plasmacytomas in these animals but the latent period was very long (150–600 days). The role of A-MuLV in this system is still not completely clear, although the dramatic reduction in the latent period — with some of the tumors appearing as early as 20 days post-infection — favors a direct transforming role for the virus. Regardless of the mechanism involved, these experiments showed that A-MuLV could interact with highly differentiated B-lymphocytes.

A relationship of Abelson lymphosarcoma cells to cells of the B-lymphocyte pathway was initially suggested by both the absence of thymic pathology in A-MuLV-induced disease [1–4] and the observation that athymic nude mice were susceptible to Ableson disease [28]. Synthesis of small amounts of both IgM and IgG was detected in some of the tumor cells [29] and none of the tumor cells expressed thy or TL antigens [30, 31].

To determine the frequency with which lymphoid-specific cell markers were expressed by A-MuLV-transformed cells, we took advantage of an in vitro hematopoietic cell transformation system that allows isolation of single foci of A-MuLV-transformed lymphoid cells [32]. These transformed cells resemble in vivo-derived transformants morphologically and most of them form tumors in syngeneic recipients. Using this system, large numbers of clonally-derived transformants from various mouse strains were examined for expression of lymphoid cell markers.

The in vitro-derived lymphoid cells resembled A-MuLV tumor cells in expressing Fc receptor and in lacking detectable T-lymphocyte markers, such as thy, TL and lyt1, 2, 3 antigens [31, 33].

The most diagnostic characteristic of the B-lymphocyte lineage is the presence of Ig in cells. An extensive survey of clonally-derived A-MuLV-transformed cell lines derived in our laboratory [34] revealed that 50% to 60% of the transformants synthesized Ig in the form of cytoplasmic mu chain. The mu chain was glycosylated and heavy chain dimers could be detected. when immunoprecipitated cell extracts were examined under nonreducing conditions. No light chain or 7–8S IgM was present in most of the clones. The few isolates that did synthesize small amounts of light chain were unstable, and recloning of these populations showed that most of the cells in the culture did not synthesize light chain.

The high frequency of A-MuLV isolates expressing mu chain was independent of the strain of mice from which the bone marrow cells were taken. About 60% of the transformants from C57L/J, C57B16/J, BALB/c, and NIH/Swiss bone marrow expressed mu chains [22, 34]. The level of mu in all of these isolates was low and represented about 0.1% to 1.0% of the amount synthesized by myeloma cells. The observation of mu chains in these cells clearly indicates that some A-MuLV transformants are on the pathway to B-lymphocyte but the cells may have other potentials too.

Expression of terminal deoxynucleotidyl transferase (TdT), an enzyme classically associated with T-lymphocytes [35, 36], can be detected in A-MuLV-transformed lymphoid cells [33, 37]. The enzymatic activity is ten to 100 times lower than the amount expressed by T-lymphocytes. A recent survey of bone marrow transformants demonstrated that about 90% of the clones derived from C57L/J, BALB/c and C57B16/J mice synthesized low levels of immunoprecipitable TdT [22]. While most mu-positive clones synthesized TdT, both mu-positive/TdT-negative and mu-negative/TdT-positive isolates were detected (Table I). Some isolates expressed neither marker.

One of the clones studied by Siden et al [34] synthesized large amounts of κ light chain in the absence of detectable heavy chain. This type of clone must be very rare because only one such clone has been detected in the more than 50 isolates screened. Its relationship to the mu-positive clones, if any, is unclear.

Fluorescent staining studies in which the pre-B cell was first identified [38] detected a higher number of mu-positive cells than k positive cells in fetal liver. Although Raff et al [38] did not interpret these results as indicating the existence of mu-only cells, this possibility should be considered. Studies to detect mu-only cells in the fetal liver have yielded suggestive evidence of such cells (Siden and Baltimore, unpublished results), but direct identification of a mu-only cell has not yet been provided.

Future Perspectives

The studies described here suggest that A-MuLV may have two important future aspects. One is the ability to transform cells at specific stages along the pathway of B-lymphocyte differentiation. Study of these cells should help to define with increasing precision how lymphocyte specificity is programmed. There are, at present, two major limitations to this type of analysis. One is uncertainty over how precisely transformed cells represent a counterpart to normal cells. It

TABLE I. Expression of μ Chain and TdT by A-MuLV-Transformed Bone Marrow Cells*

Source of bone marrow cells	Frequency of marker expression			
	μ+, TdT+	μ−, TdT+	μ+, TdT−	μ−, TdT−
BALB/cAn	4/6	1/6	0/6	1/6
C57L/J	10/20	9/20	1/20	0/20
C57BL6/J	5/8	1/8	2/8	0/8

*Individual foci of A-MuLV-transformed bone marrow cells were isolated using agar medium and adapted to grow in vitro. Expression of μ and TdT was determined by SDS-polyacrylamide gel analysis of immunoprecipitated, ^{35}S-methionine-labeled cell extracts [17, 39].

is entirely possible that the transformation act *per se* perturbs the normal cell program such that constellations of characteristics appear in transformed cells having no normal equivalent. Hopefully, continued comparison of normal and transformed cells will put this question into perspective. Also, with increasing understanding of the transformation event itself, it may be possible to understand whether transformation perturbs differentiation or merely leads to differentiation arrest.

A major, as yet unrealized, potential of the A-MuLV system could come if it were possible to induce further differentiation of the A-MuLV-transformed cells. The enormous utility of Friend virus-induced erythroid leukemic cells in the understanding of erythroid differentiation is a model for how the A-MuLV-transformed cell could help to understand B-lymphocyte differentiation. Unfortunately, most attempts to induce further differentiation of A-MuLV-transformed cells have failed. It has been possible to greatly increase the synthesis of immunoglobulin-related proteins using a variety of manipulations of the growth medium of the cells [34] and it has been possible to increase surface immunoglobulin expression on certain cells by using lipopolysaccharides (LPS) [37]. LPS has also been able to increase light chain expression in cells that make mainly heavy chains [39] but the clear-cut differentiation of a pre-B-lymphocyte to a B-lymphocyte under direct stimulation has not been possible using A-MuLV transformed pre-B-lymphocytes. Also, those A-MuLV-transformed cells that have no immunoglobulin expression have not been able to be induced to immunoglobulin expression. One great lack in the A-MuLV system is temperature-sensitive viral mutants. If such mutants could be developed it is possible that, by shifting cells to non-permissive temperature, further differentiation of A-MuLV-transformed cells could be induced. It might even be possible with such mutants to re-inject transformed cells into animals and elicit differentiation. Attempts to do this with wild-type A-MuLV-transformed-cells have, thus far, failed to demonstrate any further differentiation, although the tumors are so lethal that the experiment is difficult to perform in a convincing manner (Siden, unpublished observations).

A second utility of A-MuLV could come from an understanding of the properties of the A-MuLV protein. Because of the possibility that this protein may represent at least a portion of a protein that plays a role in normal lymphocyte differentiation, any understanding of the action of this protein may help to understand the program of differentiation of lymphocytes. Although we do not yet know whether this protein is expressed normally in B-lymphocytes, its highest expression appears to be in T-lymphocytes [21] and the intriguing possibility that a T-lymphocyte protein expressed in a B-lymphocyte leads to cell transformation deserves further investigation.

ACKNOWLEDGMENTS

This work was supported by grant VC-4J from the American Cancer Society and grant CA-14051 from the National Cancer Institute (core grant to Dr. S.E. Luria). DB is a Research Professor of the American Cancer Society.

REFERENCES

1. Abelson HT, Rabstein LS: Influence of prednisolone on Moloney leukemogenic virus in BALB/c mice. Cancer Res 30:2208, 1970.
2. Abelson HT, Rabstein LS: Lymphosarcoma: Virus-induced thymic-independent disease in mice. Cancer Res 30:2213, 1970.
3. Rabstein LS, Gazdar AF, Chopra HC, Abelson HT: Early morphological changes associated with infection by a murine nonthymic lymphatic tumor virus. J Natl Cancer Inst 46:481, 1971.
4. Siegler R, Zajkel S, Lane I: Pathogenesis of Abelson virus-induced murine leukemia. J Natl Cancer Inst 48:189, 1972.
5. Scher CD, Siegler R: Direct transformation of 3T3 cells by Abelson murine leukemia virus. Nature 253:729, 1975.
6. Shields AF, Rosenberg N, Baltimore D: Virus production by Abelson murine leukemia virus-transformed lymphoid cells. J Virol, 31:557, 1979.
7. Shields A, Otto G, Goff S, Paskind M, Baltimore D: Structure of the Abelson murine leukemia virus genome. Cell, 18:955, 1979.
8. Parks WP, Howk RS, Anisowica A, Scolnick EM: Deletion mapping of Moloney type-C virus: Polypeptide and nucleic acid expression in different transforming virus isolates. J Virol 18:491, 1976.
9. Fan H, Paskind M: Measurement of the complexity of cloned Moloney murine leukemia virus 60 to 70S RNA: Evidence for a haploid genome. J Virol 14:411, 1974.
10. Stehelin D, Guntaka GV, Varmus HE, Bishop JM: Purification of DNA complementary to nucleotide sequences required for neoplastic transformation of fibroblasts by avian sarcoma viruses. J Mol Biol 101:349, 1976.
11. Baltimore D: Tumor viruses: 1974. Cold Spring Harbor Symp Quant Biol 39, 1187, 1974.
12. Scolnick EM, Rands E, Williams D, Parks WP: Studies on the nucleic acid sequences of Kirsten sarcoma virus: A model for formation of a mammalian RNA-containing sarcoma virus. J Virol 12:458, 1973.
13. Frankel AE, Fischinger PJ: Rate of divergence of cellular sequences homologous to segments of Moloney sarcoma virus. J Virol 21:153, 1977.
14. Stehelin D, Varmus HE, Bishop JM, Vogt PK: DNA related to the transforming gene(s) of avian sarcoma virus is present in normal avian DNA. Nature 260:170, 1976.
15. Sheiness D, Fanshier L, Bishop JM: Identification of nucleotide sequences which may encode the oncogenic capacity of avian retrovirus MC29. J Virol 28:600, 1978.
16. Frankel AE, Gilbert JH, Porzig KJ, Scolnick EM, Aaronson SA: Nature and distribution of feline sarcoma virus nucleotide sequences. J Virol 30:821, 1979.
17. Witte ON, Rosenberg N, Paskind M, Shields A, Baltimore D: Identification of an Abelson murine leukemia virus-encoded protein present in transformed fibroblasts and lymphoid cells. Proc Natl Acad Sci USA 75:2488, 1978.
18. Reynolds FH, Sacks TL, Deobagkar DH, Stephenson JR: Cells nonproductively transformed by Abelson murine leukemia virus express a high molecular weight polyprotein containing structural and nonstructural components. Proc Natl Acad Sci USA 75:3974, 1978.
19. Witte ON, Rosenberg N, Baltimore D: Preparation of syngeneic tumor regressor serum reactive with the unique determinants of the Abelson MuLV encoded P120 protein at the cell surface. J Virol 31:776, 1979.
20. Risser R, Stockert E, Old LJ: Abelson antigen: A viral tumor antigen that is also a differentiation antigen of BALB/c mice. Proc Natl Acad Sci USA 75:3918, 1978.
21. Witte ON, Rosenberg NE, Baltimore D: A normal cell protein cross-reactive to the major Abelson murine leukemia virus gene product. Nature 281:396, 1979.
22. Rosenberg N, Witte O, Baltimore D: Characterization of A-MuLV isolates differing in P120 expression. Cold Spring Harbor Symp Quant Biol 44:859–864, 1979.

23. Collett MS, Erikson RL: Protein kinase activity associated with the avian sarcoma virus src gene product. Proc Natl Acad Sci USA 75:2021, 1978.
24. Levinson AD, Oppermann H, Levintow L, Varmus HE, Bishop JM: Evidence that the transforming gene of avian sarcoma virus encodes a protein kinase associated with a phosphoprotein. Cell 15:561, 1978.
25. Witte ON, Sun L, Rosenberg N, Baltimore D: A trans-acting protein kinase identified in cells transformed by Abelson MuLV. Cold Spring Harbor Symp Quant Biol 44:855–858, 1979.
26. Potter M, Sklar MD, Rowe WP: Rapid viral induction of plasmacytomas in pristane-primed BALB/c mice. Science 182:592, 1973.
27. Potter M, Premkumar-Reddy E, Wivel NA: Immunoglobulin production by lymphosarcomas induced by Abelson virus in mice, in Sanford KK (ed): Gene Expression and Regulation in Cultured Cells. National Cancer Inst Monograph 48, 1978, p 311.
28. Raschke WC, Ralph P, Watson J, Sklar M, Coon H: Oncogenic transformation of murine lymphoid cells by in vitro infection with Abelson leukemia virus. J Natl Cancer Inst 54:1249, 1975.
29. Premkumar E, Potter M, Singer PA, Sklar MD: Synthesis, surface deposition and secretion of immunoglobulins by Abelson virus-transformed lymphosarcoma cell lines. Cell 6:149, 1975.
30. Sklar MD, Shevach EM, Green I, Potter M: Transplantation and preliminary characterization of lymphocyte surface markers of Abelson virus-induced lymphomas. Nature 253:550, 1975.
31. Pratt DM, Strominger J, Parkman R, Kaplan D, Schwaber J, Rosenberg N, Scher CD: Abelson virus-transformed lymphocytes: Null cells that modulate H-2. Cell 12:683, 1977.
32. Rosenberg N, Baltimore D: A quantitative assay for transformation of bone marrow cells by Abelson murine leukemia virus. J Exp Med 143:1453, 1976.
33. Silverstone AE, Rosenberg N, Sato VL, Scheid MP, Boyse EA, Baltimore D: Correlating terminal deoxynucleotidyl transferase and cell surface markers in the pathway of lymphocyte ontogeny, in Clarkson B, Till JE, Marks P (eds): Differentiation of Normal and Neoplastic Hematopoietic Cells. New York, Cold Spring Harbor Press, 1978, p 432.
34. Siden EJ, Baltimore D, Clark D, Rosenberg N: Immunoglobulin synthesis by lymphoid cells transformed in vitro by Abelson murine leukemia virus. Cell 16:389, 1979.
35. Chang LMS: Development of terminal deoxynucleotidyl transferase activity in embryonic calf thymus gland. Biochem Biophys Res Comm 44:124, 1971.
36. Kung PC, Silverstone AE, McCaffrey RP, Baltimore D: Murine terminal deoxynucleotidyl transferase: Cellular distribution and response to cortisone. J Exp Med 141:855, 1975.
37. Boss M, Greaves M, Teich N: Abelson virus transformed haemopoietic cell lines with pre-B cell characteristics. Nature 278:551, 1979.
38. Raff MC, Megseon M, Owen JJT, Cooper MD: Early production of intracellular IgM by B-lymphocyte precursors in mouse. Nature 259:224, 1976.
39. Rosenberg N, Siden E, Baltimore D: Synthesis of mu chains by Abelson virus-transformed cells and induction of light chain synthesis with lipopolysaccharide, in Cooper M, Mosier D, Scher I, Vitetta E (eds): B lymphocytes and the Immune Response. New York, Elsevier/North Holland Biomedical Press, 1979, pp 379–386.

SUMMARY OVERVIEW

A Century of Mammalian Genetics and Cancer: Where Are We at Midpassage?

James F. Crow

I am somewhat reticent, being a student of Drosophila population genetics, to comment on fields as far from my own as the topics of this symposium. But Dr. Russell argues that a view from a distance may be a good thing. By stepping back — or in my case, by being far back in the first place — one gets a better perspective, she says. By that criterion I clearly qualify.

Final symposium speakers may do any of several things. Some summarize or repeat what has been said. I shall not do that. It would be fatuous, for the talks have been remarkable for their clarity. Some say what they think the speakers might have or should have said. I won't do that either. Some seize the opportunity to tell of their own latest research, however inconsequential or irrelevant to the main purpose of the symposium. I am tempted, but will resist. Some integrate the presentations, pointing out new interrelationships and new ways in which the field could advance. I wish I could. In no case has the final speaker simply said that he has nothing of significance to add and sat down. I should, but I won't.

At the time the Jackson Laboratory was getting under way I, too, was getting under way. I had heard of Mendel and Morgan, but not of C.C. Little, W.E. Castle, or Sewall Wright. In fact, in the early 1930s my heroine was Maud Slye. I did not realize how completely, and with what relish, Little had disproved her simplistic theories of cancer inheritance. Much later, I finally met C.C. Little, and was charmed as others had been.

A history of the Jackson Laboratory from its beginnings in 1929 to the present time has been written by Jean Holstein [1]. I want first to take a look backward to that time.

This is paper no. 2388 from the Laboratory of Genetics, University of Wisconsin.

WHAT WAS KNOWN WHEN THE JACKSON LABORATORY WAS BEGINNING?

What was known about genetics in the early 1930s when the Jackson Laboratory was getting started? Although the answers were not known, the basic question was. It was clearly formulated, especially by H.J. Muller. In fact, one of Muller's great prophetic papers was published in 1929, the year of the Jackson Laboratory founding [2]. He told us what the gene must do. It must replicate itself and, moreover, do so with extreme precision; the low error rate was already well known from mutation studies. Most important of all, Muller said, the gene must make mistakes. This is not hard, of course; it is expected in any biological system. But the gene must, when it makes a mistake, copy the mistake and not what was there before the mistake was made. Finally, the gene must have a heterocatalytic as well as an autocatalytic function; it must somehow influence the cell chemistry to bring about development and differentiation. All this seems so obvious now, but it wasn't in the early days.

The chromosome theory of heredity was, of course, well established, thanks especially to Calvin Bridges and his nondisjunction experiments [3]. The final doubts had been removed not long before the year of the Jackson Laboratory founding. Sewall Wright, who is here today, told me how Bridges convinced a skeptical William Bateson of the chromosomal basis of heredity by showing him his exceptional flies and their exceptional chromosomes. This was only a half-dozen years before the founding of the Jackson Laboratory although, of course, many geneticists had been convinced since the turn of the century.

In 1929 chromosome mapping was well understood and a large number of gene loci had been mapped in Drosophila and maize. A small beginning had been made in the mouse. Clearly, it was destined to grow, and this we have seen.

A great deal was known about mutation, not the chemical nature of the process but the kinetics. Its temperature dependence had been analyzed and interpreted as a chemical reaction. The induction of mutations by ionizing radiation in Drosophila by Muller and in maize by L.J. Stadler was at the time very new and very exciting. Many geneticists thought then that the road to an eventual understanding of the nature of the gene was through mutation studies. Physics seemed, to some at least, to offer more promise than chemistry. The relation of mutation to chromosome replication was not known, although at a much earlier date Wilhelm Weinberg (of Hardy-Weinberg fame) had suggested that an increased paternal age might be expected in sporadic cases of dominant diseases.

Chromosome rearrangements were quite well understood, as were the effects of polyploidy and aneuploidy, although not in mammals. Chromosomes could be broken by irradiation and there had been some crude confirmation of the position of genes in linkage maps by this means. The first radiation-induced translocation in mice was produced by George D. Snell [4], then working with Muller at the University of Texas.

Gene duplications were known and their evolutionary significance was emphasized, especially by C.B. Bridges. Unequal crossing over was postulated as a mechanism and later demonstrated for the *Bar* eye duplication in Drosophila.

Along this same line, R.A. Fisher [5] had emphasized the tendency for genes with associated functions to become more closely linked, this being opposed on the chromosome as a whole by the advantages of recombination in generating new variability. Fisher knew about super-genes, as they were then called, in mimetic butterflies and grouse locusts. It was not unlike what we now see in the H-2 and HLA regions. It is no accident, I think, that R.A. Fisher was the first to suggest a gene cluster for the Rh blood group. As Walter Bodmer was speaking about genes that switch several others on or off at one time, I wondered whether he, being a Fisher student, had been influenced in this direction by his association with Fisher.

In those early days, there was already some evidence for gene subdivision. Serebrovsky in Russia and C.P. Oliver in the United States were finding recombination and unusual complementation patterns in what had been thought to be a single gene. But the clear subdivision of several classical genes and the exquisite, fine-structure analysis of genes in microorganisms was to come later.

Half-tetrad analysis had been employed to understand some aspects of meiosis in Drosophila. Perhaps it could have been predicted that this would eventually be used in the mouse, but the predicted method would surely have been wrong. Most people would have expected that such analysis, if it were possible, would be done through analysis of nondisjunction of sex chromosomes, not the elegant means discussed by the mouse geneticists at this symposium.

It is important to remind ourselves that Griffith's [6] work on Pneumococcus transformation had been done at the time of the founding of the Jackson Laboratory. I think it is wrong to suppose that this work was not known. I believe, rather, that geneticists simply preferred other explanations. I well remember, as a graduate student, hearing discussions of possible subtle ways by which selection might produce these results.

There was little thought given to the use of microorganisms. The advantages of some liverworts, which held together the products of a single meiosis and thus facilitated tetrad analysis, were known. Work on Neurospora was in its infancy; the emphasis was on understanding meiosis, not biochemical genetics, which was yet to come.

As early as 1921 Muller had been much impressed by the possibility of learning about the gene from bacteriophages, then called "d'Hérelle substances." He speculated that these might be naked genes, subject to direct test tube study. Muller's famous quote of 1921 bears repeating. Referring to bacteriophage, he wrote: "It would be very rash to call these bodies genes, and yet at present we must confess that there is no distinction known between the genes and them. Hence we cannot categorically deny that perhaps we may be able to grind genes in a mortar and cook them in a beaker after all. Must we geneticists become bacteriologists, physiological chemists and physicists, simultaneously with being zoologists and botanists? Let us hope so" [7]. But as far as I know, neither he nor anyone else gave the idea much further thought at that time. The fruitful path, through Max Delbrück, was to come later.

Androgenic males were obtained in the wasp Habrobracon by destroying the female pronucleus by radiation. More striking was the work of Astaurov, who produced silkworms of exclusively paternal or maternal origin by destruction of one pronucleus and doubling the chromosomes in the other. I might mention that Astaurov did not choose to work with silkworms. Having fallen into disfavor, he was prohibited from working with Drosophila and assigned to a remote silkworm breeding station [8].

Although the chemistry of proteins was quite primitive, there was considerable information about gene-enzyme relations. The work of Garrod on inborn errors of metabolism was known, but not widely discussed. There had been chemical studies of genetically determined flower pigments. The most thorough analyses were of rodent skin pigments by Sewall Wright [9]. He used the theory of enzyme kinetics, as it was then understood, and the assumption of flux equilibrium to provide quantitative explanations of dominance and various epistatic combinations. He also predicted the occurrence of a class of mutant, not foreseen by Muller. This was a "mixomorph," that by producing an enzyme that could compete effectively for substrate but could not convert it efficiently into product, would compete and therefore detract from the effectiveness of more active gene products while acting as a hypomorph with an amorphic mutant. Such a mutant, the white-allele "pearl," was later found in Drosophila by Arthur Steinberg. Some of the problems raised by Wright and Muller could well be studied again by the much better chemical methods now available.

I want to mention two other genetic phenomena, both of special significance to the Jackson Laboratory. One is tissue grafts. The general rules of graft rejection were understood and it was even possible to make a crude estimate of the number of loci involved from the proportion of grafts accepted in various crosses. Little had done this in mice and Wright did similar experiments with guinea pigs. Out of this has grown the great work of Snell and others leading to our present knowledge of mouse and human graft rejection.

The other phenomenon is inbreeding. Thanks mainly to Wright, geneticists knew almost as much then as today. The summary of Wright's paper [10] in the early 1920s describing and explaining his guinea pig results could be reprinted today and would stand without correction. He noted that the decrease in vigor in inbred lines, the increased uniformity within lines, the increased variance between lines, and the immediate recovery of vigor on outcrossing were all predictable consequences of increasing homozygosity and dominance. Furthermore, Wright showed how the amount of decreased heterozygosity could be computed for any pedigree, however complex or irregular. This *quantitative* understanding of inbreeding is one of the solid foundations on which the Jackson Laboratory's program rests.

In cancer research, some things were known. It was known that there was an increased cancer risk in relatives of affected persons, although little was known about specific modes of inheritance. Some environmental factors were known. Radiation was known to be carcinogenic as well as mutagenic, adding some evidence for the mutation theory of cancer.

Viruses as causes of malignancy had been shown much earlier by Peyton Rous in chickens. One of the earliest results from the Jackson Laboratory was the clear proof of nonreciprocal inheritance of mammary cancer in C3H mice. The proof that the maternal factor was transmitted by the milk was a high point in cancer research of the early 1930s.

Perhaps most important of all, from the standpoint of future research on cancer in mice, was the establishment of inbred strains each with a characteristic, organ-specific cancer rate.

The other early discovery of great lasting significance has already been referred to — transplantable leukemia and tumors. Not only did these provide methods for experimental manipulation, but helped establish the fact that, at least in some cases, malignancy starts in a single cell.

WHAT MIGHT HAVE BEEN PREDICTED IN 1929?

Looking at the first 50 years of the Jackson Laboratory and wondering what the next 50 years have in store, we might first ask what might have been predicted in 1929 for the ensuing 50 years.

In general, further advances using the same techniques can be foreseen with some confidence. Yet the most exciting and spectacular advances have come from the employment of new techniques and in some cases totally new ideas that were not foreseen, and probably could not have been.

The only confident predictions, as David Baltimore has said, are made by extrapolating the past for short times into the future. Since transmission genetics and chromosome mechanics were understood in 1929, and since there had been substantial progress in the study of these in other plants and animals, one could expect that those aspects of genetics that could be studied by breeding methods would steadily advance.

Thus, a forecaster in 1929 would have predicted a steady increase in the number of genes assigned to linkage groups and placed in their proper sequence. He could also have predicted that, by use of radiation to induce chromosome rearrangements, it would be possible to associate linkage groups with specific chromosomes, as had been done in Drosophila. However, the mouse map was particularly unpromising because of the uniform terminal centromeres and the similarity in appearance of the chromosomes.

It could have been predicted that more mutants would be discovered. Perhaps the efficiency of radiation in producing these would have been overestimated. No one would have foreseen the rate at which they have actually been found; I doubt that anyone expected the large numbers of mice that would be processed in this and other laboratories in later years. Nor would the great increase in knowledge of human mutant genes have been foreseen. The prospects for human gene mapping looked dismal.

It might have been predicted that methods for measurement of chromosomal, as opposed to genic, rates of mutation would be developed in the mouse. Since chromosome breaks could be produced by radiation, it might have been predicted that methods analogous to the ClB and Muller-5 methods in Drosophila would be contrived. I think this may be a case where the realization, which is now beginning to happen as a result of the work of Tom Roderick, has taken longer than would have been expected in 1929. One reason has been the slow discovery of X-linked genes in the mouse; it certainly would have been predicted that these would be found more rapidly, relative to autosomal mutants, than has been observed.

It would have been a reasonable prediction that the genetics of sex determination would be worked out. Most American geneticists would have predicted, from analogy with Drosophila, that the Y chromosome in mammals would not play any role in sex determination. Indeed, there was a published pedigree of a human X-linked trait that was interpreted by assuming attached X chromosomes. A more correct prediction might have been made in Japan, where the silkworm analogy would have been as compelling as Drosophila was here.

I think that we might have expected the use of mosaics for fate mapping, although no one would have known how soon the techniques would become available. It would probably have been predicted that this would be done by nondisjunction of X-chromosomes, as had been foreshadowed by Sturtevant in Drosophila, rather than by the elegant methods now used to produce mouse chimeras. Most embryologists, I suspect, would have placed their bets on better vital stains rather than on mosaics as the way to study cell lineage.

That tissue and cell cultures would some day be practical and useful for analysis of gene action, and that clones might be possible might well have been expected. Indeed there was considerable discussion of this possibility. I suspect, however, that most people would have expected this to be done much sooner and more effectively with plant than with animal cells, and if animal cells were used, they would be from invertebrates. It is noteworthy in this connection that there was an active program in plant cell culture at the Jackson Laboratory starting in 1950 with Dr. Philip White, a rare choice for staff appointment at C.C. Little's almost exclusively mammalian laboratory. He must have thought it promising.

There was talk in the early days about embryo transplants, and this would have been expected to happen soon. It did, as we know, with the work of Elizabeth Fekete. She not only accomplished this, but helped answer the important question about the transmission of the milk factor. It could, I think, have been predicted that amniocentesis would some day be feasible and that it could possibly be used for prenatal diagnosis, but I am not aware of any serious discussion of the possibility.

One wrong prediction would almost certainly have been made; in fact, it was made. The use of identical twins separated in infancy had been used to allocate variance between genotypic and environmental causes. Wright had used his path coefficient method to analyze the differences between children reared by their biological parents and those who had been adopted. It was predicted that better controlled studies with much larger numbers would soon be done and the heredity-environment issue, at least as far as intelligence is concerned, would soon be essentially answered [11]. How different has been the actual course of events!

In 1929 there was no way to study genetic differences except by breeding methods. Genetic differences between different strains of mice could be understood, but no differences between mice and rats, to say nothing of the differences between vertebrates and invertebrates, or animals and plants. The magnificent technology of molecular and cell genetics has completely transcended this limitation that in earlier days seemed so fundamental. A geneticist in 1929 would find it incredible, I think, that within 50 years we would know how many amino acid substitutions had taken place in 100 million years of hemoglobin evolution.

The most exciting studies, those with revolutionary impact, were of course not foreseen. The greatest triumph, beside which all else is dwarfed, is molecular biology. No one in the 1930s foresaw the early use of microorganisms in genetic research. Attempts to find recombination in bacteria had failed, mainly because no biologist before Joshua Lederberg took Darwinism seriously enought to apply it to this problem. The now commonplace procedure of using enormous numbers and strongly selective environments to identify very rare events was not then practiced; if you wanted to study rare events you did it by strong-arm methods. At the same time, as fine structure analysis by recombination became possible in microorganisms, the chemistry of large molecules advanced to the point of being useful to biologists.

It is interesting to note that the Watson-Crick model for the structure of DNA came just at the halfway point between the origin of the Jackson Laboratory and this symposium.

The gradual, sometimes grudging acceptance of the fact that DNA is the genetic material has been reported too many times for me to repeat. I was interested to note that E.B. Wilson's great classic, "The Cell in Development and Heredity," suggested nucleic acid as the genetic material in the 1896 edition, but Wilson saw the error of his ways and revised this to protein in later editions.

I can recall that, in some circles at least, there was not much interest in the chemical nature of he gene. Some geneticists were not optimistic that simply knowing the chemical structure would provide much insight; after all, it has often been true that the chemical structure of a molecule has revealed little or nothing about its biological role. I was interested in the report that Watson and Crick also had doubts on this point; they wondered if they would find the structure of DNA only to discover that it didn't telegraph any message. But of course it did, in a way instantly obvious to anyone who had Muller's gene-requisites in mind.

Then came the manificent story — transcription, translation, promoters, ribosomes, start and stop sequences, the central dogma, then reverse transcriptase, restriction endonucleases, insertion sequences, DNA cloning, and the rest of the story, so richly illustrated in this symposium.

In looking back it is most impressive to see the role played by technical discoveries and inventions: The electron microscope and ways of preparing materials to identify the trees *and* the forest; Smithies's starch gel electrophoresis and its many successors; protein sequencing; even more exciting, DNA sequencing, thanks to the happy insight of Maxam and Gilbert; cell culture; cell fusion; DNA cloning; high speed computers that do much more than compute.

To capsulize: Transmission genetics was well understood, and its further development could have been predicted. Molecular mechanisms were almost totally unknown, and the whole of molecular genetics was unforeseen.

WHAT HAVE WE LEARNED IN THIS SYMPOSIUM?

So much for the past. What of the present? We have heard a succession of lucid, exciting presentations, each giving its successor a hard act to follow. The speakers have shown mercy for and sympathetic understanding of those not specialists in the several disciplines. They have followed the admonition to regard each member of the audience as a *tabula rasa*, but a quick study. I was impressed by the enormous fraction of information presented that was not merely new in the 50-year period of the Jackson Laboratory, but new within the past half-dozen years. As Philip Leder has reminded us, intervening sequences are only two years old.

Formal mouse genetics, the most classical of the work reported, has added to the store of knowledge and it is gratifying to see the steady increase in density of the chromosome map, thanks to the work of Eva Eicher and others who combine classical linkage analysis with cytogenetic trickery and inventive use of ovarian teratomas. The Jackson Laboratory geneticists are filling the niche that Calvin

Bridges occupied for Drosophila. Actually, I guess the real successor to Bridges is a computer data-bank, for the information has become too vast to store in a single mind.

Although the findings of molecular biology have proceeded in an orderly and, for the most part, seemingly logical way, recently there have been two utter surprises: One was that in some viruses the same nucleotides may be part of as many as three translated sequences. DNA appears to be such a scarce commodity that it has to be recycled; as if rapid replication necessitates the minimization of size of the molecule. The other surprise is that there are intervening sequences in the gene message that are not informational with regard to the protein product. Although surprising and puzzling, these offer yet other possibilities for the further refinement of the control of differentiation through processing of the DNA information.

It was particularly gratifying to hear the two-year progress report of Philip Leder. It was a testimony to the powerful technique brought about by the use of restriction endonucleases and DNA cloning and to Leder's experimental virtuosity. We are beginning to understand more of the diversity of roles played by DNA, with obvious implications for development and for evolution.

The development and differentiation of multicelled organisms has long been a central problem in biology. Until very recently the deepest understanding of basic mecahnisms has come from single-celled organisms and their viruses. Now the new techniques are being widely applied to eukaryotes, in particular, the mouse. Virginia Papioannou has told us how useful chimaeras can be for the study of differentiation. In Drosophila fate mapping of both morphological and behavioral raits can be studied by genetically-induced mosaics. In the mouse, delicate microsurgery accomplishes equivalent results.

Roy Stevens has shown how a seemingly routine discovery of several years ago, teratocarcinoma-producing strains, can be turned into a powerful way of studying not only tumors but also normal differentiation, all because of the remarkable property of teratomas to develop into highly differentiated tissues that are abnormal only in being in the wrong place. I particularly enjoyed hearing from him, not only of current advances, but of the way in which the original discovery was made.

Karl Illmensee has the embryological equivalent of a green thumb. He, along with Peter Hoppe, did what other people only talked about doing. The chimaeras, the parthenogenotes — both androgenic and gynogenic — along with the genetic control and transplantability that goes with well-understood inbred strains have at last made possible an attack on fundamental problems of development in a mammal. It is no longer necessary to run to plants and insects for such experiments. In the past the study of embryology and genetics, although each gave a great deal of lip-service to the other, followed rather separate courses. That time is over.

The techniques are too new to have had a major impact on a fundamental understanding of differentiation. Also the questions are not as clearly formulated as were those stated by Muller for the nature of the gene, to which I referred earlier. It is possible that the techniques are ahead of the questions. Jim Ebert, in his summary of the Differentiation session, has pointed one way to the future – the study of cell surface interactions. Differentiation involves relations among cells as well as changes within cells; clearly, the cell surface is important. I, too, can remember, as he did, the days when nucleic acids were discussed more seriously as candidates for the organizer than for the gene.

Douglas Coleman has shown us that an inquisitive and persistent biochemist can show geneticists what to do with their mutants. First, he demonstrated that two seemingly quite different mouse mutants have the same phenotype when the background genotype is controlled. He then showed that obesity and diabetes in both mutants are cuased by defects in a satiety factor, the one in its production, the other in its utilization. The Rube Goldberg contraptions which let one mouse determine the diet of another are remarkably simple in principle, yet enormously informative. The idea that diabetes represents a thrifty phenotype has long been around, but never with such convincing evidence as these mouse experiments supply. The value of such genes would be quite different for our ancestors with their near-starvation diets than for us.

A large number of anemic mutants have been identified in the mouse over a period of years and several of them appear to have human counterparts. Samuel Lux has told us, with clarity and satisfying detail, how many of these exert their mischief through deficiencies in the red cell membrane and the skeletal proteins thereof. It is an exciting new field for using mouse mutants both to understand normal membrane function and to understand human anemias.

From Charles Scriver we learned how remarkably far we have gone since Garrod first wrote of "inborn errors of metabolism." Well over 50 specific amino acid disorders are now known. Some can be effectively treated by an altered diet; particularly successful cures are possible in those where the deficiency is in a co-factor for which a vitamin can supply the needed element. With increasing understanding has come the humanitarian utilization of embryonic detection. It is interesting to learn that one of Garrod's first types, cystinuria, has tuned out to be a transport deficiency. Again, mouse models are very useful; some clear homologies have been demonstrated and others will surely be found among the numbers of mutants now existing and others yet to be found.

All human geneticists owe thanks to Victor McKusick for his continued leadership in classifying human genetic disease. His regularly revised and always enlarging catalog of human mendelian traits has become as indispensable to the geneticist as a pouch to a kangaroo. The number of known mendelian traits, mostly disease, doubled in the 13 years from the first edition in 1966 to the present. The increase continues at about the same rate; diminishing returns are not yet evident. The summer genetics course has been an integral part of the Jackson Laboratory for the past 20 years; we include the celebration of its 20th anniversary along with the 50th anniversary of its parent, which is 30 years older in good human demographic tradition.

George Klein and Walter Bodmer have led us gently by the hand through the maze of complexities of the histocompatibility complex, but I doubt that many of us could find our way through unassisted. The subject has become enormously complex, and ever more interesting as its relations to other biological phenomena become increasingly apparent. The relation with human disease conditions is only one example. I can remember when linkage disequilibrium was applied mainly to experiments designed to determine the causes of heterosis in maize. Now it is central to the discussion of linked gene clusters. Klein and Bodmer both combine an interest in the mechanisms with an attempt to understand the selective advantage of such closely linked clusters. The HLA region clearly has some important relation to human disease and disease resistance. It is typical of biologists to seek both a mechanistic and an evolutionary explanation of every phenomenon. The faith that a persisting association has some biological purpose has been heuristic over and over again in biological history. We don't have to be teleologists to be taleonomists; alternatively stated, teleology is fine if you don't inhale.

We had a chance to hear two of the pioneers in mouse genetics, both associated with the Jackson Laboratory. Walter Heston told us about the history of the inbred mice and C.C. Little's central role. Little was interested in cancer and in genetics. He thought the way to study cancer was through heredity, the way to study heredity was in the mouse, and that the way to study mice was with inbred strains. The inbred strains made possible an early understanding of transplantable tumors and the role of graft rejection. They also enabled Little to demonstrate the error of Maud Slye's simple, monogenic theories of cancer heredity. Not even Little could have foreseen the great accomplishments of this technique to immunogenetics, cancer inheritance, cancer viruses, carcinogen testing, and studies of chemotherapy — all mentioned by Heston. Let me add to this the great value of congenic lines, which enable the experimenter to study the effect of a particular gene (or chromosome region) without the otherwise inevitable complication of noise from genetic variability at other loci.

George Snell has looked into his crystal ball and given us his predictions of what will happen in immunogenetics. He noted, as have others in this symposium, that the most certain prediction is that there will be surprises. Yet some things, representing short-term extrapolations from the past, can be foreseen to some degree of approximation, the number of loci yet to be discovered, for example.

It will be interesting to see if his predictions about the role of the cell surface, the function of the system in different cells types, and possible medical outgrowths will be realized, and when.

The final session on carcinogenesis shows once again the intimate connection between virology and genetics. Wallace Rowe has shown how a viral genome is integrated into the mouse genome at a recognizable site. David Baltimore, the final speaker, and Howard Temin are, respectively, the Wallace and Darwin of reverse transcriptase. Both, incidentally, are alumni of the Jackson Laboratory summer program for high school and college students. Baltimore's review of transforming genes and the possibilities they are offering and will offer for deeper insights was both thoughtful and exciting — a fitting climax to an exciting and informative series of talks.

WHAT OF THE FUTURE?

We have heard a magnificent series of reviews of the latest results in mammalian genetics and cancer. What can we expect in the next 50 years?

As David Baltimore has just said, we can be confident only of rather short range predictions based on extrapolations from the recent past. I would add that such extrapolations can be made meaningfully only by those deeply immersed in each field, those who are aware of the possibilities and limitations of the techniques. There is another feature that makes predictions uncertain. This is the increasing pace of new knowledge. It was probably easier to make a 20-year prediction in 1929 than it is to make a 5-year prediction now; such is the accelerated rate of acquisition of new information.

Even our greatest scientists can make outrageously bad predictions. No less a figure than Alfred Russell Wallace opined in 1900 that in the twentieth century phrenology would finally come into its own [12, p 151].

Some things can be reasonably expected. We can expect solid and detailed information about rates of evolution at the DNA level. Not only will we know the rates, but the chemical nature of the changes and the functions affected. This is particularly welcome to the population geneticists, since it provides the opportunity to study evolution at its most basic level and to test theory against observation.

We will soon have a detailed comparison of the genetics of different species. Comparative genetics may become as popular as comparative anatomy, and perhaps as much resented by students. We will no longer have to resort to speculation about genetic differences between species that cannot be mated. We shall soon know much more about the amount of variability in the population, not just at the phenotypic and protein level, but at the DNA level.

There is a reasonable expectation of a practical solution to transplantation difficulties. Then we shall have increasingly difficult decisions about the allocation of scarce resources.

We certainly can predict the greater use of chimeras in mice as a standard means of developmental analysis, as is now the case in Drosophila. For example, if it were only easier to do (as it will be) one could presumably use this technique to find the source of the satiety factor in Coleman's diabetic mice.

I hope, and expect, that we shall see behavioral genetics at a new level. Early work at the Jackson Laboratory has clearly shown the role of genetic factor in behavior, and at the same time the role of environmental factors. Behavioral mutants have been useful and will be increasingly so. But we can also hope for a more fundamental understanding of learning and memory. Recent research has shown that learning is not as exclusively a property of vertebrates as was once thought. Learning experiments in insects and nematodes are now very sophisticated. In Drosophila there are mutants that cause retarded learning or inability to learn, as well as mutants affecting the memory of learned behavior. It has been said that the more fundamental distinction between higher organisms, such as ourselves, and other animals is not in the ability to learn, which is very widespread, but in the ability to teach. What is the genetic and molecular basis for this? I think we can predict a greater understanding of and, I hope, a greater appreciation of human individuality. We already note these in biochemical differences, immunological patterns, and disease susceptibilities. The extent of individual differences will surely be understood in much more detail, and our treatment of ourselves and of each other can become increasingly individualized. Haldane once said that liberty is the practical recognition of human polymorphism. Shaw may have said it better, when he said not to treat your neighbor as you would be treated; his tastes may differ.

The most exciting possibilities in the immediately foreseeable future lie in the use of recombinant DNA. This work will surely impinge on all aspects of genetic knowledge. We are beginning to see the results and I am impressed not only by the depth of knowledge thereby generated but by the rapidly accelerating rate of technical advance.

We can expect more understanding and control of the mutation process. We shall certainly learn a great deal about the environmental causes of mutation, and will be able to use this information practically for the benefit of our descendants. It is likely that some of the causes of "spontaneous" mutation will be learned. So, I would welcome knowledge as to how to reduce the spontaneous rate, and such knowledge may be forthcoming. The coupling between mutagenesis and carcinogenesis seems close enough that such knowledge would very likely lead to reduced cancer as well.

R.A. Fisher was the first to point out the great capacity of mendelian inheritance to conserve variability. Only in very small or inbred populations is reduced variability a problem (or, as in the case of inbred mice, a useful experimental tool). Genetic variance decays at a rate of $1/2N$ per generation, where N is the effective population number. To take an extreme example, a population that is reduced for one generation to a size of two would lose only 25% of its variance. So an extremely low rate of mutation is sufficient to balance any decay of genetic variability from random drift in any sizable population, and the human population in particular. For this reason I would argue that the human mutation rate is already too high and that our descendants for the foreseeable future would be better off it were lowered.

In the past, the major beneficiary of genetical research has been agriculture. The accomplishments are too well known to need mention. I think we can foresee that in the decades ahead human benefits will be much more direct — through medicine and through a better understanding of ourselves.

More striking, as we look back through the past 50 years, is the acceleration of knowledge. I have lived through the first 50 years of the Jackson Laboratory, and shan't live through the next 50 (unless the practical results of aging research are far more spectacular than now seems at all likely.) But I think the actuarial odds are good that I'll live through a period that will see an increase of knowledge equal to that in the Laboratory's first half century, and considerably greater if current rates of progress continue.

As regards carcinogenesis, we can confidently expect much better information about the extent to which environmental chemicals are causative. We can also expect to know about the extent of virus causation. This symposium has clearly shown the exciting interrrelationship between viral and eukaryotic genomes. But where this will lead, theoretically and practically, is nothing that I would want to guess.

Can we count on a continuing increase of knowledge? It depends strongly on the scientific ambiance. If, back in 1929, one had tried to predict not only *what* the most important advances in genetics would be, but also *where* they would be made, surely the United States and Britain would have been mentioned. But I think there would also have been the prediction that Germany would be a leader. There were Weinberg, Bauer, Vershuer, Nachtsheim, Goldschmidt, Stern, Timofeeff-Ressovsky, and Delbrück. Physicists like Heisenberg were actively interested in genetic problems. Then came Hitler. Another certain prediction would have been Russia. There were Karpenshenko, Serebrovsky, Levitt, Tschetverikov, Astaurov, and Vavilov. Bridges had been there and Muller was soon to go to start an outstanding Drosophila program. Then came Lysenko. The science was destroyed for a generation, not to mention the people. Genetics, it seems, is particularly fragile and susceptible to political manipulation. Do we dare predict that studies of recombinant DNA and human genetics will be able to continue their accelerating rate of progress, free of dampening influences of special interest pressure groups, excessive government regulations, and self-appointed guardians of the public interest? I hope so, but cannot predict this with any confidence.

In 1929 it might have been said that the next 25 years would be the time when the mouse became the major research species for genetics. Drosophila research was then getting a bit in a rut; but then came salivary gland chromosomes and the study of natural populations. Next came the study of microorganisms and the whole excitement of molecular biology. The mouse was upstaged once again.

Now the time for research in mammalian genetics gain seems ripe. The major unsolved problems are in development and in the nervous system. The classical techniques can be combined with those of molecular and cell genetics for a fundamental attack on these problems in mammals. So I fearlessly predict a rich future for mammalian genetics and an exciting second half century for the Jackson Laboratory.

We are here to honor C.C. Little for his persistence and faith in mouse research, for the friendly and cooperative atmosphere that he fostered, for carrying the Laboratory through depression, war, and fire, and for his foresight in developing inbred mouse strains. I once read an anthology of Sherlock Holmes stories in which the collector thanked the would-be patients of young Dr. A. Conan Doyle for *not* seeking his services, thereby converting him into a writer. In the same vein, I should like to thank those people at the University of Michigan whose opposition to Little's personal and administrative decisions led to his resignation as President and to the founding of the Jackson Laboratory.

REFERENCES

1. Holstein, J: The First Fifty Years at the Jackson Laboratory. The Jackson Lab, 1979.
2. Muller, HJ: The gene as the basis of life. Proc Int Cong Plant Sci 1:897, 1929.
3. Bridges, CB: Nondisjunction as proof of the chromosome theory of heredity. Genetics 1:1, 1916.
4. Snell, GD: X-ray sterility in the male house mouse. J Exp Zool 65:421, 1933.
5. Fisher, RA: The Genetical Theory of Natural Selection. Oxford, Clarendon Press, 1929.
6. Griffith, F: The significance of pneumococcal types. J Hygiene 27:113, 1928.
7. Muller, HJ: Variation due to change in the individual gene. Amer Nat 56:32, 1922.
8. Berg, R: Boris L. Astaurov. His life and research. Quart Rev Biol 54:397, 1979.
9. Wright, S: The physiology of the gene. Physiol Rev 21:487, 1941.
10. Wright, S: The effects of inbreeding and crossbreeding on guinea pigs. Bull US Dept Agric 1121:1, 1922.
11. Woodworth, RS: Heredity and environment: A critical survey of recently published material on twins and foster children. Soc Sci Res Counc Bull No 47, 1941.
12. Hutchings, E and E: Scientific Progress and Human Values. New York, American Elsevier Pub Co, 1967.

Index

Agouti locus, 86, 147–148
Androgenesis, 105, 312, 314
Anemias, 160–161, 318
Antigens, 199–203
   differentiation, 247
   and viruses, 252–253

Bacteriophage, 311–312
Biochemical genetics, 130–133, 187
Brush border membrane, 176, 184–185

Cancer, 284–289, 320–322; see also Oncogenesis; Tumor
Carcinogens, 287–289
Cell membrane
   alloantigens, 242–246, 247
   immunogenetics, 248–251
   see also Membrane
Cellular immunity, 251–253, 258–263
Chemotherapy, 289
Chimeras, 78–88, 98, 101, 102, 111, 113, 317
   and early differentiation, 79–88
   and mutants, 86–87
   and normal genetic variants, 81–85
   producing, 78–79
   and teratocarcinoma cells, 87–88
Cholesterol, 177
Chromosomal aberrations, 129, 133, 134
Clinical genetics, 137–140
Cloning, 51, 52
Coimmunization, 246

Complement, 219–222
Cystinuria, 174–175
Cytogenetics, 129

Diabetes mutant, 147–152, 155–157; see also Obesity-diabetes syndrome, murine, 145–157, 318
Differentiation, 77–88, 247–248, 318
Disease, 130–133, 217–219, 253
DNA, 24, 315–316, 317, 321; see also Recombinant DNA

Elliptocytosis, hereditary, 162–163
Embryo, 77–88
   blastocyst, 78, 81, 83–85, 87, 98, 101, 102, 105–106, 108–109, 115
   inner cell mass, 78, 83–85, 87, 115
   morula, 82, 83, 115
   trophectoderm, 78, 83–84, 87, 115
Embryo-derived teratoma, 93, 97
Embryoid bodies, 93
Embryonal carcinoma, 93, 102
Endocytosis, 177–178
Enzymatic deficiencies, 130–132
Evolution
   and immune system, 234
   mouse genes, 56–62
   tumor, 209

Fat mouse, 148, 151

Galactokinase, 111
Gene(s)
  clusters, 213–237
  immune response, 219–223
  intervening sequence, 53–64, 234–236
  mapping, 134–136
  and mouse hemoglobin, 51, 53–56
Genetic variants, 81–85
Genital ridge, 95–97
Germ cells, 96–97, 100
Gynogenesis, 105

Hemolytic anemia, 159–165
Histidinemia, 173
Histocompatibility, 197–209, 213–238, 248–253, 256–263, 319
HLA, 134, 139, 259–260, 263
  and homology with H2, 213–216
  and monoclonal antibodies, 223–229
  population distribution of, 217–219
  structure of, 219–223
  region, 234–236
H-2, 213–216, 243–251
  and histocompatibility complex, 251–253, 256–257
  and tumor, 197–202
Hyperprolinemia, 174, 181
Hypophosphatemia, 174, 183–187
Hypoxanthine phosphoribosyltransferase, 109, 111

Immunogenetics, 133–134, 241–263, 273–274, 319
  and cell membrane, 248–251
  and differentiation, 247–248
  and inbred mice strains, 282–284
Inborn errors of metabolism, 130, 141, 169–188, 312, 318

Inbred strains, murine, 279–289, 315, 319
  and cancer, 284–285
  and cancer virology, 285–287
  and chemotherapy, 289
  and immunogenetics, 282–284
  carcinogen testing on, 287–289
  development of, 280–282
  recombinant, 99, 246

Leukemia, murine, 291–294, 297–306
Lipids, 175–176, 248
Little, Clarence Cook, 279–283, 319, 323

Medical genetics, 127
Membrane
  red cell, 159–165
  transport, 174–188, 248
  see also Cell membrane
Mendelian disease, 130–133
Monoclonal antibodies, 223–234
Mouse genetics, 7–43
  anchoring linkage maps, 27–30
  band-to-gene relationships, 30–37
  future problems of, 25, 37, 40
  linkage maps, 7–24, 25–27, 38–43
  special approach to, 22–37
Mouse globin, 51–65
Mouse models
  His, 171–173
  Hyp, 174, 183–187
  PRO/Re, 174, 181–183
  Spf, 173–174
Mouse mutants, 160–161

Natural killer cell, 253, 258–262
Nuclear transplantation, 113–117

Obesity-diabetes syndrome, murine, 145–157, 318
Obesity mutant, 146–157

Oncogenesis, 203–208
Ovarian teratoma, 97–100

Parthenogenesis, 93–102, 105, 108
Paternal age effect, 134
Prenatal diagnosis, 137–140
Proline oxidase, 174, 182–183
Pyropoikilocytosis, 163–164

Radiation, 313
Receptor disease, 133
Recombinant DNA, 43, 51–65
Recombinant inbred strains, 99, 246
Red cell membrane, 159–165
Renal phosphate transport, 184–186
Rickets, 174
RNA
 splicing, 62–65
 viruses, 254–255, 298

Somatic cell hybridization, 110–113 113
Spectrin, 159–165
Spherocytosis, 160–161, 164–165

Supergenes, 311

Teratocarcinoma, 87–88, 108–109, 111, 317
Teratogenesis, 93–102
Testicular teratoma, 93–97
Thriftiness, 156
Thymidine kinase, 109–111
T-locus, 86, 134
Tubby mouse, 148, 151
Tumor
 biology, 273–274
 evolution, 209
 and immune response, 203–209, 258–262
 transplant, 95–97, 101–102, 197

Uniparental mice, 105–108

Viruses, 291–294, 297–306, 313
 and immune system, 252–263
 oncogenic, 203–208, 285–287

X-chromosome, 99–100, 108, 118, 134, 136, 142, 314
Xenogeneic gene expression, 108, 113

8104492

3 1378 00810 4492